Auditory Effects of Microwave Radiation

James C. Lin

Auditory Effects
of Microwave Radiation

 Springer

James C. Lin
University of Illinois at Chicago
Chicago, IL, USA

ISBN 978-3-030-64546-5   ISBN 978-3-030-64544-1   (eBook)
https://doi.org/10.1007/978-3-030-64544-1

This Springer imprint is published by the registered company Springer Nature Switzerland AG
The registered company address is: Gewerbestrasse 11, 6330 Cham, Switzerland

Preface

The unique microwave auditory effect has been widely recognized as one of the most interesting and significant biological phenomena from microwave exposure. Its potential applications are just now beginning to be seriously explored.

The microwave auditory effect is defined as the auditory perception of microwave radiation or simply hearing microwaves. This description may seem surprising, and one might even question whether such sensation could exist. It is indeed a unique exception to the acoustic energy normally encountered by humans in auditory perception of sound. Although mammalian hearing apparatus responds to acoustic or sound pressure waves in the audio-frequency range (up to 20 kHz for humans), frequencies above 35 kHz are considered as ultrasound for humans (beyond the range of audible frequencies). The hearing of microwave pulses involves electromagnetic radiation whose frequency is in the much higher megahertz (MHz) to gigahertz (GHz) range. As electromagnetic waves (such as sunlight and optical radiation) can normally be seen but not heard, the auditory sensation of microwave pulses is obviously surprising, and initially, its authenticity was widely questioned.

The microwave auditory effect involves a cascade of events. Minuscule but rapid rise in tissue temperature ($\sim 10^{-6}$ °C) resulting from the absorption of short microsecond (µs) wide pulses of microwave energy creates a thermoelastic expansion of brain matter. This small theoretical temperature elevation is undetectable by any currently available temperature sensors, and at threshold levels, it cannot be felt as a thermal sensation or heat. Nevertheless, it can launch an acoustic wave of pressure that travels inside the head to the inner ear. It then activates the nerve cells in the cochlea and relays it to the central auditory nervous system for perception via the same process involved for normal hearing. Thus, the discovery of microwave thermoelastic pressure wave generation in biological tissues by deposition of short microwave pulses in tissue came about as the result of an intense effort in search of a mechanism to help understand the observed auditory response to microwave radiation. Furthermore, identification of the propagation nature of the acoustic wave of pressure in biological tissues has prompted the exploration of its potential for applications in biomedical imaging, specifically a new dual modality diagnostic imaging system – microwave thermoacoustic tomography (MTT).

It is interesting to note the U.S. State Department's disclosure that Havana-based US diplomats were experiencing health issues associated with hearing loud buzzing or what was described as bursts of sound in 2017. A similar announcement was made by the Government of Canada. The staff of both embassies had reported symptoms of hearing loss, ringing in the ears, headaches, nausea, and problems with balance or vertigo, which are suggestive of a connection to the inner ear apparatus within the human head, where the cochlea and vestibular organs are located.

Government officials had difficulty pinning down the source of sound. There are speculations that the diplomats may have been attacked with an advanced sound weapon. Assuming reported accounts are reliable, the microwave auditory effect may be the scientific explanation. It is plausible that the loud buzzing, burst of sound, or acoustic pressure waves could have been covertly delivered using high-power pulsed microwave radiation, rather than blasting the subjects with conventional sonic sources. Indeed, many have come to believe that the microwave auditory effect – induced by a targeted beam of high peak-power pulsed microwave radiation – may be the most likely scientific explanation for the sonic attack. Of course, until the truth is revealed, this specific matter will remain somewhat of a mystery.

The objective of this volume is to bring together in a comprehensive book the multidisciplinary research investigation leading to a scientific understanding of the microwave auditory effect and related applications, especially the emerging microwave thermoacoustic tomography (MTT) imaging modality. The analysis and discussions in the chapters of this book pertain to relevant physical laws, exposure and dosimetry, anatomy and physiology, psychophysics and behavior, theories and models, mechanisms of interaction, computer analysis and simulation, applications in biology and medicine, and other applied aspects. Another purpose of this volume is to expand and update existing knowledge and understanding of microwave auditory effects and applications, which was the title of an earlier research monograph. A considerable amount of scientific knowledge, data, and information have been generated since then through theoretical study, laboratory experimentation, numerical analysis, and computer simulation.

As this may be the readers' first encounter to this material and the treatment is multidisciplinary, after the Introduction, the four chapters that follow each begins with basic background information that may appear as elementary to some readers but is essential to understanding the discussions on microwave auditory effect and applications for those from a different discipline. There is a chapter on the principles of microwave and RF exposure and one on brain anatomy and auditory physiology. The succeeding two chapters present the microwave property of biological materials and its influence on dosimetry and microwave absorption in biological tissues. They are intended to facilitate a fuller understanding of discussions in the ensuing five chapters about microwave auditory effect and applications.

The focus of Chaps. 6, 7, 8, and 9 is all on the microwave auditory effect. In Chap. 6, neurophysiological evidence and psychophysical and behavioral observations from laboratory studies involving animals and humans as experimental subjects are discussed in detail. The objective is to provide a complete account of what is scientifically known about the microwave auditory effect. The possible mechanisms that have been suggested whereby auditory responses might be induced by pulse-modulated microwave radiation are analyzed in Chap. 7 with the conclusion of the microwave thermoelastic theory being the favored.

Solving partial differential equations may be mathematically fun and satisfying. However, some of the ramifications of mathematical solutions can be hidden if there is little knowledge of their parametric dependence, such as on peak power or pulse width or the values they may take at various frequencies. Also, because long uninterrupted strings of formulas tend to become dull, computational and experimental results are interspersed throughout these chapters when appropriate.

Chapter 8 presents rigorous multidisciplinary, mathematical analyses of the thermoelastic pressure waves generated in canonical or spherical human and animal head models exposed to pulsed microwave radiation. The results include variations of induced sound pressure frequency and strength on microwave pulse characteristics. They also correctly predict the attributes of sound waves generated in the head as perceived by humans. More precise computer simulations of the properties of microwave-pulse-induced sound pressure waves using realistic anatomic human head structures are presented in Chap. 9. The simulation confirmed by experimental results clearly indicates that the microwave auditory effect or the hearing of microwave pulse-induced sound involves a cascade of events that start from microwave absorption in the brain, where it is converted into an acoustic pressure wave.

Chapter 10 discusses the important diagnostic imaging application of microwave thermoelastic pressure wave interaction including a summary of early investigations and current developments in microwave thermoacoustic tomography (MTT) and imaging. It also describes some applied aspects of the microwave auditory effects in directed messaging, mind control, and its possible role in the recently reported covert personnel attacks at some diplomatic missions (the Havana syndrome) and fighter pilot disorientations.

An extensive list of references is provided in this book at the end of each chapter to furnish the knowledge base to put the materials in proper perspective, and to overcome potential misunderstanding. The guiding principles throughout the book are that any description or conclusion must be consistent and compatible with physical laws, biological evidence, and experimental observations based on laboratory data and findings.

The many graduate students and colleagues whose contributions to various aspects covered in this book are acknowledged with appreciation. Instead of repeating their names that are included in the reference citations, I would like to direct the readers to the references. It is with gratitude and love that I thank my family for their support and encouragement and especially my wife, Fei Mei, for her patience and forbearance during the writing of this book.

Chicago, USA   James C. Lin

Contents

Chapter 1
Introduction

Humans have been living in a milieu of natural electromagnetic radiation for millennia. The natural electromagnetic radiation originates from terrestrial and extraterrestrial sources such as lightning and electrical discharges in the Earth's atmosphere and radiant energies from the solar system and outer space. Indeed, the discovery and measurement of 3 K microwave background radiation are the crucial steps leading to the standard "Big Bang" model of universe [Grandin, 2007; Lundqvist, 1992]. Furthermore, scientific and technological advances have ushered in a myriad of artificial electromagnetic sources through semiconductor and vacuum tube processes that are unique in each case and have given rise to devices and systems that enable applications to benefit human endeavors and embellish our daily lives. Examples are found in a wide variety of commercial, communication, industrial, scientific, residential, and medical applications.

1.1 Electromagnetic Radiation and Spectrum

The wide-ranging spectra of electromagnetic radiation span from cosmic gamma rays to static electric and magnetic fields. In between these are the well-known X-ray, ultraviolet (UV), visible light, infrared, microwave, and radio frequency (RF) waves (Fig. 1.1). Electromagnetic radiation may be described in terms of its wavelength, frequency, and energy, each with a different set of units of measures. The speed (v) of propagation of electromagnetic radiation in a material medium depends on the permittivity and permeability properties of the medium. It is related to the product of frequency (f) and wavelength (λ), such that

$$v = f \times \lambda \tag{1.1}$$

The highest speed of electromagnetic radiation can travel is 2.998×10^8 m/s in vacuum or free space.

© Springer Nature Switzerland AG 2021
J. C. Lin, *Auditory Effects of Microwave Radiation*,
https://doi.org/10.1007/978-3-030-64544-1_1

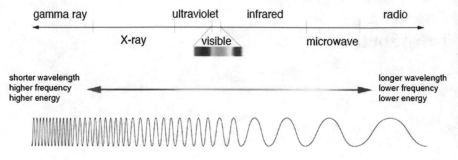

Fig. 1.1 A portion of the electromagnetic spectrum including microwave and radio frequency radiation (NASA)

Wavelength measures the closeness between any two successive peaks or valleys of the wave variations shown in Fig. 1.1 and uses meter (m) as the standard unit of measure. Frequency is measured in cycles per second (s) or Hertz (Hz). Energy is measured in Joules (J) or electron volts (eV). One eV is equal to 1.602×10^{-19} J. Gamma rays have shorter wavelength, higher frequency, and higher energy per quantum or photon, while microwave and RF radiation have longer wavelength, lower frequency, and lower energy per quantum or photon energy.

At shorter wavelength, electromagnetic radiation may be conceptualized as massless wave packets, referred to as photons, traveling at the speed of light in vacuum or free space with a finite amount of energy per packet or photon. Thus, the term photon energy is often used to describe the energy for various regions of electromagnetic radiation. The relationship among these quantities is specified by the equation

$$\varepsilon = h \times f \tag{1.2}$$

where ε is energy per photon (J), h is Planck's constant (6.625×10^{-34} J-s), and f is frequency (Hz).

A list of the photon or quantum energy of common forms of electromagnetic radiation is provided in Table 1.1. The photon energy of Gamma ray is greater than 2×10^{-14} J, ultraviolet radiation has photon energy varying between 5×10^{-19} and 2×10^{-17} J, and those of optical radiation and microwaves are less than 5×10^{-19} J.

Depending on the photon energy carried by the electromagnetic radiation, some of them may be capable of ejection or promotion of orbital electrons from the atoms of materials through which the electromagnetic radiation travels and creating ions in the process. During ionization, the impinging electromagnetic radiation imparts a finite amount of photon energy to each ejected or promoted electron.

The minimum amount of photon energies required to produce ionization in water and atomic carbon, hydrogen, nitrogen, and oxygen is between 10 and 25 eV. Inasmuch as these atoms constitute the basic elements of living organisms, 10 eV is considered as the lower limit of ionization in biological systems. Although weak hydrogen bonds in biomolecules may involve energies less than 10 eV, photon energies below this value can generally be considered, biologically, as nonionizing.

Table 1.1 Approximate wavelength, frequency, and photon energy limits of the various regions of the electromagnetic spectrum

	Wavelength (m)	Frequency (Hz)	Energy (J)	Energy (eV)
Radio frequency (RF)	$>1 \times 10^{-1}$	$<3 \times 10^{9}$	$<2 \times 10^{-24}$	$<1.2 \times 10^{-5}$
Microwave	$1 \times 10^{-3} - 1 \times 10^{-1}$	$3 \times 10^{9} - 3 \times 10^{11}$	$2 \times 10^{-24} - 2 \times 10^{-22}$	$1.2 \times 10^{-5} - 1.2 \times 10^{-3}$
Infrared	$7 \times 10^{-7} - 1 \times 10^{-3}$	$3 \times 10^{11} - 4 \times 10^{14}$	$2 \times 10^{-22} - 3 \times 10^{-19}$	$1.2 \times 10^{-3} - 1.8$
Optical	$4 \times 10^{-7} - 7 \times 10^{-7}$	$4 \times 10^{14} - 7.5 \times 10^{14}$	$3 \times 10^{-19} - 5 \times 10^{-19}$	$1.8 - 3.1$
Ultraviolet (UV)	$1 \times 10^{-8} - 4 \times 10^{-7}$	$7.5 \times 10^{14} - 3 \times 10^{16}$	$5 \times 10^{-19} - 2 \times 10^{-17}$	$3.1 - 1.2 \times 10^{2}$
X-ray	$1 \times 10^{-11} - 1 \times 10^{-8}$	$3 \times 10^{16} - 3 \times 10^{19}$	$2 \times 10^{-17} - 2 \times 10^{-14}$	$1.2 \times 10^{2} - 1.2 \times 10^{5}$
Gamma ray	$<1 \times 10^{-11}$	$>3 \times 10^{19}$	$>2 \times 10^{-14}$	$>1.2 \times 10^{5}$

$1 \text{ eV} = 1.602 \times 10^{-19} \text{ J}; 1 \text{ J} = 6.242 \times 10^{18} \text{ eV}; 1 \text{ nm} = 10^{-9} \text{ m}$

X-rays and Gamma rays have photon energies higher than 100 eV. The principal biological effects of the ionizing X-ray and Gamma radiation are therefore largely the result of ionization they produce, including their effect on deoxyribonucleic acid (DNA). It is important to know that ionizing radiation is also associated with thermal effects in biological systems through heat generation.

Terrestrial solar UV consists mainly of UV-A radiation (315–400 nm) and the balance UV-B (280–315 nm). Only artificial UV sources emit radiant energy within the UV-C spectral band (100–280 nm). Ultraviolet radiation is important for several biological processes and has been shown to have deleterious effects on certain biological systems, including carcinogenesis. One commonly known effect of UV radiation is sunburn. Aside from beneficial medical treatments, they are used in cellular and microbiology laboratories to kill bacteria and prevent bacterial growth and for public health applications as antiviral agents.

Ultraviolet radiation at wavelengths of less than 320 nm, especially the UV-B and UV-C bands, transfers their energies to atoms or molecules almost entirely by excitation, that is, by promotion of orbital electrons to some higher energy levels. Consequently, some of the effects produced by UV photons may resemble the changes resulting from ionizing radiation. In fact, the effects of UV share some aspects of effects of ionizing radiation such as effects on DNA. Also, since the advent of high-power lasers, some of the principal hazard recognized in the use of UV sources has been the potential for injury of the skin and eye from exposure to UV radiation at wavelengths of less than 320 nm. The widespread use of high-power UV in industry has been the cause of many corneal injuries. The UV rays can circumvent the natural defense of the body by allowing direct exposure of the cornea at normal angles of incidence, unshielded by the brow or eyelids.

Although the photons of visible light with relatively low energy levels, 1.8–3.1 eV, are not capable of ionization or excitation, they have the unique ability of producing photochemical effects or photobiological reactions. Through a series of biochemical reactions, green plants, for example, can use optical energy to fix carbon dioxide and split water such that carbohydrates and other chemicals are synthesized.

Because visible light is not very penetrating, the eye and the skin are the major organs of concern. Optical radiation is transmitted through the eye media without appreciable attenuation before reaching the retina. There it is absorbed by light-sensitive cells that initiate biochemical reactions whose end result is the sensation of vision. The leading acute effects are thermal and photochemical retinal injury for the eye and erythema and burns for the skin. Retinal injury and transient loss of vision may occur as a result of exposure to intense visible light. Delayed effects include cataract formation in the lens, retinal degeneration for the eye, and accelerated aging and cancer for the skin.

Infrared radiation of the sun is the major source of heat on the Earth and has wavelength varying from approximately 700 nm to 1 mm. It is also emitted by all hot bodies. With photon energies of 0.002–1.8 eV, it does not produce ionization. There is little evidence that photons in the infrared region can initiate photochemical reaction in biological systems. It is known that changes in vibrational modes are responsible for absorption in infrared region. The absorbed energy increases the kinetic energy in the system, which dissipates in heat. Thus, the main response of biological systems to an exposure to infrared radiation is thermal.

Microwave and RF radiation increases the kinetic energy of the system when it is absorbed by the biological media. In this case, the increase in kinetic energy is due to induced translation of electronic particles and change of rotational or vibrational energy levels; the latter ends up primarily in the form of increased system temperature while it may not be exclusively so. Microwave and RF radiation has low-energy photons; therefore, under ordinary circumstances, the photon energy is too low to affect excitation or ionization of a biological atom or molecule. However, the situation may be different in some cases, if multiple microwave photons imping on a single atom or molecule, or under intense bombardment, ionization may still occur. Simultaneous absorption of strong RF radiation, 8×10^5 greater than the low-energy microwave photons, could potentially produce ionization in biological materials. The point is that RF radiation has low energy photons, therefore under ordinary circumstances, RF radiation is too weak to affect ionization or cause significant damage to biological molecules such as DNA, which is especially renowned for its repair mechanism. Additional information and description on microwave and RF radiation are given in the following section and subsequent chapters.

1.2   Microwave Technology and Applications

In scientific, industrial, commercial, communication, medical, and security applications, common names such as RF, microwave, millimeter wave (mm-W), and Terahertz (THz) or T-wave are used to describe specific regions of the electromagnetic spectrum between 30 kHz and 3 THz (Table 1.2), using manufactured artificial sources. Also, for RF region, the LF, MF, HF, and VHF band designations are often employed to denote the low, medium, high, and very high frequency bands, respectively. Likewise, microwave radiation may be further divided into UHF (ultrahigh frequency) and SHF (super high frequency) bands.

Table 1.2 Commonly used frequency band designations of RF and microwave region of the electromagnetic spectrum

Common name	Band designation	Frequency
RF	Low frequency (LF)	30–300 kHz
	Medium frequency (MF)	300 kHz–3 MHz
	High frequency (HF)	3–30 MHz
	Very high frequency (VHF)	30–300 MHz
Microwave	Ultrahigh frequency (UHF)	300 MHz–3 GHz
	Super high frequency (SHF)	3–30 GHz
Millimeter wave (mm-W)	Extremely high frequency (EHF)	30–300 GHz
Terahertz wave (T-wave)	Terahertz (THz)	300 GHz–3 THz

Microwaves are generated as continuous-wave (CW) sinusoids or pulses of various forms and are transmitted in CW, baseband, amplitude- or frequency-modulated carrier waves, or pulse-amplitude modulated waves. In some high-power microwave applications, magnetron tubes are the generator of choice, although in some cases the multiple-beam klystron is favorite. However, in many current wireless communication and data transmission systems, solid-state semiconductor devices have replaced magnetron and klystron tubes.

Scenarios with sources of high levels of microwave and RF radiation are typically found in medical facilities such as magnetic resonance imaging (MRI) scanners, or specialized establishments with high-power radars. Some medical diagnostic imaging procedures may involve high levels of microwave and RF radiation at the patient's location or even inside a patient's body. Situations associated with personal use by the general public such as for wireless communication, data transmission, security operations, or food processing produce comparably much lower exposures at the position of the user.

In open medium or free space, these waves propagate or travel in straight line-of-sight path in the absence of obstacles. They may be reflected or refracted upon encountering any material bodies. It is interesting to note in this regard the moon's far side is continuously shielded from radio transmissions from the Earth. The recent touch down of China's Chang'e-4 lander on the lunar surface's far side in January 2019 involved the support of new technology-based relay platform at a location well beyond the moon [Greshko, 2019].

As a powerful but relatively compact microwave source for practical use, the magnetron tube was developed in the early 1940s under the stimulus of an intense war-time effort for military radars. It was reported that these microwave radars caused warmth sensation for radar operators in certain work environment. Furthermore, anecdotally, some radar personnel had stated that they were able to auditorily sense when radars were in active operation. However, microwave sources were unavailable for biological and medical research since they were reserved solely for military use during the war. Nonetheless, in response to the concerns over the biological effects and potential hazards of microwave radiation, the United States military services conducted some human studies to determine whether or not rumors such as sterilizing effects in males might have some factual ground. The available reports [Daily, 1943; Lidman and Cohn, 1945] failed to record any clinical (or hematological) changes resulting from exposure to radar microwave radiation.

Shortly after the war, magnetron sources capable of generating high power levels of microwaves were introduced into industrial and medical applications. Indeed, industrial microwave heating, drying, and curing systems are currently in widespread use in many industrial processes. They are found in agricultural, food processing, material treatment, pharmaceutics, waste management, and manufacturing facilities, to name a few. The following are some familiar applications of microwave and RF radiation beginning with microwave magnetrons operating under the CW mode.

1.2.1   Microwave Diathermy

In 1946, Mayo Clinic in Minnesota received two magnetron microwave generators for biomedical research, almost a decade after they had first become interested in the medical use of microwave radiation. They immediately embarked on their first studies of microwave diathermy, which means deep heating of tissues through the skin. Studies on animals showed that microwave diathermy could cause severe tissue burns within a short time of application. Nevertheless, with the proper selection of power and treatment duration, it was found that deep tissues could indeed be heated to therapeutically effective levels [Krusen et al. 1947]. It was noted that the initial temperature rise was greater in the skin and subcutaneous fat than in deeper musculatures. The final temperature in the muscle, however, was higher [Licht, 1965].

Research conducted at the Mayo Clinic had helped to establish the initial uses and limitations of microwave diathermy and led to its acceptance as a therapeutic instrument by the American Medical Association in 1947. Some early reports showed that microwave diathermy was effective in complete relieve of the pain of osteoarthritis in most cases. Microwave diathermy has been described as valuable in the management of disorders of the shoulder [Rae, et al., 1950]. Clinical studies found that microwave diathermy can be used to advantage in traumatic conditions such as sprains and strains, especially when combined with massage. Readers interested in this topic are referred to [Kotttke and Lehmann, 1990] for more detailed descriptions, indications, and contraindications and its use in the current practice of physical and sports medicine.

There are two related topics in medical applications of RF and microwave heating: ablation and hyperthermia treatments.

1.2.2   Microwave Ablation Therapy

Since 1987, percutaneous transluminal microwave [Beckmann, et al., 1987; Lin and Wang, 1987] and RF [Huang and Wilber, 2000] catheter technologies have been developed for cardiac ablation treatment of arrhythmia or abnormal heart rhythm [Lin, 1999, 2004]. Within a few years, it has become the method of choice for treatment of cardiac arrhythmias. Catheter ablation for atrial fibrillation and ventricular tachycardia is expanding; it stems in part from the nonpharmacological approach and the minimally invasive nature of these procedures. In cardiac ablation, endocardia conducting tissue responsible for causing arrhythmia is destroyed by a burst of microwave or RF energy applied through a catheter to the cardiac tissue to effect thermal coagulation. The field of catheter cardiac ablation has progressed with the development of new methods and tools [Lopresto et al., 2017] and with the publication of large clinical trials [Calkins et al. 2018; Cronin et al., 2019]. Catheter ablation of atrial fibrillation is widely available and is now the most performed catheter ablation procedure.

For ablation of atrioventricular junction to treat supraventricular arrhythmia, the catheter antenna or electrode is inserted percutaneously through the femoral vein, first to record electrograms from the His bundle using the catheter as electrode and then to apply heating energy to destroy the atrioventricular conduction tissue. For microwave catheters, (2-mm diam and operating at 2450 MHz), an average energy of 200 joules is required for tissue temperatures of 65 °C [Lin, 1999; Lin et al., 1996]. In the RF technique, a 500-kHz current is induced to flow from an electrode inside the heart to a large patch reference electrode on the body surface. The resulting resistive heating, which is the highest at the electrode-tissue interface, produces desiccation of the cardiac tissue to aid ablation.

The development of coagulative ablation therapy over the past decades has revolutionized not only the practice of cardiac electrophysiology but also oncology, especially in using microwave or RF energy to ablate tumors. For many of the diseases, surgical intervention has been the principal method of treatment, although alternatives to surgery have been sought to reduce the cost and morbidity of treatment. Minimally invasive catheter ablation offers several potential benefits: long incisions are replaced with a puncture wound, and the need for postoperative intensive care is significantly reduced and, in many cases, offers a complete or lasting cure. It also has important advantages over drugs that are merely palliative. It avoids the side effects, expense, and inconvenience of chronic drug therapy, often with only partial success.

Percutaneous microwave and RF ablation is considered a prime therapy for early hepatocellular carcinomas with improved rates of local tumor control and survival. However, RF current relies on a conductive or resistive heating process. It is limited to contact heating and is not successful in treating large tumors and tumors in regions of tissue with high blood perfusion. As a result, RF end up with an undesirably high rate of local tumor progress in larger tumors. In contrast, microwave energy at 915 or 2450 MHz is radiated from the catheter antenna into the tissue volume surrounding the antenna and can reach a larger tumor mass. Also, microwave technology affords better power control and has greater capacity to overcome perfusion-mediated tissue cooling associated with large vessels. A contraindication is the increased power can result in vascular thrombosis of the portal vein in patients with reduced portal venous flow rate. Note that blood flow velocity within the inferior vena cava, hepatic arteries, and major hepatic veins is usually sufficient to prevent significant vascular thrombosis [Wells et al., 2015].

1.2.3   Hyperthermia Treatment of Cancer

Hyperthermia cancer therapy is a treatment in which tumor temperatures are elevated to the range of 40–45 °C. The cytotoxic effects of some antitumor drugs are enhanced, and the cell-killing ability of ionizing radiation is potentiated by hyperthermia serving as a sensitizing agent. Moreover, the synergism between the beneficial effects of local or regional hyperthermia combined with chemotherapy and radiotherapy has been postulated for some time, and it has been well documented [Dewhirst et al., 2018; Issels, 2008; van der Zee et al., 2000]. Clinical and laboratory results from various investigators have indicated a promising future for hyperthermia. Its efficacy depends on the induction of a sufficient temperature rise throughout the tumor volume. Many external and implanted antennas and applicators have been designed to produce therapeutic heating of localized tumors of different volumes in a variety of anatomical sites [Burfeindt et al., 2011; Hand 1989; Lin et al., 2000; Nguyen et al., 2017]. Clearly, each modality has its own advantages and liabilities. Monitoring and control of tumor temperature in real time during hyperthermia treatment is essential for effective treatment. While progress in temperature sensing in vivo has been dramatic, considerable advance is needed along with appropriate treatment planning to support routine clinical application of hyperthermia for cancer. Nevertheless, hyperthermia is gaining wider use in clinical practice in treating a variety of malignant tumors [Cheng et al., 2019; Dooley et al., 2010; Yamamoto et al., 2014]. Also, hyperthermia as a multifaceted therapeutic modality represents a potent radiosensitizer, interacts favorably with a host of chemotherapeutic agents, and, in combination with radiotherapy, enforces immunomodulation akin to in situ tumor vaccination [Datta et al., 2020].

Whole-body hyperthermia has been employed to enhance the effectiveness of chemotherapy for patients with systemic metastatic cancer. A variety of conductive and convective heating techniques such as warm air, water, wax, radiant heating techniques, and heating the blood via extracorporeal circulation have been applied. Mild whole-body hyperthermia at 40 °C for as long as 10 hours in rats has shown promising therapeutic potentials on a metastatic primary tumor. Studies in pigs have shown body-core temperature elevations to 41.5–42.0 °C for 60 min are safe and associated with acceptable toxicity rates using radiant heat sources [Hildebrandt et al., 2015]. While clinical results remain guarded, a recent pig study reported on thermal distribution, pathophysiological effects, and safety of whole-body hyperthermia [Lassche et al., 2020].

1.2.4   Microwave Ovens

The search for new applications for the magnetron microwave-generating technology alighted on the idea of heating food with microwaves [Osepchuk, 1984, 2002]. Concerted efforts were directed toward a microwave oven for commercial and residential use, perhaps aided by prior accounts of warmth sensations close to radiating microwave radars, as mentioned above. In 1946, Raytheon in Massachusetts unveiled its new "Radarange" microwave oven for heating and cooking food, introducing a new use for the company's magnetron tubes and allowing it to avoid obsolescence instead, for a bright future.

The successful marketing of countertop microwave ovens for consumer use, in subsequent years, had not only launched an economically important technology-based business, but it enabled magnetrons to become one of the best-known sources of microwave power. It also assured the future of magnetrons in helping to fulfill the "promise" of a microwave oven in every kitchen. Microwave oven operates at the industrial, scientific, and medical (ISM) band frequency of 2450 MHz for heating and cooking food in the household or commercial establishment. It remains an essential appliance in state-of-the-art kitchens. Some of the technological and economic accomplishments, along the way, include achievement of close to 100% efficiency at 1 kilowatt (kW) of microwave power and cost and weight reductions by a factor of 100 or more since 1946.

It is noteworthy that a Radiation Control for Health and Safety Act (PL90-602) was adopted by the United States Congress in October 1968, to protect the public from unnecessary exposure to potentially harmful radiation, which includes microwaves emitted by electronic products. The Act prescribes different and individual performance standards, to the extent appropriate and feasible, for different electronic products, to recognize their different operating characteristics and uses. For microwave ovens, the performance standard limits radiation at any point 5 cm from the oven surface to 1 mW/cm^2 prior to purchase and 5 mW/cm^2 throughout their useful life (1 mW/cm^2 = 10 W/m^2).

A primary source for microwave oven emission comes from the leakage of microwave radiation from the door. There had been several designs to effectively limit microwave leakage from the door and seal to the specified levels. In fact, most of the microwave ovens currently on the market are based on designs developed, tested, and evaluated decades ago for their effectiveness in minimizing leakage radiation.

1.2.5   Magnetic Resonance Imaging

Advantages of magnetic resonance imaging (MRI) have made it the radiological modality of choice for many diagnostic medical procedures. MRI has become perhaps the most successful application of electromagnetic fields and waves in biology and medicine. MRI systems use strong static, spatially varying gradient, and RF magnetic fields to make images. The clinical successes have also heightened the desire for increased spatial resolution from MRI systems. This demand has prompted the exploration of ever higher strength static magnetic fields and the associated use of higher RF spectra and powers. Pulsed RF magnetic fields are used to elicit magnetic resonance signals from tissues, principally from the hydrogen protons in fat and water molecules. Many open designs for interventional and intraoperative procedures are increasingly being installed in healthcare sites and radiological imaging centers.

In a typical MRI procedure, the patient is exposed to numerous pulses of RF radiation. For the common 1.5 tesla (1.5 T) clinical MRI scanner, the associated RF frequency for proton imaging at 1.5 T is about 64 MHz. The new 3 T MRI scanners use 128-MHz RF energy for operation. In fact, higher field MRI systems are becoming more common. Furthermore, at present, several experimental ultra-high-field strength and ultra-fast MRI systems operate at 7.0 T, 9.4 T, or higher, with a corresponding 300-MHz, 400-MHz, or higher RF frequencies for proton imaging (see more discussions in Chap. 9.).

1.2.6   Modern Microwave Radars

While radars based on magnetrons were initially developed for military purposes, current uses of microwave radars have expanded to numerous civilian applications where information on relative location of objects of interest is crucial. Examples include air traffic control, weather monitoring, marine navigation, and autonomous vehicles (self-driving cars). Many of these radar applications involve high-power microwave radiation ranging from a few kilowatt (kW) to several megawatt (MW). There are installations operating at gigawatt (GW), but other radar devices use power that are at or below the milliwatt (mW) level. Moreover, many modern radar applications are based on solid-state systems for high and low power applications [Balanis, 2008]. Also, most of these operations involve use of pulse or frequency modulation techniques to acquire both location and speed information.

1.3   Auditory Effects from Pulsed Microwave Exposure

Exposure to microwave and RF radiation is comprised of electric field, magnetic field, and the electromagnetic power carried by microwave and RF radiation in these fields. The exposures generally vary with time and space (or location) because the electric and magnetic fields change with time and space and the relative position (and distance) of the exposed subject and source of microwave and RF radiation. Of concern here are principally the macroscopic phenomena in which bodies or volumes whose dimensions are larger than atomic dimensions. Moreover, the time intervals of observation are assumed to be long enough to allow for an averaging of atomic fluctuations.

The microwave and RF radiation exposure may be measured using appropriate instruments and systems or quantified through theoretical analysis and computer simulation based on Maxwell's equations which form the classic electromagnetics theory. Indeed, solution of Maxwell's equations leads to descriptions with mathematical vigor and computational exactness of the macroscopic characteristics of radio and microwave radiation and their interaction with biological systems.

For exposure durations that are suitably long and average applied field strengths or incident power densities of microwave radiation that are sufficiently high, appreciable temperature elevation including heating may occur in biological materials. The heating effects have been extensively studied, documented, and served as the foundation for therapeutic treatment in physical medicine and rehabilitation, in minimally invasive ablative interventions in cardiology and oncology, and in the use of microwave oven for heating and cooking food stuff.

The unique microwave auditory effect has been widely recognized as one of the most interesting and significant biological phenomena from microwave exposure. Its potential applications are just now beginning to be seriously explored. In contrast to heating using high average-power CW microwaves in the medical applications and industrial microwave heating, drying, and curing process described above, the report of microwave auditory effect was striking since it occurred for exposures to very low average-power levels of pulsed microwave radiation [Lin, 1978, 1980, 1990; Chou et al., 1982; Elder and Chou, 2003; Lin and Wang, 2007].

The microwave auditory effect is defined as the auditory perception of microwave radiation or simply hearing microwaves. This definition may seem surprising, and one might even question whether such sensation could exist. It is indeed a unique exception to the acoustic energy normally encountered by humans in auditory perception of sound. Although mammalian hearing apparatus responds to acoustic or sound pressure waves in the audio-frequency range of 0 to 20 kHz for humans, which could be up to 35 kHz or higher for some animals (cats and rats for example), frequencies above 35 kHz are considered as ultrasound for humans (beyond the range of audible frequencies). The hearing of microwave pulses involved electromagnetic waves whose frequency are much higher, ranging from 300 MHz to tens of gigahertz (GHz). Since electromagnetic waves (such as sun lights and optical radiation) are seen but not heard, the report of auditory sensation of microwave pulses was quite surprising, and initially its authenticity was widely questioned.

The earliest reports of the auditory perception of microwave pulses were provided anecdotally by radar operators during World War II [Airborne Instruments Laboratory, 1956]. They described an audible sound, a zip, click, or buzz that occurred at the repetition rate of radar when standing in front of radar antennas. These reports of the microwave auditory effect were initially documented in a technical report, in which several persons who had reported the sensation were interviewed and tested under field conditions [Frey, 1961]. It had been hypothesized that the microwave auditory effect involved direct electrical stimulation of the cochlear nerves or neurons at more central sites along the auditory nervous system pathway. However, skepticism of the microwave auditory effect and a responsible mechanism of interaction remained obscure for more than a decade.

Research conducted since has shown that a cascade of events take place when a beam of microwave pulses is aimed at a human or animal subject's head [Guy et al., 1975; Lin, 1978, 1980, 1990; Elder and Chou, 2003; Lin and Wang, 2007].

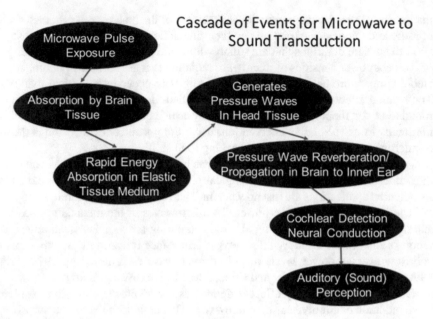

Fig. 1.2 The cascade of events illustrating the thermoelastic theory of auditory perception of pulsed microwaves

Absorption of pulsed microwave energy creates a rapid expansion of brain matter and launches an acoustic wave of pressure that travels inside the head to the inner ear. There, it activates the nerve cells in the cochlea, and the neural signals are then relayed through the central auditory system to the cerebral cortex for perception (Fig. 1.2). The center frequency of microwave pulse-induced acoustic pressure wave is about 8 kHz for human adults – well within the range of human auditory response.

However, if the average power and intensity of the impinging pulsed microwave is sufficiently high, the level of induced sound pressure could be considerably above the threshold of human auditory perception – to approaching or exceeding levels of discomfort and even causing potential brain tissue injury including structures in the cerebrum and cerebellum. It is significant to note that the microwave pulses may be targeted remotely, so that only the intended subject would perceive the microwave pulse-generated sound in the subject's head.

1.4   A Diplomatic Affair – the Havana Syndrome

It is interesting to note the US State Department disclosure that Havana-based US diplomats were experiencing health issues associated with hearing loud buzzing or what was described as bursts of sound in 2017 [Gearan, 2017; Harris and Goldman, 2017]. A similar announcement was made by the Government of Canada. Staff of both embassies had reported symptoms of hearing loss, ringing in the ears, headaches, nausea, and problems with balance or vertigo, which are suggestive of a connection to the inner ear apparatus within the human head.

Government officials had difficulty to pin down the source of sound. There were speculations that the diplomats may have been attacked with an advanced sound weapon. Assuming reported accounts are reliable, there may be a scientific explanation. It is plausible that the loud buzzing, burst of sound, or acoustic pressure waves could have been covertly delivered using high-power pulsed microwave radiation, rather than blasting the subjects with conventional sonic sources [Lin, 2017, 2018]. Many have come to believe that the microwave auditory effect – induced by a targeted beam of high peak-power pulsed microwave radiation – may be the most likely scientific explanation for the sonic attack [Best, 2017; Hambling, 2017; Hignett, 2017; Broad, 2018; Deng, 2018]. Of course, until the truth is revealed, this specific matter will remain somewhat of a mystery.

It is well known that robust audible sound could damage hearing and vestibular sensory systems and can alter human emotions and moods. However, it is not clear whether a weapon that covertly uses sonic energy to injure people exists today. Nonetheless, this event is calling attention to the reality that pulsed microwave radiation-induced sonic pressure waves in human heads could potentially be weaponized for health attacks. Furthermore, the required technology is mature and for the most part, commercially available or readily adaptable from existing microwave radar systems. However, existing hardware may need to be optimized to meet some specific requirements in specific operations including adaptive beam formation, steering, and focusing. Nevertheless, they can be used in nonlethal mode to transmit targeted messages, to cause task disruption and personnel disorientation, and even to apply the microwave auditory effect remotely as a lethal or nonlethal weapon. Additional discussions on this topic which is being referred to as the Havana Syndrome are included in Chap. 10.

1.5   Organizing Principles of the Book

The objective of this book is to bring together in a comprehensive book the multidisciplinary research investigation leading to a scientific understanding of the microwave auditory effect and related applications, especially the emerging microwave thermoacoustic tomography (MTT) imaging modality. The analysis and discussions in the chapters of this book pertain to relevant physical laws, exposure and dosimetry, anatomy and physiology, psychophysics and behavior, theories and models, mechanisms of interaction, computer analysis and simulation, applications in biology and medicine, and other applied aspects. Another purpose of this volume is to expand and update existing knowledge and understanding of microwave auditory effects and applications, which was the title of an earlier research monograph [Lin, 1978]. Considerable amount of scientific knowledge, data, and information have been generated since then through theoretical study, laboratory experimentation, numerical analysis, and computer modeling.

As this may be the reader's first encounter to the subject material, and since the treatment is multidisciplinary, after the Introduction, the four chapters that follow each begins with basic background information that may appear as elementary to some readers but is essential to understanding the discussions on microwave auditory effect and applications for those from a different discipline. There is a chapter on principles of microwave and RF exposure and one on brain anatomy and auditory physiology. The succeeding two chapters present the microwave property of biological materials and its influence on dosimetry and microwave absorption in biological tissues. They are intended to facilitate a fuller understanding of discussions in the ensuring five chapters about microwave auditory effect and applications.

An extensive list of references is provided in this book at the end of each chapter to furnish the knowledge base, to put the materials in proper perspective, and to overcome potential misunderstanding. The guiding principles throughout the book are that any description or conclusion must be consistent and compatible with physical laws, biological evidence, and experimental observations based on laboratory data and findings.

References

Airborne Instruments Laboratory (1956) An observation on the detection by the ear of microwave signals. Proceedings of the IRE 44:10 (Oct), p. 2A

Balanis C (ed) (2008) Modern antenna handbook. Wiley, Hoboken

Beckman KJ, Lin JC, Wang Y, Illes RW, Papp MA, Hariman RJ (1987) "Production of reversible and irreversible atrio-ventricular block by microwave energy," The 60th Scientific Sessions, American Heart Association, Anaheim, CA November 1987; also in Circulation 76: 1612

Best S Brain abnormalities in US and Canadian diplomats living in Havana were caused by a mystery microwave weapon and not 'noisy crickets', claims scientist. https://www.dailymail.co.uk/sciencetech/article-5174763/Mysterious-sonic-weapon-Cuba-microwaves.html. 2017/12/17

Broad W (2018) Microwave weapons are prime suspect in Ills of U.S. embassy workers https://www.nytimes.com/2018/09/01/science/sonic-attack-cuba-microwave.html

Burfeindt MJ, Zastrow E, Hagness SC, Van Veen BD, Medow JE (2011) Microwave beamforming for non-invasive patient-specific hyperthermia treatment of pediatric brain cancer. Phys Med Biol 56(9):2743–2754

Calkins H, Hindricks G, Cappato R et al (2018) 2017 HRS/EHRA/ECAS/APHRS/SOLAECE expert consensus statement on catheter and surgical ablation of atrial fibrillation. Europace 20(1):e1–e160

Cheng Y, Weng S, Yu L, Zhu N, Yang M, Yuan Y (2019) The Role of Hyperthermia in the Multidisciplinary Treatment of Malignant Tumors. Integr Cancer Ther 18:1534735419876345

Chou CK, Guy AW, Galambos R (1982) Auditory perception of radio-frequency electromagnetic fields. J Acoust Soc Am 71(6):1321–1334

Cronin EM, Bogun FM, Maury P et al (2019) HRS/EHRA/APHRS/LAHRS expert consensus statement on catheter ablation of ventricular arrhythmias [published online ahead of print, 2020 Jan 27]. J Interv Card Electrophysiol 2020:1–154. https://doi.org/10.1007/s10840-019-00663-3

Daily LE (1943) A clinical study of the results of exposure of laboratory personnel to radar and high frequency radio. US Navy Med Bull 41:1052–1056

Datta NR, Kok HP, Crezee H, Gaipl US, Bodis S (2020) Integrating loco-regional hyperthermia into the current oncology practice: SWOT and TOWS analyses. Front Oncol 10:819. Published Jun 12. https://doi.org/10.3389/fonc.2020.00819

Deng B (2018) Mystery noises and illnesses evoke terrors of cold war. https://www.thetimes.co.uk/article/mystery-noises-and-illnesses-evoke-terrors-of-cold-war-htvj63t58. 2018/6/15

Dewhirst MW, Kirsch D (2018) Technological advances, biologic rationales, and the associated success of chemotherapy with hyperthermia in improved outcomes in patients with sarcoma. JAMA Oncol 4:493–494

Dooley WC, Vargas HI, Fenn AJ, Tomaselli MB, Harness JK (2010) Focused microwave thermotherapy for preoperative treatment of invasive breast cancer: a review of clinical studies. Ann Surg Oncol 17(4):1076–1093

Elder J, Chou C (2003) Auditory response to pulsed radiofrequency energy. Bioelectromagnetics 24:S162–S173

Frey AH (1961) Human auditory system response to RF energy. Aerospace Med 32:1140–1142

Gearan A (2017) State Department reports new instance of American diplomats harmed in Cuba. The Washington Post. [Online]. Available: www.washingtonpost.com/world/national-security/american-diplomats-sufferedtraumatic-brain-injuries-in-mystery-attack-incuba-union-says/2017/09/01/9e02d280-8f2f-11e7-91d5-ab4e4bb76a3a_story.html?utm_term=.1f1d5aee26f4

Grandin K (ed) (2007) Les Prix Nobel. The Nobel Prizes 2006. Nobel Foundation, Stockholm

Greshko M (2019) China just landed on the far side of the moon: What comes next? https://www.nationalgeographic.com/science/2019/01/china-change-4-historic-landing-moon-far-side-explained/

Guy AW, Chou CK, Lin JC, Christensen D (1975) Microwave induced acoustic effects in mammalian auditory systems and physical materials. Ann NY Acad Sci 247:194–218

Hambling D (2017) Weaponised microwave may be behind alleged sonic attacks in Cuba: https://www.newscientist.com/article/2156164-weaponised-microwave-may-be-behind-alleged-sonic-attacks-in-cuba/#ixzz5v1xYNXup. 2017/12/12

Hand JW (1989) Technical and clinical advances in hyperthermia treatment of cancer. In: Lin JC (ed) Electromagnetic interaction with biological systems. Springer, Boston, pp 59–80

Harris G, Goldman A (2017) Illnesses at U.S. embassy in Havana prompt evacuation of more diplomats. [Online]. Available: www.nytimes.com/2017/09/29/us/politics/usembassy-cuba-attacks.html?mcubz=0

Hignett K Mass hysteria or microwave weapons—what's behind the 'sonic attacks' https://www.newsweek.com/mass-hysteria-microwave-weapons-sonic-attacks-cuba-747107, 2017/12/16

Hildebrandt B, Hegewisch-Becker S, Kerner T et al (2015) Current status of radiant whole-body hyperthermia at temperatures >41.5 degrees C and practical guidelines for the treatment of adults. The German 'Interdisciplinary Working Group on Hyperthermia'. Int J Hyperth 21(2):169–183

Huang SKS, Wilber DJ (eds) (2000) Radiofrequency catheter ablation of cardiac arrhythmias: basic concepts and clinical applications, 2nd edn. Futura, Armonk/New York

Issels RD (2008) Hyperthermia adds to chemotherapy. Eur J Cancer 44:2546–2554

Kottke FJ, Lehmann JF (1990) Handbook of physical medicine and rehabilitation, 4th edn. Saunders

Krusen FH, Herrick JF, Leden U, Wakim KG (1947) Microkymato therapy-preliminary report of experimental study of the heating effect of microwaves (radar) in living tissue. Proc. Staff Meeting, Mayo Clinic 22:201–224

Lassche G, Frenzel T, Mignot MH et al (2020) Thermal distribution, physiological effects and toxicities of extracorporeally induced whole-body hyperthermia in a pig model. Physiol Rep 8(4):e14366. https://doi.org/10.14814/phy2.14366

Licht S (ed) (1965) Therapeutic heat and cold, New Haven

Lidman BI, Cohn C (1945) Effects of radar emanations on the hematopoietic system. Air Surg Bull 2:448–449

Lin JC (1978) Microwave auditory effects and applications. CC Thomas, Springfield

Lin JC (1980) The microwave auditory phenomenon. Proc IEEE 68:67–73

Lin JC (1990) Chapter 12: Auditory perception of pulsed microwave radiation. In: Gandhi OP (ed) Biological effects and medical applications of electromagnetic fields. Prentice-Hall, New York, pp 277–318

Lin JC (1999) Catheter microwave ablation therapy for cardiac arrhythmias. Bioelectromagnetics 20(Supplement #4):120–132

Lin JC (2004) Studies on microwaves in medicine and biology: from snails to humans. Bioelectromagnetics 25:146–159

Lin JC (2017) Mystery of sonic health attacks on havana-based diplomats. URSI Radio Science Bulletin No 362:102–103

Lin JC (2018) Strange reports of weaponized sound in cuba. IEEE Microwave Magazine 19/1:18–19

Lin JC, Wang YJ (1987) Interstitial microwave antennas for thermal therapy. Int J Hyperth 3:37–47

Lin JC, Wang ZW (2007) Hearing of microwave pulses by humans and animals: effects, mechanism, and thresholds. Health Phys 92(6):621–628

Lin JC, Hariman RJ, Wang YG, Wang YJ (1996) Microwave catheter ablation of the atrioventricular junction in closed-chest dogs. Med Biol Eng Comput 34:295–298

Lin JC, Hirai S, Chiang CL, Hsu WL, Su JL, Wang YJ (2000) Computer simulation and experimental studies of SAR distributions of interstitial arrays of sleeved-slot microwave antennas for hyperthermia treatment of brain tumors. IEEE Trans Microwave Theory Tech 48:2191–2197

Lopresto V, Pinto R, Farina L, Cavagnaro M (2017) Treatment planning in microwave thermal ablation: clinical gaps and recent research advances. Int J Hyperth 33(1):83–100

Lundqvist S (ed) (1992) Nobel lectures, physics 1971–1980. World Scientific Publishing, Singapore

Nguyen PT, Abbosh AM, Crozier S (2017) 3-D focused microwave hyperthermia for breast cancer treatment with experimental validation. IEEE Trans Antennas Propag 65(7):3489–3500

Osepchuk JM (1984) A history of microwave heating applications. IEEE Trans Microwave Theory Tech 32:1200–1224

Osepchuk JM (2002) Microwave power applications. IEEE Trans Microwave Theory Tech 50(3):975–985

Rae JW, Martin GM, Treanor WJ, Krusen FH (1950) Clinical experience with microwave diathermy, Proceedings of Staff Meeting, Mayo Clinic

van der Zee J, Gonzalez Gonzalez D, van Rhoon GC, van Dijk JD, van Putten WL, Hart AA (2000) Comparison of radiotherapy alone with radiotherapy plus hyperthermia in locally advanced pelvic tumours: a prospective, randomised, multicentre trial. Dutch Deep Hyperthermia Group Lancet 355:1119–1125

Wells SA, Hinshaw JL, Lubner MG, Ziemlewicz TJ, Brace CL, Lee FT Jr (2015) Liver Ablation: Best Practice. Radiol Clin N Am 53(5):933–971

Yamamoto C, Yamamoto D, Tsubota Y, Sueoka N, Kawakami K, Yamamoto M (2014) The synergistic effect of local microwave hyperthermia and chemotherapy for advanced or recurrent breast cancer. Cancer Chemo 41(12):1921–1923

Chapter 2
Principles of Microwave and RF Exposure

The physical fundamentals of microwave and RF radiation are presented in this chapter for an understanding of the essential elements of exposure of biological systems to RF and microwave radiation. These exposures are functions of the source frequency and configuration, shape and size of the exposed subjects, and orientation and location of the subject with respect to the source, among others.

The approach of this chapter is to develop but confine the coverage to the most relevant topics, rather than a comprehensive discussion of bioelectromagnetics describable by physical laws of electromagnetic theory. The reader who is interested in a broader physical and engineering account of microwave and RF radiation is referred to some readily available texts devoted entirely to the subject of electromagnetics and microwave engineering.

An understanding of the interaction of microwave and RF radiation with biological systems is facilitated through knowledge of the physical laws describing the behavior and characteristics of microwave and RF radiation in space and time. The physical principles of microwave and RF radiation are prescribed by physical laws referred to as the Maxwell's equations of electromagnetics theory. These equations represent mathematical expressions of experimentally validated observations. They are applicable for linear or nonlinear, isotropic or anisotropic, and homogeneous or heterogeneous medium. Maxwell's equations are macroscopic laws that define the relationship between space- and time-averaged electric and magnetic fields. They apply to regions or volumes whose dimensions are larger than atomic dimensions. Time intervals of observation are assumed to be long enough to allow for an averaging of atomic fluctuations.

From these four laws of electromagnetics, one may deduce all macroscopic electromagnetic characteristics and behavior including microwave and RF propagation from source to target or object, exposure of biological subjects, reflection and transmission at tissue interfaces, and dosimetry – the quantification of microwave and RF radiation's distribution and absorption in biological bodies and materials. This knowledge is important for understanding and interpretation of the effects in

© Springer Nature Switzerland AG 2021
J. C. Lin, *Auditory Effects of Microwave Radiation*,
https://doi.org/10.1007/978-3-030-64544-1_2

biological systems exposed to microwave and RF radiation as well as to applying them for health and safety assessment and for biomedical applications.

Microwave and RF radiation consists of oscillating electric and magnetic fields propagating through free space at the speed of light, 2.998×10^8 m/s. This speed varies depending on the material medium through which the wave travels. It is a function of the permittivity and permeability of the material media. In addition to being radiated through a transmitting antenna, microwave and RF radiation also may be conducted from the source by coaxial transmission lines or waveguides. Microwave and RF radiation may be detected and measured by diodes or similar devices and their associated instruments and systems.

2.1 The Maxwell Equations

Maxwell's four mathematical equations may be stated in either integral or differential equation form. Each formulation provides a unique description of electromagnetic radiation in space. In integral forms, they specify them along lines, through surfaces and over volumes, and lend the equations to easy physical interpretation. In contrast, the differential equations depict their behavior at points in space. These formulations are mathematically equivalent since they may be derived from each other using Stokes' and divergence theorems from vector analysis.

The integral forms of Maxwell's equations are given by

$$\oint E \cdot dl = -\int_s \frac{\partial B}{\partial t} \cdot ds \quad \left(\text{Faraday's law}\right) \tag{2.1}$$

$$\oint H \cdot dl = \int_s \left(J + \frac{\partial D}{\partial t} \right) \cdot ds \left(\text{Ampere-Maxwell's law}\right) \tag{2.2}$$

$$\oint D \cdot ds = \int_v \rho dv \left(\text{Gauss electric law}\right) \tag{2.3}$$

$$\oint B \cdot ds = 0 \left(\text{Gauss magnetic law}\right) \tag{2.4}$$

where
  E = Electric field strength in volt/meter (V/m)
  H = Magnetic field strength in ampere/meter (A/m)
  D = Electric flux density in coulomb/square meter (C/m^2)
  B = Magnetic flux density in weber/square meter (Wb/m^2)
  J = Conduction current density in ampere/square meter (A/m^2)
  ρ = Charge density in coulomb/cubic meter (C/m^3)

According to Faraday's law (2.1), the total voltage induced in an arbitrary closed path is equal to the total time rate of decrease of magnetic flux through the area bounded by the closed path. Therefore, a time-varying magnetic field generates an electric field. There is no restriction on the nature of the medium.

Ampere's law (2.2) states that the line integral or sum of the magnetic field strength around a closed path or loop is equal to the total current enclosed by the path or loop. The total current may consist of two parts: a conduction current with density J and a displacement or dielectric current with density $\partial D/\partial t$. Thus, Ampere's law implies that a magnetic field can only be produced by currents or movement of charges. The displacement current was first introduced by Maxwell and allowed him to unify the separate laws governing electricity and magnetism into a unified electromagnetic theory. It also led to the postulate of electromagnetic waves that can transport energy and the concept that light is an electromagnetic wave.

Equation (2.3) is Gauss's electric law, which states that the net outward flow of electric flux through a closed surface is equal to the charge contained in the volume enclosed by the surface. Gauss's law for the magnetic field (2.4) states that the net outward flow of magnetic flux through a closed surface is zero. Therefore, magnetic flux lines are always continuous and form closed loops.

Although Eqs. (2.1) through (2.4) lend to ready physical interpretation of electromagnetic phenomena associated with macroscopic surface and volumes, they are not in forms most suitable for the mathematical analysis of physical events taking effect at or surrounding a point in space. It is often expedient to solve differential equations to obtain field quantities. The desired differential equations may be derived by using Stokes's and divergence theorems from vector analysis. As a result, Maxwell's equations in differential form are

$$\nabla \times E = -\frac{\partial B}{\partial t} \ \left(\text{Faraday's law}\right) \tag{2.5}$$

$$\nabla \times H = J + \frac{\partial D}{\partial t} \ \left(\text{Ampere-Maxwell's law}\right) \tag{2.6}$$

$$\nabla \cdot D = \rho \ \left(\text{Gauss electric law}\right) \tag{2.7}$$

$$\nabla \cdot B = 0 \ \left(\text{Gauss magnetic law}\right) \tag{2.8}$$

Equation (2.5) says that the curl or circulation of electric field around a point in space equals to the time rate of decrease of magnetic flux density at that point. Again, a time-varying magnetic field generates an electric field around it.

Equation (2.6) declares that the curl or circulation of magnetic field around a point in space equals to the total current passing through that point. The total current may consist of two parts: a conduction current with density J and a displacement current with density $\partial D/\partial t$. Again, Ampere's law holds and a magnetic field is produced by currents or movement of charges.

The Gauss electric law of Eq. (2.7) states that the divergence or outward flow of electric flux from a point in space is equal to the charges located at that point, thus indicating that electric fields are associated with electric charges.

The companion Gauss law for the magnetic field (2.8) states that the net outward flow of magnetic flux at any point is zero. Therefore, magnetic flux lines converging at a point is always equal to the number of fluxes leading away from the same point in space. This is consistent with the statement of Eq. (2.4) that magnetic flux lines are always continuous and formed closed loops. It suggests that there is no net accumulation of magnetic charges at any point in space and allows the well-known conclusion that, unlike electric charges, magnetic charges do not exist.

It is seen from the right-hand sides of Eqs. (2.6) and (2.7) that the sources of electromagnetic fields and waves are electrical charges and current flow. In fact, the conservation of charge principle gives rise to an equation of continuity for current flow in differential form

$$\nabla \cdot \boldsymbol{J} = -\frac{\partial \rho}{\partial t} \tag{2.9}$$

which indicates that the outward flow of electric current is equal to the time rate of decrease of electric charge density at that point.

When the electromagnetic fields are harmonically oscillating functions with a single frequency, f, all field quantities may be assumed to have a time variation represented by $e^{j\omega t}$, where $\omega = 2\pi f$ in radians is the angular frequency. Under this assumption, the $j\omega t$ derivatives in Maxwell's equations may be replaced by a multiplier $j\omega$, and the common factor $e^{j\omega t}$ may be omitted from these equations. Maxwell's equations can be reduced and written as

$$\nabla \times \boldsymbol{E} = -j\omega \boldsymbol{B} \tag{2.10}$$

$$\nabla \times \boldsymbol{H} = \boldsymbol{J} + j\omega \boldsymbol{D} \tag{2.11}$$

$$\nabla \cdot \boldsymbol{D} = \rho \tag{2.12}$$

$$\nabla \cdot \boldsymbol{B} = 0 \tag{2.13}$$

Note the use of boldface letters for the vectors that are complex functions of space coordinates.

A set of auxiliary equations relating the fields and flux densities are required to determine the electric and magnetic fields produced by a given current or charge distribution. For a linear, isotropic, and homogeneous medium, the field quantities are related as follows:

$$\boldsymbol{D} = \varepsilon \boldsymbol{E} \tag{2.14}$$

$$\boldsymbol{B} = \mu \boldsymbol{H} \tag{2.15}$$

$$\boldsymbol{J} = \sigma \boldsymbol{E} \tag{2.16}$$

The free space or vacuum is such a medium in which the permittivity is given by

$$\varepsilon = \varepsilon_0 = 8.854 \times 10^{-12} \, farad \, / \, m \, (F \, / \, m) \tag{2.17}$$

permeability is given by

$$\mu = \mu_0 = 4\pi \times 10^{-7} \, henry \, / \, m \, (H \, / \, m) \tag{2.18}$$

and conductivity, $\sigma = 0$. For other media, it is conventional to introduce the dimensionless ratios,

$$\varepsilon_r = \frac{\varepsilon}{\varepsilon_0} \tag{2.19}$$

$$\mu_r = \frac{\mu}{\mu_0} \tag{2.20}$$

which are usually labeled the relative dielectric constant and relative permeability, respectively. Biological materials generally have relative permeabilities close to that of free space and relative dielectric constants that show characteristic dependence on material and frequency.

2.2   The Wave Equation

Equations (2.10) and (2.11) are a set of coupled equations both containing electric and magnetic field quantities. If we assume that the conduction current density is zero in the region of interest, these two equations may be combined to give two second-order differential equations, one containing the electric field strength and the other containing the magnetic field strength:

$$\nabla^2 E + \omega^2 \mu\varepsilon E = 0 \tag{2.21}$$

$$\nabla^2 H + \omega^2 \mu\varepsilon H = 0 \tag{2.22}$$

where we also have taken the charge density as zero for a source-free region. It is customary to let $k^2 = \omega^2\mu\varepsilon$, where k is the wave number and has the following relations:

$$k = \omega\sqrt{\mu\varepsilon} = \frac{\omega}{v} = \frac{2\pi}{\lambda} \tag{2.23}$$

Equations (2.21) and (2.22) are wave equations in harmonic form. Their solutions represent electromagnetic waves propagating with velocity v equal to $(\mu\varepsilon)^{-1/2}$. The wavelength λ is equal to $\dfrac{v}{f}$, where f is the frequency. In free space, v is equal to the speed of light, $c = 2.998 \times 10^8$ m/s, as mentioned previously.

In a medium with finite electrical conductivity σ, a conduction current density $\boldsymbol{J} = \sigma\boldsymbol{E}$ will exist, and this will give rise to energy loss to Joule heating. The wave equations in media of this type have a loss term given by $j\omega\mu\sigma$, such that

$$\left(\nabla^2 + \omega^2\mu\varepsilon - j\omega\mu\sigma\right)\boldsymbol{E} = 0 \tag{2.24}$$

$$\left(\nabla^2 + \omega^2\mu\varepsilon - j\omega\mu\sigma\right)\boldsymbol{H} = 0 \tag{2.25}$$

It should be mentioned that a finite conductivity is equivalent to an imaginary component in the permittivity ε. Comparing Eqs. (2.21 and 2.22) with (2.24 and 2.25), it is seen that the equivalent permittivity is given by $\left(\varepsilon - \dfrac{j\sigma}{\omega}\right)$.

2.3 Boundary Condition at Material Interfaces

The behavior of electric and magnetic fields in situations where the physical or biological properties of the medium change abruptly across one or several interfaces is governed by certain boundary conditions to be satisfied at the interfaces. These conditions must be satisfied by any solution to the electric and magnetic fields and the wave equation. They may be derived by applying Maxwell's Eqs. (2.1, 2.2, 2.3 and 2.4) to infinitesimal regions containing these interfaces (Fig. 2.1).

If medium 1 and medium 2 are separated by a common boundary, the boundary conditions at the interface may be summarized as follows:

1. The tangential components of the electric field strengths are continuous across the boundary, such that

$$E_{t1} = E_{t2} \tag{2.26}$$

2. The normal components of the electric flux densities differ by an amount equal to the surface charge density, such that

$$D_{n1} - D_{n2} = \rho_s \tag{2.27}$$

If there is no surface charge on the boundary, $\rho_s = 0$, which is the usual case of dielectric materials, then

$$D_{n1} = D_{n2} \tag{2.28}$$

Fig. 2.1 Boundary conditions and schematics for evaluating boundary conditions at material interface between two different media for tangential and normal electric and magnetic field components. (**a**) Tangential components of electric field; (**b**) Normal components of electric field; (**c**) Tangential components of magnetic field; (**d**) Normal component of magnetic field

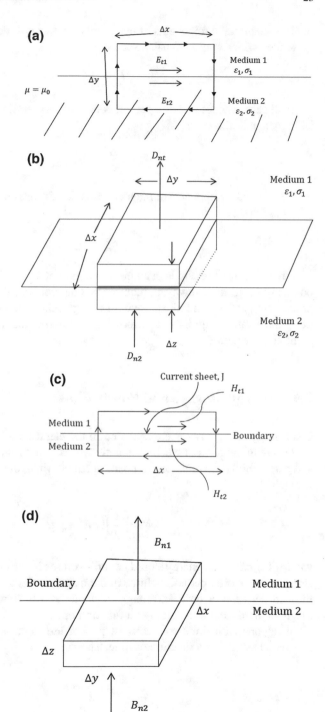

3. The tangential components of magnetic field strength differ by an amount equal to the surface current density, such that

$$H_{t1} - H_{t2} = J_s \tag{2.29}$$

In all cases except that of a perfect conductor, the surface current density is vanishingly small; then

$$H_{t1} = H_{t2} \tag{2.30}$$

4. The normal components of magnetic flux density are continuous across a boundary, such that

$$B_{n1} = B_{n2} \tag{2.31}$$

When dealing with fields in an infinite region of space, the radiation condition, a boundary condition equivalent, requires that the field at infinity must be an outward propagating wave with finite amount of energy. Alternatively, electric and magnetic fields must vanish rapidly so that the energy stored in the fields and the energy flow at infinity are zero.

2.4   Energy Storage and Power Flow

Electromagnetic energy is either stored in the electric and magnetic fields or radiated away in the form of electromagnetic waves. For a region in which permittivity and permeability are functions of position but not time, the energy density at a point is given by

$$W = \left(\frac{1}{2}\right)\left(\varepsilon E^2 + \mu H^2\right) \tag{2.32}$$

where $(1/2)\varepsilon E^2$ is the energy stored per unit volume in the electric field and $(1/2)\mu H^2$ is the energy stored per unit volume in the magnetic field. The stored energy is partitioned between electric and magnetic fields and is transferred from electric to magnetic fields and back again as a function of time.

In regions with finite conductivity, part of the electromagnetic energy may be dissipated as heat as a function of time. The term

$$P_a = \left(\frac{1}{2}\right)\sigma E^2 \tag{2.33}$$

represents the absorbed power density or dissipated power; the absorbed power may be converted to heat energy over time. The absorbed power density also equals the specific rate of energy absorption or specific absorption rate (SAR), which is generally accepted as the metric for quantifying microwave and RF energy deposition in biological systems. In this case, $SAR = P_a$, in units of watts per kilogram or W/kg. Note that SAR is independent of time.

The determination of SAR and its distribution in biological tissues is commonly referred to as dosimetry.

The instantaneous density of power that flows across a surface bounding a given region is given by Poynting vector $\mathbf{P}$,

$$P = E \times H \tag{2.34}$$

Thus, the direction of power flow is perpendicular to $\mathbf{E}$ and $\mathbf{H}$ and is in the direction of Poynting vector $\mathbf{P}$. It can be shown by an application of the Poynting theorem that the total power flowing into a region is equal to the total power dissipated within the region plus the time rate of increase of energy stored within the region. The power density that impinges on a surface area normal to the direction of propagation is proportional to the square of electric or magnetic field and is expressed in watts per square meter (W/m^2) or milliwatt per square centimeter (mW/cm^2). Note that 1 mW/cm^2 = 10 W/m^2.

For electric and magnetic fields that vary sinusoidally with time, the time-average power flow or rate of energy flow per unit area is

$$P_d = \left(\frac{1}{2}\right) \mathrm{Re}\left(E \times H^*\right) \tag{2.35}$$

where H^* is the conjugate of $\mathbf{H}$ and P_d is given by the real (Re) part of the product of two sinusoidal quantities.

2.5   Plane Waves and Far-Zone Field

The radiated energy of a small antenna (a point source) in free space takes the form of a spherical wave in which the wave fronts are concentric spherical shells. The spherical wave fronts expand as the wave propagates outward from the source. At points far from the source, the wave front would essentially appear as a plane. This is analogous to the situation of an observer on Earth who sees the Earth's surface as a plane (flat earth), since the person can view only a small portion of the global surface of the Earth. Both electric and magnetic fields of the wave lie in the plane of the wave front. They vary only in the direction of propagation. Such a wave is called a plane wave. It is a very important practical case since fields radiated by any transmitting antenna appear as a plane wave at distances far from the source. Moreover, through Fourier analysis a suitable combination of plane waves may be made to represent a wave of any desired form in space and time.

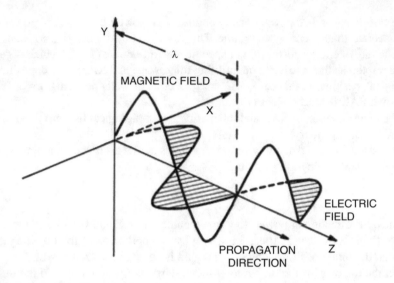

Fig. 2.2 The direction of plane wave propagation, electric field and magnetic field orientation in space at a given moment in time

At distances far from the source or transmitting antenna (the far zone), typically 10 wavelengths or more, RF and microwaves may be considered as plane waves whose electric and magnetic fields are perpendicular from each other and both are perpendicular to the direction of wave propagation. Moreover, the electric and magnetic field maximum occurs at the same location in space at any given moment in time, as depicted in Fig. 2.2. In this case, the electric field strength in volts per meter (V/m) is related to the magnetic field strength in amperes per meter (A/m) through a constant known as intrinsic impedance, which is medium dependent and is approximate 376.7 ohms for air, vacuum, or free space. In all other dielectric media including biological materials, the intrinsic impedance is always smaller than that of free space.

For distances less than 10 wavelength from the transmitting antenna (the near zone), the maxima and minima of electric and magnetic fields do not occur at the same location along the direction of propagation. That is, the electric and magnetic fields are out of time phase. The ratio of electric and magnetic field strengths is no longer constant; it varies from point to point. The direction of propagation is also not uniquely defined as in the far zone case, making the situation complicated (more details are given below). It should be noted that various field regions generally do not affect the basic mechanisms by which RF and microwaves act on biological system, although the quantitative aspects of the interaction may differ due to changes in energy coupling. In general, when plane-wave microwave and RF radiation impinge from air on a planar biological structure, a large fraction of the incident energy may be reflected. The transmitted fraction is attenuated exponentially as it propagates in the tissue.

2.6   Polarization and Propagation of Plane Waves

The polarization of a plane wave refers to the time-varying characters of the electric field at a given location in space. To specify the orientation of microwaves and RF radiation in space, it is necessary to specify the orientation of one of the field vectors. Since the magnetic field vector is always perpendicular to the electric field vector, a knowledge of the orientation of the latter is sufficient to describe the orientation of the wave. If the electric field of the waves always lies in a specific location, the wave is said to be linearly polarized. An example would be a linearly polarized plane wave in the x direction of a rectangular coordinate system. If both E_x and E_y are present and are in time phase, the resultant electric field has a direction dependence on the relative magnitude of E_x and E_y. The angle that this direction makes with the x axis, however, is constant with time. The wave is therefore also linearly polarized (Fig. 2.3a).

If E_x and E_y are equal in magnitude and E_y leads E_x

$$E = E_x \mathbf{x} + jE_y \mathbf{y} \tag{2.36}$$

E_x and E_y reach their maximum values at different instants of time; the direction of the resultant electric field will vary with time. It can be shown that the locus of

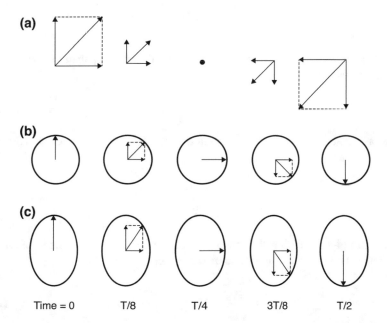

Fig. 2.3 Polarization of electromagnetic planes: a. linear polarization; b. circular polarization; c. elliptical polarization. Arrows indicate directions of electric field and its coordinate components

the endpoint of the resultant electric field will be a circle. The wave is said to be circularly polarized. Moreover, it may be seen that looking in the direction of propagation, the rotation of the electric field vector is that of a left-handed screw advancing in the direction of propagation. The wave is therefore also said to be left circularly polarized (Fig. 2.3b). If the x component of the electric field leads the y component instead, a reversal of the direction of rotation is obtained.

The most general form of polarization is elliptical polarization. This happens when E_x and E_y differ in magnitude as well as time phase. Assuming again E_y leads E_x by 90°, since E_x and E_y are not equal in magnitude, the endpoint of the resultant field traces out an ellipse in the plane normal to the direction of propagation (see Fig. 2.3c). The circular polarization is in fact a special case of elliptical polarization.

A wave propagating in any direction has no field variations in planes normal to the direction of propagation and is a linearly polarized plane wave. In this section, we shall describe the properties of plane wave propagation in free space, lossy (biological) media, and media involving plane boundaries.

2.6.1 Plane Waves in Free Space

Assume that the electric field is polarized in the x direction and the wave propagating in the z direction, the wave Eq. (2.24) reduces to

$$\frac{d^2 E_x}{dz^2} + k_0^2 E_x = 0 \tag{2.37}$$

where $k_0 = \omega(\mu_0 \varepsilon_0)^{1/2}$ is the free space wave number or propagation constant. The solution to this equation is

$$E_x = E_0 e^{-jk_0 z} \tag{2.38}$$

The associated magnetic field is related to the electric field through Maxwell's Eq. (2.10), thus

$$H_y = \sqrt{\frac{\varepsilon_0}{\mu_0}} E_x \tag{2.39}$$

The ratio of E_x to H_y has the dimension of impedance and is called the intrinsic impedance. The intrinsic impedance of free space is

$$\eta_0 = \sqrt{\frac{\mu_0}{\varepsilon_0}} = 120\pi \simeq 377 \, ohms \tag{2.40}$$

Figure 2.2 shows the electric and magnetic fields as a function of distance at some instant of time. The wavelength λ or λ_0 in free space is defined as the distance over which the sinusoidal waveform passes through a full cycle of 2π radians. Note that the electric and magnetic fields are in time phase but in space quadrature.

The time-average power flow or average power density associated with this wave is

$$P_d = \left(\frac{1}{2}\right)\left(\frac{E_0^2}{\eta_0}\right) \tag{2.41}$$

The direction of power flow is normal to both electric and magnetic field vectors and is in the direction of wave propagation.

2.6.2 Plane Waves in Lossy or Biological Media

In the case of a plane wave propagating through a homogeneous isotropic medium with dielectric loss or finite conductivity such as biological materials, the governing Eq. (2.24) becomes

$$\frac{d^2 E_x}{dz^2} - \gamma^2 E_x = 0 \tag{2.42}$$

where γ is the propagation factor given by

$$\gamma = \sqrt{j\omega\mu(\sigma + j\omega\varepsilon)} \tag{2.43}$$

The solution of Eq. (2.42) for a wave propagating along the positive z direction is

$$E_x = Ee^{-\gamma z} \tag{2.44}$$

The corresponding solution for the magnetic field is

$$H_y = \left(\frac{\gamma}{j\omega\mu}\right)Ee^{-\gamma z} \tag{2.45}$$

The intrinsic impedance of a medium was previously defined as the ratio of electric field to magnetic field. Thus,

$$\eta = \frac{E_x}{H_y} = \frac{j\omega\mu}{\gamma} = \sqrt{\frac{j\omega\mu}{\sigma + j\omega\varepsilon}} \tag{2.46}$$

For a lossless dielectric medium, $\sigma = 0$, the intrinsic impedance reduces to

$$\eta = \sqrt{\frac{\mu}{\varepsilon}} \tag{2.47}$$

As shown earlier, η has a value of 377 ohms for free space. Since dielectric media have approximately the same permeability of free space and have permittivity greater than free space, it follows that the intrinsic impedance of free space is the upper limit of attainable value for dielectric materials.

The real and imaginary parts of complex propagation factor γ may be represented by α and β such that

$$\gamma = \alpha + j\beta \tag{2.48}$$

and

$$\alpha = \omega\sqrt{\frac{\mu\varepsilon}{2}\left(\sqrt{1+\frac{\sigma^2}{\omega^2\varepsilon^2}}-1\right)} \tag{2.49}$$

$$\beta = \omega\sqrt{\frac{\mu\varepsilon}{2}\left(\sqrt{1+\frac{\sigma^2}{\omega^2\varepsilon^2}}+1\right)} \tag{2.50}$$

where α and β are referred to as attenuation coefficient and propagation coefficient, respectively. It is clear from Eqs. (2.44), (2.45), and (2.48) that the amplitude of the wave decreases as it advances in the lossy medium and the reduction is exponential in nature. Furthermore, since α is related to ω and σ, the rate of attenuation is proportional to frequency and conductivity. The factor, $\dfrac{\sigma}{\omega\varepsilon}$, is known as the loss tangent. It is the ratio of the magnitude of conduction current density to the magnitude of displacement current density in a material medium. It is a useful way to classify the transitional behavior of materials at different frequencies by the loss tangent being unity; the medium acts as a dielectric material for smaller loss tangents, and conversely, it behaves more like a conducting material. When $\sigma \gg \omega\varepsilon$, such as for a good conductor, α and β may be simplified to yield

$$\alpha = \beta = \sqrt{\frac{\omega\mu\sigma}{2}} \tag{2.51}$$

The time average power density associated with a plane wave propagating through a lossy media is

$$P_d = \left(\frac{1}{2}\right)\mathrm{Re}\left[\frac{E^2 e^{-2\alpha z}}{\eta}\right] \tag{2.52}$$

It is easily seen from Eq. (2.52) that the average power density decreases according to $e^{-2\alpha z}$ as the wave propagates in the lossy medium. This is as expected since the field decreases exponentially ($e^{-\alpha z}$) as it travels in the medium. At $z = \delta$, both electric and magnetic field strengths decrease to $1/e$, 36.8% of their value at the surface or the point of entry into the lossy medium (the term conducting medium is also used to mean lossy medium).

The quantity δ is known as the depth of penetration or skin depth and is given by

$$\delta = \frac{1}{\alpha} \tag{2.53}$$

Therefore, the depth of penetration is inversely proportional to the conductivity and the frequency. It should be mentioned that the fields do not fail to penetrate beyond the depth of δ; this is merely the point at which they have decreased to 37% of their initial value. The power density will decrease to about 14% accordingly.

The concept as presented here applies strictly to planar media or planar-like structures. It may be extended, however, to bodies of other geometry so long as the depth of penetration is much smaller than the radius of curvature of the body surfaces.

2.7   Reflection and Transmission at Interfaces

When microwave and RF radiation propagating in one medium impinges on a second medium having different dielectric properties, partial reflection occurs at the boundary between the two media. A portion of the incident radiation may also be transmitted into the second medium. If intrinsic impedances of the two media are approximately equal, most of the energy is transmitted into the second medium, and the reflected radiation is relatively small. Conversely, if intrinsic impedances differ greatly, the transmitted radiation is small, and the reflected radiation is relatively large. If the wave impinges normally on the boundary surface, the resulting reflected wave propagating back toward the source combines with the incident wave to form a standing wave in the first medium. An example of the standing wave created by a perfect conductor is shown in Fig. 2.4. The term standing wave ratio (SWR) is defined as the ratio of maximum to minimum electric field strength in a standing wave. It is used as a measure of the degree of impedance mismatch or of differences between the electromagnetic properties of the two media.

The essential features of the behavior of RF and microwaves at the surface between two media may be deduced from an analysis of the simple equation of a plane wave incident normally upon a plane surface between two media (see Fig. 2.5). Both media are assumed to be infinite in extent in all directions except at the boundary ($z = 0$). At any plane z we may define a wave impedance, $Z(z)$ as the ratio of total electric to total magnetic fields at that plane:

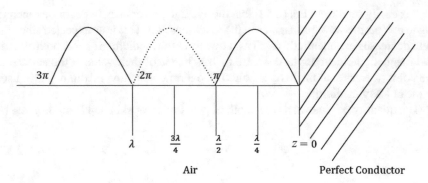

Fig. 2.4 A standing wave created when a plane wave impinges normally on a perfect conductor. The fields are shown at time, t = T/8, where T is the period of the wave

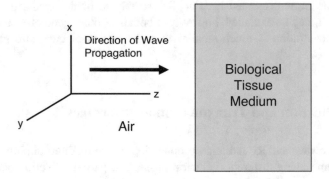

Fig. 2.5 Plane wave impinging on two layered media: medium 1 or air and medium 2 is a tissue layer

$$Z(z) = \frac{E_x(z)}{H_y(z)} \quad (2.54)$$

For the wave propagating along the positive z direction in medium 2 or a tissue layer, $Z = \eta_2$ is the intrinsic impedance. Similarly, for the incident plane wave traveling positively in medium 1, $Z = \eta_1$; however, there is also a negatively traveling wave having a $Z = -\eta_1$. The combination of the incident and reflected components gives rise to a wave impedance that varies with z in medium 1, such that

$$Z(z) = \eta_1 \left(\frac{\eta_2 - \eta_1 \tanh \gamma_1 z}{\eta_1 - \eta_2 \tanh \gamma_1 z} \right) \quad (2.55)$$

where γ_1 is the propagation factor in medium 1 or air.

The reflected radiation is characterized by the reflection coefficient R, which is defined as the ratio of the reflected electric field strength to the incident field strength at the boundary and is given by

$$R = \frac{\eta_2 - \eta_1}{\eta_2 + \eta_1} \qquad (2.56)$$

In a similar manner, the transmission coefficient T is defined as the ratio of the transmitted electric field strength to that of the incident field at the boundary,

$$T = \frac{2\eta_2}{\eta_2 + \eta_1} \qquad (2.57)$$

It is seen from Eqs. (2.56) and (2.57) that when $\eta_2 = \eta_1$, i.e., the electrical properties of the media are approximately equal, there is no reflection and transmission is maximal. On the other hand, there is complete reflection when η_2 is zero. As shall be seen later, for biological materials, microwave reflection and transmission behaviors fall between the two extremes. The latter of the two extremes may be encountered when metallic components are involved in research on biomedical aspects of radio and microwave radiation.

The SWR may be expressed in terms of the magnitude of the reflection coefficient as

$$SWR = \frac{1 + |R|}{1 - |R|} \qquad (2.58)$$

It may be shown using the results of the last paragraph that the SWR = 1 when there is matching equality of media. Any mismatch of the two media will result in an SWR greater than unity.

For the situation illustrated in Fig. 2.4, the electric and magnetic fields may be expressed as

$$E_{1x} = E_i \left(e^{-\gamma_1 z} + R e^{\gamma_1 z} \right) \qquad (2.59)$$

$$H_{1y} = \left(\frac{E_i}{\eta_1} \right) \left(e^{-\gamma_1 z} - R e^{\gamma_1 z} \right) \qquad (2.60)$$

$$E_{2x} = E_i T e^{-\gamma_2 z} \qquad (2.61)$$

$$H_{2y} = \left(\frac{E_i}{\eta_2} \right) T e^{-\gamma_2 z} \qquad (2.62)$$

where the reflection coefficient R and transmission coefficient T are as defined previously. E_i is the incident electric field strength.

The time-average density of power transmitted across the interface is

$$P_t = \frac{1}{2}\frac{E_i^2}{\eta_1}\left(1-|R|^2\right)$$ (2.63)

The difference between the incident and transmitted power must be that reflected, or

$$P_r = \frac{1}{2}\frac{E_i^2}{\eta_1}\left(|R|^2\right)$$ (2.64)

Let us consider, for example, the case of a metal conductor ($\eta_2 = 0$); according to (2.56), the reflection coefficient R = −1. There will not be any transmitted energy. The incident and reflected components of the electric and magnetic fields will combine in medium 1, as indicated in Eqs. (2.59) and (2.60), such that

$$E_x = -2jE_i sin\beta z$$ (2.65)

$$H_y = \left(\frac{2E_i}{\eta}\right)cos\beta z$$ (2.66)

These equations represent a wave that is stationary in space. The values of E and H are sine and cosine functions of z, respectively. The maxima and minima do not move in the z direction but remain at a fixed position as time passes (Fig. 2.4). Note also that E_x is 90° apart in time phase from H_y, and the peak values of E do not occur at the same point in space as H field. In other words, the electric and magnetic energies of a standing wave are in space and time quadrature. The energy is not transferred but oscillates back and forth from the electric field to the magnetic field over a distance of $\lambda/4$.

The situation involving a plane wave incident upon several parallel layers of dielectric materials is also of practical interest. The problem may be treated by considering quantities in each medium and use of an impedance formulation like Eq. (2.55). Examples of wave propagation in multiple layers of biological material are given in Chap. 5.

2.8   Refraction of Microwave and RF Radiation

The refraction and transmission of a plane wave at a plane interface depend on the frequency, polarization, and angle of incidence of the wave and on the dielectric constant and conductivity of the medium. A wave of general polarization usually is decomposed into its orthogonal linearly polarized components whose electric and magnetic fields are parallel to the interface. These components are called E and H

Fig. 2.6 Plane wave
incident upon a boundary
surface at an angle of
incidence θ

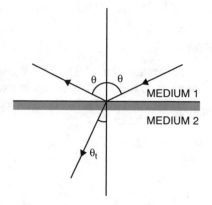

polarizations, respectively, and can be treated separately and combined afterward.
The reflection coefficients for H and E polarizations are given by

$$R_h = \frac{-\left(\dfrac{\varepsilon_2}{\varepsilon_1}\right)\cos\theta + \sqrt{\left(\dfrac{\varepsilon_2}{\varepsilon_1}\right) - \sin^2\theta}}{\left(\dfrac{\varepsilon_2}{\varepsilon_1}\right)\cos\theta + \sqrt{\left(\dfrac{\varepsilon_2}{\varepsilon_1}\right) - \sin^2\theta}} \tag{2.67}$$

$$R_e = \frac{\cos\theta - \sqrt{\left(\dfrac{\varepsilon_2}{\varepsilon_1}\right) - \sin^2\theta}}{\cos\theta + \sqrt{\left(\dfrac{\varepsilon_2}{\varepsilon_1}\right) - \sin^2\theta}} \tag{2.68}$$

and the transmission coefficients are given by

$$T_h = \left(1 + R_h\right)\frac{\cos\theta}{\sqrt{1 - \left(\dfrac{\varepsilon_1}{\varepsilon_2}\right)\sin^2\theta}} \tag{2.69}$$

$$T_e = 1 + R_e \tag{2.70}$$

where θ is the angle of incidence (Fig. 2.6) and ε_1 *and* ε_2 are the complex permittivity of the medium in front of and behind the interface, respectively. In particular,
$\varepsilon = \varepsilon_0\left[\varepsilon_r - j\left(\dfrac{\sigma}{\omega\varepsilon_0}\right)\right]$ with free-space permittivity ε_0 and radian frequency $\omega = 2\pi f$.
It is noted that the angle of reflection is equal to the angle of incidence, while the angle of transmission, θ_t is given by Snell's law of refraction.

$$\sin \theta_t = \sqrt{\frac{\varepsilon_1}{\varepsilon_2}} \sin\theta \tag{2.71}$$

An examination of Eq. (2.67) shows that, for H polarization, it is possible to find an angle so that $R_h = 0$ and the wave is totally transmitted. This angle, referred to as the Brewster angle θ_B, can be obtained by setting the numerator of Eq. (2.67) equal to zero such that

$$\theta_B = \tan^{-1} \sqrt{\frac{\varepsilon_2}{\varepsilon_1}} \tag{2.72}$$

Note that a Brewster angle exists for either $\varepsilon_1 > \varepsilon_2$ or $\varepsilon_1 < \varepsilon_2$. Thus, a circularly polarized wave incident at θ_B becomes linearly polarized upon reflection since there will be no reflection for the H-polarized component.

There is a second phenomenon that applies to both polarizations. For a wave that is incident from a medium with a higher permittivity onto a medium with a lower permittivity, total reflection can take place at the interface between the two dielectric media. This incident angle is called the critical angle

$$\theta_c = \sin^{-1} \sqrt{\frac{\varepsilon_2}{\varepsilon_1}} \tag{2.73}$$

Under these conditions, the incident wave is totally internally reflected. The wave in the medium with smaller permittivity will decay exponentially away from the interface.

2.9   Radiation of Electromagnetic Energy

RF and microwaves are radiated into space or material media through antennas or radiators that serve as transitions between the transmission system and space or material medium. Furthermore, an antenna can also be used as a device to receive microwave and RF radiation. The distribution of radiated energy from an antenna as a function of direction or orientation is given by the antenna radiation pattern. The pattern usually consists of several lobes or sectors. The lobe with the largest maximum is referred to as the main lobe, while the smaller lobes are called minor or side lobes. If the pattern is measured sufficiently far from the antenna so that there is no change in pattern with distance, the pattern is called the far-zone or radiation pattern. Measurement at lesser distances gives near-zone patterns, which are functions of both angle and distance.

In generally, antennas involved in biomedical applications are of the order of one wavelength in size and include such diverse types as dipoles, slots, horns, and apertures. Many of these are designed for near-zone operations. Therefore, these

Fig. 2.7 An elementary
dipole antenna of length ℓ
which is short compared with
a wavelength, $(\lambda \gg \ell)$, and
the width or diameter is small
compared with its length

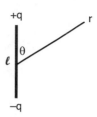

antennas are referred to as broad-beam antenna. Consequently, their far-zone radiated energy may be distributed broadly in space if they reach that far. Another class
of antennas, called narrow-beam antennas such as arrays, horns, and apertures that
focus radiated energy into confined regions and are mostly used in communication
and target acquisition situations such as radars, but also for biomedical applications.

Another important parameter of an antenna is the power gain, or simply the gain,
G of an antenna; this may be defined as the ratio of the maximum radiation intensity
to the radiation intensity from a lossless isotropic antenna radiating the same amount
of total power. In this case, the radiation intensity is the average power radiated per
unit solid angle. An isotropic antenna is one that radiates uniformly in all directions,
and is commonly referred to as an omnidirectional antenna. It should be noted that
the gains of an antenna may also be defined with respect to any reference antenna.
Furthermore, the gain of an antenna is applicable to all antenna operations regardless of its function. Specifically, the gain of an antenna when it is used for transmitting is the same as its gain when used for receiving.

The near- and far-zone characteristics of microwave and RF antennas will be
briefly discussed using the example of a short dipole. The short dipole is an elementary antenna but a very important one both practically and theoretically. For example, any linear antenna may be regarded as a series of short dipoles, and large
antennas of other shapes may be regarded as being composed of many short dipoles.
Thus, knowledge of the properties of the short dipole is useful in understanding and determining the properties of large antennas of complex shape.

2.9.1   The Short Dipole Antenna

For the short dipole illustrated in Fig. 2.7, the length l is short compared with a
wavelength $(l \ll \lambda)$ and the width or diameter is small compared with its length. It
is energized by a transmission line that does not radiate. Hence, for purposes of
analysis, the short dipole may be considered simply as a thin conductor of length l
carrying a uniform current I. It can be shown [Jordan and Balmain, 1968] that the
electric and magnetic fields from the dipole have only three components E_θ, E_r and
H_ϕ in the spherical coordinate system and they are given by

$$E_r = \frac{nIle^{j(\omega t - \beta r)}\cos\theta}{2\pi}\left(\frac{1}{r^2} + \frac{1}{j\beta r^3}\right)$$ (2.74)

$$E_\theta = \frac{nIle^{j(\omega t - \beta r)} \sin\theta}{4\pi} \left(\frac{j\beta}{r} + \frac{1}{r^2} + \frac{1}{j\beta r^3} \right) \tag{2.75}$$

$$H_\phi = \frac{Ile^{j(\omega t - \beta r)} \sin\theta}{4\pi} \left(\frac{j\beta}{r} + \frac{1}{r^2} \right) \tag{2.76}$$

Also, the components E_ϕ, H_r and H_θ are zero at all points. At points far from the dipole antenna, r is large, the terms involving $\frac{1}{r^2}$ and $\frac{1}{r^3}$ in Eqs. (2.74, 2.75 and 2.76) can be neglected in comparison with terms involving $\frac{1}{r}$. Thus, in the far field, there are only two field components given by

$$E_\theta = j\frac{\eta I \beta l}{4\pi r} e^{j(\omega t - \beta r)} \sin\theta \tag{2.77}$$

$$H_\phi = j\frac{I \beta l}{4\pi r} e^{j(\omega t - \beta r)} \sin\theta \tag{2.78}$$

The wave impedance in the far field, i.e., the ratio of E_θ to H_ϕ as given by Eqs. (2.77) and (2.78), is the same as the intrinsic impedance of the medium. Also, E_θ and H_ϕ in the far field are in time phase and at right angles to each other. Thus, the electric and magnetic fields in the far field of a short dipole are related in the same fashion as in a plane wave. Furthermore, the direction and time-average flow of energy per unit area are given by Poynting vector (Eqs. 2.34 and 2.35),

$$P = \eta \left(\frac{I\beta l}{4\pi r} \right)^2 \sin^2\theta \hat{r} \tag{2.79}$$

Clearly, energy flow in the far field is real, outgoing, and entirely in the radial direction. The energy is hence radiated, and the term radiation field is synonymous with far field. In the far field, the intensity of radiated energy decreases as $\frac{1}{r^2}$ with increasing distance.

It is instructive to take apart the far-field expression of Eqs. (2.77) and (2.78) into seven basic physical quantities. For example, E_θ may be written as

$$E_\theta = \frac{1}{2} I \frac{l}{\lambda} \frac{1}{r} \eta j e^{j(\omega t - \beta r)} \sin\theta \tag{2.80}$$

where $\frac{1}{2}$ is a constant (magnitude) factor, I is the dipole current, $\frac{l}{\lambda}$ is the dipole length expressed in wavelengths, $\frac{1}{r}$ is the distance factor, η is the intrinsic impedance of the medium, $je^{j(\omega t - \beta r)}$ is the phase factor, and $sin\theta$ is the pattern factor specifying the field variation with angle. Apparently, the radiated electric field behavior or performance characteristics of the simple dipole antenna is a function of the seven physical parameters. Indeed, in general, the far-field description of any antenna will involve all seven of these factors, regardless how complex the antenna structure may assume.

2.9.2 Near-Zone Radiation

Examination of the expressions for E_r, E_θ and H_ϕ shows that, at points close to the short dipole antenna where r is small, the $\dfrac{1}{r^2}$ and $\dfrac{1}{r^3}$ terms become predominant and Eqs. (2.74, 2.75 and 2.76) are reduced to

$$E_r = -j\frac{\eta Il}{2\pi\beta r^3}e^{j(\omega t - \beta r)}cos\theta \tag{2.81}$$

$$E_\theta = -j\frac{\eta Il}{4\pi\beta r^3}e^{j(\omega t - \beta r)}sin\theta \tag{2.82}$$

$$H_\phi = \frac{Il}{4\pi r^2}e^{j(\omega t - \beta r)}sin\theta \tag{2.83}$$

Consequently, both components of the electric field are in time quadrature with the magnetic field. Thus, the electric and magnetic fields in the near field are related as in a standing wave. The maxima and minima of electric and magnetic fields do not occur at the same point in space. The ratio of electric to magnetic field strength varies from point to point, giving rise to widely divergent field impedance.

Furthermore, the terms that vary as $\dfrac{1}{r^3}$ in the expression for E_r and E_θ correspond exactly to the field of an oscillatory electrostatic dipole; these $\dfrac{1}{r^3}$ terms are referred to as electrostatic fields. The terms that vary inversely as r^2 are just the fields that would be obtained by a direct application of Ampere's law. Thus, the field represented by the $\dfrac{1}{r^2}$ term is called the induction field and becomes predominant at points close to the dipole.

If the Poynting vector is invoked, it will be clear that the electrostatic and induction terms contribute to energy that is stored in the field during one-quarter of a cycle and returned to the dipole during the next without any net or average outward energy flow. In the near field, the energy flow is largely reactive; only the $\dfrac{1}{r}$ terms

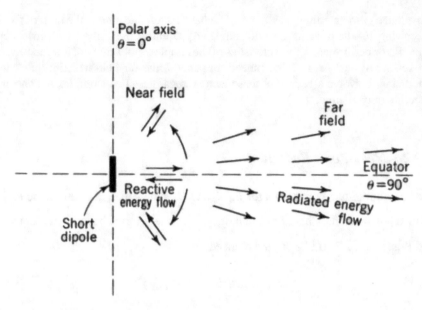

Fig. 2.8 Flow of electromagnetic energy from a dipole antenna where the arrows represent the direction of energy flow at successive instants of time. The same flow pattern is on both sides of the dipole antenna

contribute to an average outward flow of energy. The energy transfer characteristics are illustrated in Fig. 2.8 where the arrows represent the direction of energy flow at successive instants of time [Kraus and Carver, 1973].

The criterion of distance most used to demarcate near and far fields is that the phase variation of the field from the antenna does not exceed $\frac{\lambda}{16}$ [Silver, 1949]. This boundary occurs at a conservative distance of

$$R = \frac{2D^2}{\lambda} \tag{2.84}$$

where D represents the largest dimension of the antenna aperture. In the far field, the field strengths decrease as $\frac{1}{r}$ and only transverse field components appear.

The near field can be divided into two subregions: the radiative near-field region and the reactive near-field region. In the radiative near-field region, the region closer than $\frac{2D^2}{\lambda}$ the radiation pattern varies with the distance from the antenna. The region of space surrounding the antenna in which the reactive components predominate is known as reactive near-field region. The extent of this region varies for different antennas. For most antennas, however, the outer limit is of the order of a few wave-lengths or less. For the special case of the short dipole, the reactive components predominate to approximately $\dfrac{\lambda}{2\pi}$, where the radiating and reacting components are equal.

For the three field regions illustrated in Fig. 2.8, the boundary between the radiative and reactive near-field regions has been conservatively assumed to be 1λ away from the antenna. It should be noted that at low frequencies, the wavelengths are long, and the induction field may extend to very large distances from the source. The corresponding wavelengths at high frequencies are quite short, and the induction field may not exist at all.

In the far-field region, the electric and magnetic fields are outgoing waves with plane wave fronts, and the power density along the axis of the antenna in free space is given by

$$P_c = \frac{PG}{4\pi r^2} \qquad (2.85)$$

where P is the total radiated power, G is the gain of the antenna, and r is the distance from the antenna [Balanis, 2008]. Clearly, in the far field, the power density along the beam axis falls off inversely with the square of the distance.

Power density in the near field is not as uniquely defined as in the far field, since the electric and magnetic fields and their ratio vary from point to point. Furthermore, the angular distribution is dependent on the distance from the antenna. It is therefore necessary to individually arrive at a quantitative estimate of the power density even along the axis. In general, the near-field power density depends on the antenna shape and aperture field distribution. Analyses have been reported for several types of antennas.

The calculated on-axis power density for a square antenna is shown in Fig. 2.9, where the dashed line indicates the envelope of maximum power density obtained [Hansen, 1964]. It is seen that at points close to the antenna ($\dfrac{2D^2}{\lambda} < 0.1$), the power density oscillates about a normalized value of 4.5. It reaches a peak normalized value of 13.3 at $\dfrac{0.18D^2}{\lambda}$ in the radiative near-field region. However, the power density falls below the $\dfrac{1}{R^2}$ value for distances less than $\dfrac{D^2}{\lambda}$. The on-axis power density for a circular aperture is given in Fig. 2.10. The normalized value at a point close to the antenna is about 16. The peak power density occurs at about $\dfrac{0.2D^2}{\lambda}$ and is nearly 42 times the value of $\dfrac{2D^2}{\lambda}$.

These results indicate that the transition point between reactive and radiative near-field regions occurs from 0.2 to $\dfrac{0.4D^2}{\lambda}$. Moreover, they give average and maximum near-field power densities as follows:

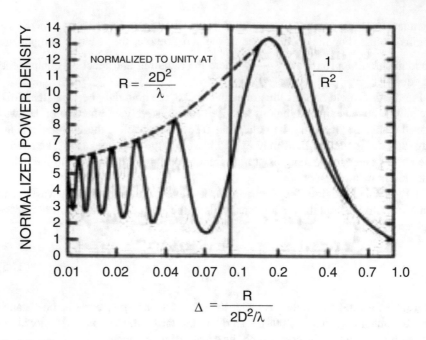

Fig. 2.9 The on-axis power density of a uniform square antenna

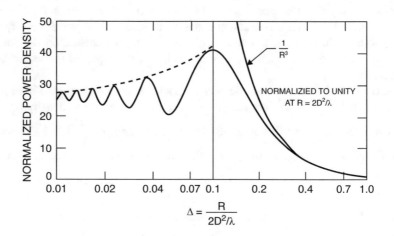

Fig. 2.10 The on-axis power density of a tapered circular aperture antenna

1. Square antenna with uniform field distribution:

$$P_d = \frac{0.88P}{A}, \text{ average} \qquad (2.86)$$

$$P_d = \frac{2.61P}{A}, \text{ maximum} \tag{2.87}$$

2. Circular antenna with uniform field distribution:

$$P_d = \frac{3.01P}{A}, \text{ average} \tag{2.88}$$

$$P_d = \frac{4.86P}{A}, \text{ maximum} \tag{2.89}$$

where P is the total radiated power and A is the area of the antenna. Thus, the maximum near-field power density that can exist on the axis of a practical aperture antenna is about $\frac{5P}{A}$. It should be noted that these formulas do not include the effect of ground reflections, which could cause a value of power density that is four times the free-space volume [Mumford, 1961]. Thus, the electromagnetic radiation in the far field is quite different from that in the near field although there is a smooth transition from one to the other. It is important to note that radiation is largely confined to within a cylinder whose cross section is the antenna aperture until the distance from the antenna approaches the transition range. This characteristic is also indicated by the on-axis power density, which at large distances varies inversely as the square of the distance but in the near field oscillates about a constant value.

2.9.3   Antenna Receiving Characteristics

Thus far, we have been mainly concerned with transmitting antennas that radiate energy. As mentioned earlier, an antenna can also be used to receive RF and microwave radiation. This is because the performance of an antenna when used for transmitting is the same as its performance when used for receiving according to the reciprocity theorem. However, a more direct characterization of receiving antenna performance is the effective cross section or effective area. The effective cross section is defined as the area of an ideal antenna that absorbs the same amount of energy from an incident plane wave as the actual antenna. The effective cross section has significance when applied to horns and reflector antennas that have well-defined physical apertures. For these antennas, the ratio of the effective cross section to the actual aperture is a direct measure of the antennas effectiveness in radiating or receiving the energy to or from the desired direction. Normal values of this ratio for reflector antennas vary from 45% to 75%, depending on antenna type and design, with 65% considered rather good for the commonly used parabolic reflector antenna [Jordan and Balmain, 1968].

The total power or rate of energy extracted by a receiving antenna with the effective cross section, S from an incident plane wave, is therefore given by

$$P = P_d S \tag{2.90}$$

Since the effective cross section is related to the gain of an antenna through

$$S = \frac{\lambda^2}{4\pi} G \tag{2.91}$$

the total power received in the field of impinging plane wave for any antenna is

$$P = \frac{\lambda^2}{4\pi} P_d G \tag{2.92}$$

The rate of energy extraction for the short dipole antenna illustrated in Fig. 2.7 is

$$P = 1.5 \frac{\lambda^2}{4\pi} P_d \tag{2.93}$$

For the special case of a thin dipole antenna of finite length, l, the maximum gain is [Harrington, 1961]

$$G = \frac{\eta}{\pi R_r} \left(1 - \cos \frac{\beta l}{2} \right)^2 \tag{2.94}$$

where η is the intrinsic impedance of the medium and R_r is the radiation resistance of an antenna defined by analogy to Ohm's law as

$$R_r = \frac{P}{|I|^2} \tag{2.95}$$

where I is the antenna current. The total received power is found using (2.92) as

$$P = \frac{\eta \lambda^2 P_d}{4\pi^2 R_r} \left(1 - \cos \frac{\beta l}{2} \right)^2 \tag{2.96}$$

Because of the cosine term and the R_r term in the denominator of Eq. (2.96), the total power received or rate of energy extracted by a thin dipole antenna reaches a peak as a function of the antenna length expressed in terms of wavelength. That is, the antenna exhibits resonant receiving characteristics (Fig. 2.11). The total energy extracted from an impinging plane wave peaks for selected antenna lengths. The first of the resonant lengths occurs at a length l and equals 0.9λ.

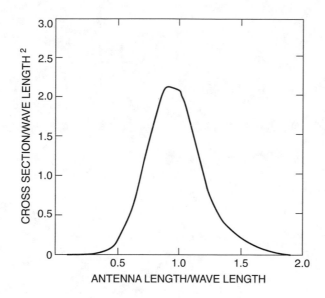

Fig. 2.11 Resonant length absorption characteristics of elementary dipole antennas

References

Balanis C (ed) (2008) Modern antenna handbook. Wiley, Hoboken

Hansen RC (ed) (1964) Microwave scanning antennas, vol I. Academic Press, New York

Harrington RF (1961) Time harmonic electromagnetic fields. McGraw- Hill, New York

Jordan EC, Balmain KG (1968) Electromagnetic waves and radiating systems. Prentice-Hall, Englewood Cliffs

Kraus JF, Carver KR (1973) Electromagnetics. McGraw-Hill, New York

Mumford WW (1961) Some technical aspects of microwave radiation hazards. Proc IRE 49:427

Silver S (ed) (1949) Microwave antenna theory and design. McGraw-Hill, New York

Chapter 3
Brain Anatomy and Auditory Physiology

This chapter briefly describes the anatomy and physiology of the mammalian brain with a concentration on the auditory system. The intent is to provide an appreciation for the central nervous system structure and function as related to one sensory system. The concepts introduced here are aimed to support later discussions on experimental and theoretical research, and numerical modeling of pulsed microwave interaction with the brain and the central nervous system. The materials presented are an important step in understanding the auditory system and how its functions are influenced by pulsed microwave radiation.

3.1 Anatomy and Physiology of the Human Brain

The mammalian brains including that of humans have the same essential structure and function. The human brain is distinguishable from other animals by virtue of having a considerably greater size to total-body mass ratio in comparison. The brain is surrounded by a rigid skull, suspended in cerebrospinal fluid, and is nourished via blood through the blood-brain barrier system serving as a differentially selective filter to maintain its homeostatic physiochemical environment.

A conservative estimate of the human brain is that it contains several billions of cells apportioned into thousands of different types, consisting mainly of neurons and the neuroglial cells. The brain is organized into complex structures of similar appearances of gross tissue masses. It is comprised of three major anatomical parts, the cerebrum, cerebellum, and brainstem (Fig. 3.1). The cerebrum is the largest part of the human brain. It is divided into two cerebral hemispheres, the outermost of which, the cerebral cortex is a layer of gray matter, covering a white matter core. The brainstem consists of the midbrain, the pons, and the medulla oblongata and is structurally continuous with the spinal cord. The cerebellum occupies the lower part of the brain, behind the brainstem; it is also known as the little brain. The cerebellum is divided into two hemispheres and is surrounded by a large mass of cortical

© Springer Nature Switzerland AG 2021
J. C. Lin, *Auditory Effects of Microwave Radiation*,
https://doi.org/10.1007/978-3-030-64544-1_3

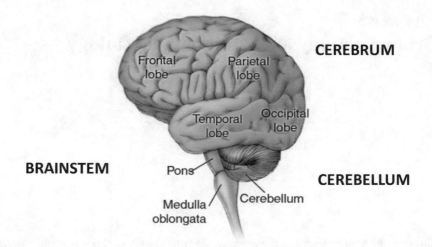

Fig. 3.1 The three major anatomical parts of the human brain: the cerebrum, cerebellum, and brainstem

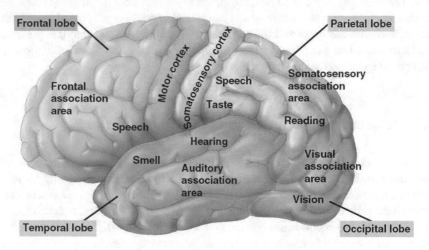

Fig. 3.2 The four lobes of the cerebrum and their associated functions and processes

materials. As the central nervous system for the body, the function of the brain is to process all incoming neurosensory signals and effect outgoing neural signals that activate and exert control over all other cells, tissues, and organs in the body.

The cerebrum is divided into two cerebral hemispheres. The hemispheres are connected by commissural nerve tracts. While some functions are associated uniquely with one side, certain functions are shared. The left and right hemispheres are generally similar in shape and anatomy. The cortex of each hemisphere is divided into four lobes – the frontal, parietal, temporal, and occipital lobes (Fig. 3.2). It is important to note that the cerebral cortex in these lobes works in concert with other areas of the brain to perform various tasks.

The frontal lobes are located at the front of the cerebrum and are associated with cognitive processes such as attention, behavior, working memory, planning, reasoning, and abstract thought. The frontal lobes are the control center for emotion and site for personality. They play a comprehensive role in all voluntary movement and host the primary motor cortex which regulates such activities as arm and leg movements.

The parietal lobes are anterior to the occipital lobes and above the temporal lobes. The cerebral cortex in these lobes is responsible for processing and integration of such sensory information as taste, temperature, touch, pain, awareness, spatial orientation, object identification, reading, writing, mathematical manipulation, visuomotor, and other somatosensory functions involved in analysis, specification, and activation.

The temporal lobes, which sit on the lateral side of the hemisphere, are located below the frontal and parietal lobes and are anterior to the occipital lobes. The auditory cortex and olfactory cortex are found in the temporal lobes. They have important functions in auditory perception, sound processing, language comprehension, speech production, and recognition of words, faces and objects, emotion response, as well as memory formation and association. In general, the left side is the dominant temporal lobe in most people and is central in understanding language and learning and remembering verbal information. Typically, the non-dominant right lobe is involved in acquiring and remembering music and visuospatial subjects.

The occipital lobes are the smallest of the paired lobes of the cerebral cortex. They are located at the back portion of the cerebrum, posterior to the parietal and temporal lobes. The dura mater separates the cerebrum from the cerebellum. The occipital lobes, comprising most of the anatomical areas of the visual cortex, are the center for visual information processing of the mammalian brain. They are dedicated to understanding visual inputs and stimuli and visual interpretation through such functions as sensor, motor, and association. The optic nerve is the second cranial nerve from the retina of the eye.

The brainstem borders the frontal, parietal, and temporal lobes of the cerebral cortex and connects the cerebrum with the spinal cord in the brain. It is structurally continuous with the spinal cord but separate from the cerebrum (Fig. 3.3). It consists of the midbrain, the pons, and the medulla oblongata. A foremost function of the brainstem is to connect the major brain regions. The brainstem connects the motor and sensory nervous systems from the brain to the rest of the body. It is involved in many essential nervous system functions of the brain such as cardiac and respiratory function, motor and somatosensory system including sound and vibration, temperature, pain, tactile, and proprioception. The brainstem also relays neural signals from the cranial nerves to the cerebral cortex including the eighth cranial nerve, which is known as the auditory nerve leading from the inner ear. In addition to connecting the cerebrum and spinal cord, the brainstem also links the cerebrum to the cerebellum.

The cerebellum is located behind or below the occipital lobe. A major function of the cerebellum is to receive and organize complex information, for example, from sensory nerves and the auditory and visual systems, among other daily tasks

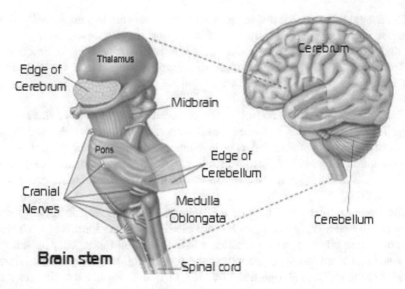

Fig. 3.3 Structure of the brainstem and its nerve connections [facebook.com]

and routines, as well as the basic memory and learning processes. The cerebellum is also notable for its regulatory functions such as balance, equilibrium, and movement coordination. Thus, a major function of the cerebellum is to coordinate the timing and force exertion of the different muscle groups to produce smooth limb or body movements.

3.2   The Human Auditory System

The ear is the organ of hearing in humans and mammals. It receives the sound pressure waves, or vibrations, in its surroundings and conducts the sound waves through the external ear canal to the ear drum, which initiates the amplification of sound pressure and sends the sound waves to the inner ear and to the fluid-filled cochlea, the hearing apparatus. The information carried by the mechanical sound waves are converted to electrical impulses and then transmitted to the central nervous system for processing and recognition. It is convenient to divide the auditory system into two parts according to their anatomic and functional characteristics. The peripheral portion consists of the external ear, the middle ear, and the cochlea of the inner ear. The central portion is made up of the auditory nerve and its various relay pathways to central neural structures.

3.2.1   External and Middle Ears

The external ear consists of the pinna or auricle, the external auditory canal or meatus, and the tympanic membrane or eardrum (Fig. 3.4). The function of the auricle is to collect airborne sound waves and channel it into the external auditory meatus; however, it is only marginally effective in human. The external auditory meatus is about 2.5–4 cm in length and 7.5–10 mm in diameter. Sound waves entering the external meatus are amplified by it much the same way as a tubal sound resonator, so that the sound pressure at the tympanic membrane is higher than the pressure at the entrance of the auditory meatus. A frequency response curve for the auditory meatus may be obtained by plotting the pressure difference between the tympanic membrane position and the center of the entrance of the auditory meatus against the sound frequency. An average frequency response curve in sound pressure level (see Sect. 3.3.6) is shown in Fig. 3.5. This curve is based on measurements by Wiener and Ross [1946] up to 8 kHz and by Djupesland and Zwislocki [1972] up to 10 kHz. The extrapolation to 12 kHz was inferred from measurements on a human ear replica and a model ear [Shaw, 1974]. The maximum increase in sound pressure occurs first near 4 kHz, falls off on both sides of this resonant frequency, and peaks again near 12 kHz. The peaks are broad and round, indicating that the walls of the auditory meatus and the tympanic membrane are not rigid. The sound pressure wave impinging on the tympanic membrane is partially reflected into the air. Some of the incoming sound energy is also lost to the walls of the external auditory meatus.

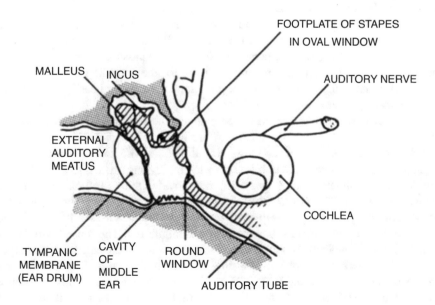

Fig. 3.4 Anatomical features of the human ear

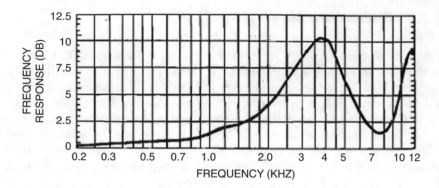

Fig. 3.5 Frequency response showing the ratio of the sound pressure level in dB at the tympanic membrane to the entrance of external auditory meatus

The obliquely positioned tympanic membrane completely separates the external ear from the middle ear or tympanic cavity. The tympanic membrane is shaped like a shallow cone with its apex directed inward and somewhat below the center. Its anatomic area is about 65 mm² [Moller, 1974]. The entire tympanic membrane vibrates in response to the impinging sound waves. The mode of vibration depends upon the sound frequency. At the threshold of hearing in humans, the membrane displacement ranges from 10^{-7} m for low frequencies to 10^{-11} m at 3 kHz. [Bekesy, 1957]. The middle ear is an air-filled cavity in the temporal bone. It is separated from the external ear by the tympanic membrane and from the inner ear by the oval and round windows. The middle ear is connected with the nasopharynx by the eustachian tube or auditory tube. The tube is normally closed, but it opens during chewing, swallowing, and yawning, keeping the air pressure within the middle ear equalized with the atmospheric pressure. Note that the passage between the middle ear and the nasal pharynx is a natural pathway for the spreading of infections of nose and throat to the middle ear. Such infections may impair hearing temporarily or permanently unless properly cared for.

The three small auditory ossicles – the malleus, incus, and stapes – are housed in the middle ear (Fig. 3.4). The handle of the malleus is directed downward and attached to the upper part of the tympanic membrane. The head of the malleus is attached to the incus which in turn is connected by its long process to the stapes (or stirrup). The footplate of the stapes rests in the oval window. The malleus and the incus vibrate as a unit; movement of the tympanic membrane therefore causes the stapes also to move back and forth against the oval window. Two small muscles, the tensor tympani and the stapedius, are in the middle ear. The tensor tympani is attached to the handle of the malleus, and the stapedius is connected to the neck of the stapes. When the tensor tympani contracts, it moves the malleus inward and increases the tension on the tympanic membrane. The stapedius pulls the stapes outward upon contraction. Contraction of either or both muscles will therefore increase the stiffness of the middle ear mechanism and thereby decreases the low-frequency energy transmission. These reflex muscle contractions are initiated only by relatively loud sounds and perform a limited protective function against them.

The most important function of the middle ear is to transform the sound pressure from a gaseous air to a fluid medium without significant loss of energy. At an air-water interface, only 0.1% of the sound energy is transmitted into water; the other 99.9% is reflected to the air. The middle ear has two arrangements that practically eliminate this potential loss. The area of the tympanic membrane is approximately 65 mm^2, and the stapedial footplate has an anatomic area of about 3.2 mm^2. Since the mode of vibrations of the tympanic membrane is not simple, the ratio of the effective areas is around 14 to 1. In addition, the pressure exerted on the stapes is amplified by the lever action of the ossicles by a factor of 1.3 to 1 [Wever and Lawrence, 1954]. Thus, there is a total gain factor of 18 between the pressure at the tympanic membrane and at the oval window.

The frequency response of the middle ear is not flat over the audible frequency range (up to 20 kHz) for humans. The mass of the middle ear ossicles, and the elasticity of the muscles influence the transmission of sound through the middle ear in different ways for different frequencies.

The elastic property predominates at high frequencies, and the mass prevails at lower frequencies. Moreover, the mode of vibrations of the tympanic membrane is also frequency-dependent. Figure 3.6 is a plot of the frequency response of the middle ear of the cat as determined by measuring the stapedial displacement in response to sound pressures at the tympanic membrane [Guinan and Peake, 1967]. It is seen that the ear is most sensitive in the region of 1 kHz for the cat. It is important to note that the middle ears of man and cat are not the same, although they are qualitatively similar in their functions.

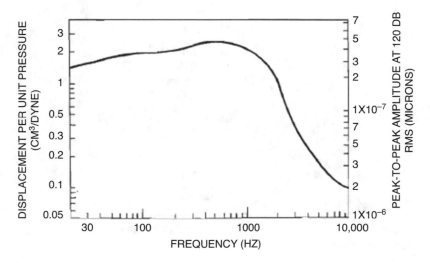

Fig. 3.6 Displacement amplitude of stapedial vibrations in cats

3.2.2   The Inner Ear

The inner ear or labyrinth consists of an osseous or bony labyrinth and a membranous labyrinth. The bony labyrinth is a series of canals and chambers in the petrous portion of the temporal bone. The membranous labyrinth lies within the bony labyrinth and is surrounded by the perilymph. Its inside is filled with endolymph. The labyrinth is divided into three parts: the vestibule, the semicircular canals, and the cochlea. The semicircular canals contain part of the sensory organ for balance. The vestibule is a chamber separated from the tympanic cavity by a thin partition of bone which is found in the oval window.

The cochlea is shaped like a snail shell which spirals for about two and three-quarter turns and has a volume of about 0.2 milliliter. The base of the cochlea is broad and tapers as it spirals to a narrow apex. The cochlea is divided by the basilar and Reissner's membranes into three chambers or scalae (Fig. 3.7). The upper scala vestibuli ends at the oval window. The lower scala tympani ends at the round window which is closed by the secondary tympanic membrane. Both of these chambers are filled with perilymph, and they are separated by the scala media except at the apex of the cochlea where they are continuous. The scala media contains

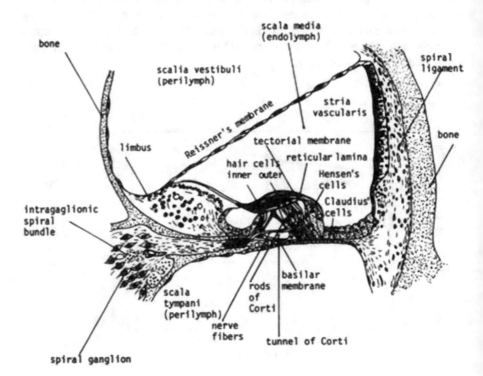

Fig. 3.7   A cross section of the cochlea

endolymph and is continuous with the membranous labyrinth. It is separated from the scala vestibuli by the Reissner's membrane and is cut off from the scala tympani by the basilar membrane. The essential organ of hearing, the organ of Corti, is in the scala media.

The organ of Corti extends from the apex to the base of the cochlea and consists of a series of epithelial structures located on the basilar membrane, which is narrow and stiff near the oval window and comparatively wide at the apex of the cochlea. The cross section of a single turn of the cochlea is shown in Fig. 3.7. The auditory receptor hair cells are arranged in row. There are about 5000 inner hair cells placed in a single row alone the entire length of the cochlea, and there are about 20,000 outer hair cells arranged in three to four rows in the basal and apical turns of the cochlea. These cells have long processes (cilia) at one end and large basal nuclei at the other. The hair cells are covered by a thin but elastic tectorial membrane which makes contact with the cilia of the hair cells. (The hair cells derive their name from the tiny little hair-like structures at the free ends of cilia.) Approximately 20,000 fibers of the cochlear branch of the auditory nerve arborize around the hair cells. The cell bodies of these afferent neurons make up the spiral ganglia. The axons leaving the spiral ganglia form the auditory portion of the eighth cranial nerve which enters the dorsal and ventral cochlear nuclei of the medulla oblongata.

3.2.3   Cochlear Mechanical Activity and Transduction

The Reissner's membrane is so thin and delicate that the scala vestibuli and the scala media function as a single unit in the passage of sound pressure waves. On the other hand, the basilar membrane is stiff and reacts in a characteristic manner to sound waves. When a sound pressure is transferred from the stapedial footplate to the cochlea, the oval window moves inward and pushes the perilymph of the scala vestibuli up toward the apex of the cochlea (Fig. 3.8). The sudden increase in pressure in the scala vestibuli forces the basilar membrane to bend toward the scala tympani, causing the round window to bulge outward. When the stapedial footplate is pulled backward, the process reverses. The vibrations of the basilar membrane are transmitted to the hair cells via the supporting cell structures of the organ of Corti and the tectorial membrane and cause the hair cells to activate the neural endings, which results in cochlear electrical potentials. However, the manner and mechanism by which movement of the basilar membrane converts sound energy into electrical nerve impulses was not completely understood for some time [Eldredge, 1974].

Since then significant advancements have been made in the identification of proteins linked to the function of the mechanotransduction in hair cells, and the kinetics of ion channel activation have been measured. However, the specific function of these proteins still needs to be defined [Cunningham and Müller, 2019]. This much is now known. Cochlear hair cells utilize mechanically gated ion channels, located in cilia that open in response to sound wave-induced motion of the basilar membrane, and convert mechanical stimulation to changes in hair cell membrane

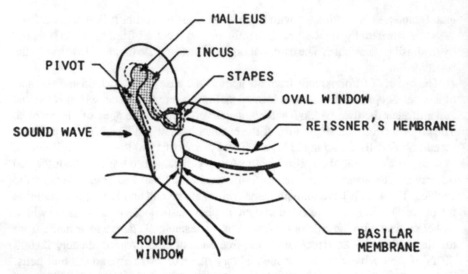

Fig. 3.8 The auditory ossicles and the way their movement translates movements of the tympanic membrane into a wave in the cochlear fluid

electrical potential. The potential changes in hair cells then cause neurotransmitters to release from hair cells that initiate electrical signals in the nerve fiber terminals of spiral ganglion neurons.

The basilar membrane responds in resonance to highest frequencies at the basal end near the oval window and to progressively lower frequencies as it continues toward the apical end. At the apical end, the basilar membrane responds resonantly to the lowest frequencies of audible sound. The cochlea functions as a very efficient frequency analyzer. A summary of the classic observations on the frequency analysis and traveling wave nature of basilar membrane movement is given below.

When a sinusoidal pressure wave is applied to the oval window, it initiates a traveling wave that propagates from the oval widow to the apex of the basilar membrane. The amplitude of the traveling wave grows slowly from the base of the membrane to a place of maximum and decays abruptly [Bekesy, 1960]. The place of maximum varies systematically with the frequency of applied pressure. The relative amplitude, as a function of distance along the basilar membrane for difference frequencies, are shown in Fig. 3.9. Each curve is for a single frequency at a constant amplitude of stapes vibration. The solid lines were obtained from measurements made when all the portion of basilar membrane vibrating at that frequency was still intact. The dotted lines were observations made at the same stapes amplitude after the more apical portion of the basilar membrane had been removed. It is seen that the higher the sound frequency, the closer the maximum to the base of the membrane. These curves represent the envelope of the peak amplitude of vibration of the basilar membrane at different instance of time (Fig. 3.10).

Fig. 3.9 Displacement amplitude along basilar membrane for different frequency at constant amplitude of stapes vibration

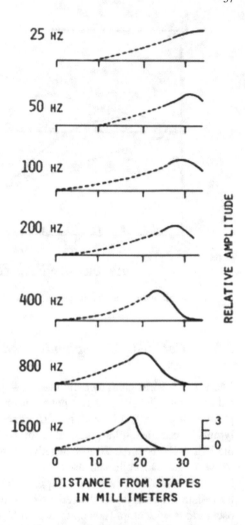

It is noteworthy that as a frequency analyzing instrument, the basilar membrane's performance reveals some very interesting features that display or account for various known biophysical principles and phenomena. As mentioned, a sound wave launched at the oval window by motion of the stapes is transmitted along the basilar membrane as a traveling wave. Each position along the length of the basilar membrane is endowed with its own specific mechanical filter or sink of characteristic frequency; filters or sinks have the highest frequency response at the basal end and lowest frequency response at the apical end. So, the higher-frequency components would be extracted first and lower frequencies last. In a traveling sound wave with multiple frequencies, the higher-frequency components attenuate faster and travel shorter compared to low-frequency components that decay slower and would propagate further to correspond to the structure and function of the basilar membrane.

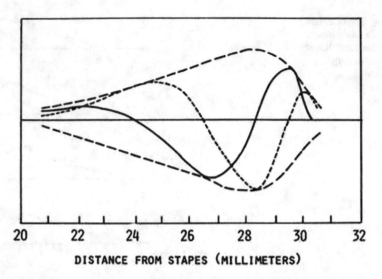

DISTANCE FROM STAPES (MILLIMETERS)

Fig. 3.10 Amplitude of basilar membrane vibration at different instances of time

3.2.4   Cochlear Microphonics and Electrical Potentials

There are several characteristic electrical potentials in the cochlea. The endocochlear potential (EP) is a DC potential existing between the endolymph and the perilymph. At rest, this potential difference is about +80 mV relative to the perilymph. The intracellular potential of the large cells in the organ of Corti, including the hair cells, is some 70 mV negative to the perilymph. The potential difference between the hair cells and the endolymph is therefore 150 mV. This potential is highly dependent on the oxygen supply. The 150 mV DC potential when modulated by resistance changes at the reticular laminar can produce cochlear microphonic oscillations with amplitude up to 3–10 mV.

The cochlear microphonic is a potential that faithfully duplicates the waveform of the applied sound stimulus. It may be recorded from within or near the cochlea, and a popular recording site is the round window. The cochlear microphonic appears without threshold and has negligible latency. It is stable over long periods of time. It increases linearly with an increase in the pressure of the applied sound wave until the potential reaches 1 mV, and it then decreases with further increase in sound pressure. It is highly resistant to such changes in the physiologic state of the test animal as cold, fatigue, and drug administration. At death, the cochlear microphonic drops to a low level, but it persists at this level for up to 30 minutes or longer. Its existence, however, depends on the presence of normal hair cells.

Some examples of cochlear microphonics recorded from three sites along the cochlea of a guinea pig are shown in Fig. 3.11. The waveforms illustrated in Fig. 3.11a and b are typical for acoustic transients, and those shown in c and d are typical for acoustic tones. The cochlear microphonic responses to acoustic tones

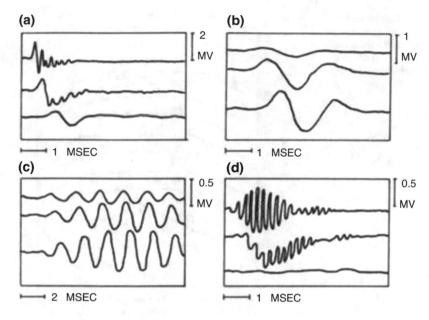

Fig. 3.11 Cochlear microphonics recorded from the first (top), second (middle), and third (bottom) turns of *the* guinea pig cochlea in response to four different acoustic stimuli. (**a**) Wide band click. (**b**) 650 Hz click. (**c**) 500 Hz pip. (**d**) 4000 Hz burst

correspond closely to the waveform of applied sound pressure. The microphonic shows increasing latency with distance from the oval window, consistent with the traveling waves described by Bekesy. The responses to bursts of tone at low frequencies are the largest at the apical tum but spread out over the entire cochlear duct. The cochlear microphonic generated is maximum in the basal turn when a burst of high-frequency tone is used. Moreover, it is distorted and shows a strong asymmetrical nonlinearity in the second turn.

The peak-to-peak potentials for the cochlear microphonic responses to tones are shown in Fig. 3.12 as a function of frequency using the applied sound pressure level at the tympanic membrane as a parameter [Engebretson, 1970]. (Sound pressure level is described in Sect. 3.3.5 of this chapter.) The solid curves are measurements made with the auditory bulla (tympanic bone) opened. The increased stiffness due to the compliance of the small volume of air enclosed behind the tympanic membrane would change the slopes of each curve by the difference between the solid curve and broken curve shown for the 30 dB case. It is interesting to note that the cochlear microphonic response is almost frequency independent. The cochlear microphonic potential increases linearly as a function of applied sound pressure up to 80 dB at any given frequency. At higher pressures, the cochlear microphonic response becomes nonlinear and the deviations from linearity increase as a function of both frequency and sound pressure [Eldredge, 1974]. Cochlear microphonics up to 100 kHz have been recorded from bats, cats, rats, and guinea pigs [Vernon and

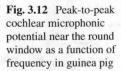
Fig. 3.12 Peak-to-peak cochlear microphonic potential near the round window as a function of frequency in guinea pig

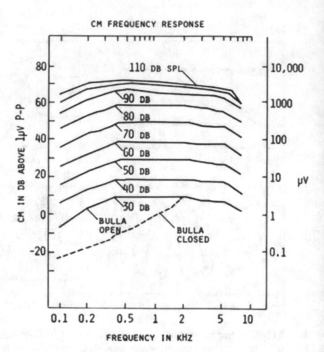

Meikle, 1974]. It is noted that the presence of the cochlear microphonics is a determining clinical finding in the differential diagnosis of auditory neuropathy or dyssynchrony – a disorder characterized by the presence of otoacoustic emissions [Hood, 2015].

There are two additional cochlear potentials generated when sound impinges on the ear. Moderate to strong sound pressure decreases the potential difference between the scala media and the scala vestibuli, and this decrease is maintained as long as the applied sound pressure persists. Similar to the cochlear microphonic, this negative summating potential shows no threshold and negligible latency. Unlike the microphonic, its amplitude continues to increase with increasing sound pressure. It is generally more resistant to drugs and anoxia and depends on the integrity of the inner ear hair cells. Under certain circumstances (namely, in fresh animal preparations and low sound pressures), the direction of change of the potential in the scala media is positive with respect to the scala tympani: it is then called the positive summating potential. The summating potential recorded in the basal turn when low-frequency sound is used is usually small and positive.

3.2.5 Action Potentials of the Auditory Nerve

Although the mechanism by which movement of the basilar membrane converts sound energy into nerve impulses was not completely known for some time, it is understood that cochlear hair cells utilize mechanically gated ion channels located in the cilia that open in response to sound wave-induced motion of the basilar membrane. Conversion of mechanical stimulation to hair cell membrane electrical potential causes neurotransmitters to release from hair cells, which then initiate electrical signals in the nerve fiber terminals of spiral ganglion neurons arborizing around the hair cells [Cunningham and Müller, 2019]. From these nerve fibers, the action potential passes through the auditory nerve into the central nervous system. The action potential of the auditory nerve as a whole can best be recorded following stimulation by an acoustic click. It consists of two distinct components, N_1 and N_2, each about one ms in duration (Fig. 3.13). The latency of the action potential relative to cochlear microphonic is a function of sound pressure amplitude, of the rate of rise of sound pressure, and of the frequency. The minimum latency for N_1 is about 0.55 ms, and the maximum is about 2.3 ms. The amplitudes of N_1 and N_2 are nonlinear functions of sound pressure. N_1 grows slowly at first and then suddenly becomes more rapid and N_2 appears. The discontinuity indicates the existence of two different sets of excitable elements with different thresholds of excitation.

The nerve response is vulnerable to almost all adverse conditions. It is more sensitive to anoxia than cochlear microphonics and recovers less readily. Quinine has been shown to abolish nerve responses selectively. The latency of N_1 is increased by cold. The nerve response can also be reduced by the activity of the efferent inhibitory fibers. A slowing of the nerve discharge in single fibers slows during constant stimulation. The neural components of the round window response are also known to decrease because of either simultaneous or previous stimulation. The masking effect is particularly sensitive if the frequency spectrum of the masking noise overlaps that of the stimulus. It is interesting to note that the polarity of the neural components of the round window response remains the same when the phase of the stimulus is reversed. The cochlear microphonic potential, on the other hand, reverses polarity with the change of stimulus phase. The same observation is true when the cochlear location of the recording electrode is changed.

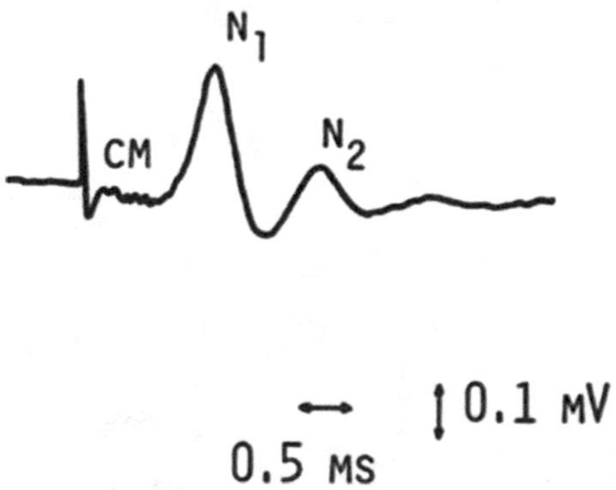

Fig. 3.13 Auditory nerve response in cats following an acoustic click stimulation. CM, cochlear microphonics. N_1 and N_2, nerve responses

3.2.6   Central Auditory Nuclei and Pathways

The auditory action potentials generated in the nerve fibers ascend from the spiral ganglia via the eighth cranial nerve (auditory nerve) to the dorsal and ventral cochlear nuclei. These nuclei project both to the superior olivary complex unilaterally and bilaterally through the trapezoid body and the superior olivary nuclei and to the lateral lemniscus nuclei (Fig. 3.14). The superior olivary complex also sends fibers to the lateral lemniscus. The inferior colliculus receives axons from the cochlear nucleus, the superior olivary complex, and the lemniscus. At this level, the axons may cross over to the contralateral inferior colliculus nucleus via the commissure. The major ascending connection runs, bilaterally, from the inferior colliculus to the ventral division (principal nucleus) of the medial geniculate body of the thalamus in the brainstem via the brachia. Lemnici axons do not contribute to the medial geniculate body since lesions of the lemniscus do not produce degeneration in the brachia or geniculate body.

After forming synapses in the medial geniculate body, the ascending axons radiate in a diffused fashion to the cerebral cortex and project to the transverse temporal gyri and insular cortex located in the superior portion of the temporal lobe, near the floor of the lateral cerebral fissure. The crossings at the levels of the superior olivary

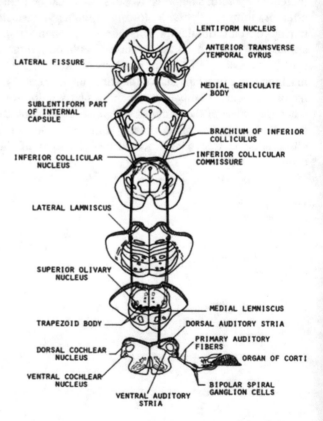

Fig. 3.14 Schematic representation of the central auditory neural pathways

complex, lateral lemniscus nuclei, and inferior colliculi are responsible for the bilateral representation which allows auditory impulses arising in either ear to be projected on both sides of the auditory cortex. Therefore, as auditory signals pass up each side of the brain from both ears, unilateral hearing loss cannot be caused by a localized brain lesion.

The olivocochlear bundle, or the bundle of Rasmussen, is a prominent bundle of efferent (descending) auditory nerve fibers that originate in the superior olivary complex. These axons cross the brainstem to reach the hair cells of the organ of Corti of the opposite ear. Stimulation of this olivocochlear bundle of Rasmussen produces an inhibitory effect on the action potential response to click. The cochlear microphonic is unaffected by the procedure, but the auditory nerve response is greatly reduced. This efferent inhibitory action is an expression of the central nervous system's regulation of the sensitivity of hearing mechanisms.

3.3 Perception of Sound and Pressure

When a sound pressure wave impinges on the ear, it is funneled into and amplified by the external auditory meatus which causes the tympanic membrane to vibrate in a characteristic manner. This vibration is transformed by the middle ear ossicles into movements of the stapedial footplate. These movements create pressure waves in the fluids of the inner ear which displace the basilar membrane of the cochlear duct and cause the hair cells located on top of the basilar membrane to generate electrical potentials. The endocochlear potential elicits impulses in the auditory nerve. After the auditory nerve, the nerve impulses are transmitted through the cochlear nuclei, the trapezoid body, the superior olivary complex, the inferior colliculus, the medial geniculate body, and finally to the auditory cortex. The primary auditory cortex receives the nerve impulses and interprets them as different sounds. The impulses are also conveyed to the surrounding auditory associative areas for recognition.

3.3.1 Transmission of Sound Pressure

The audible sound frequency range of humans is between 16 Hz and 20 kHz (approximately 10 octaves). The sensitivity is lower at the two ends but becomes much more sensitive above 128 Hz up to about 4 kHz. A sound wave in the surrounding air is collected by the outer ear and directed into the air-filled ear canal. The geometry and dimensions of the ear canal lets it to act as a resonating sound tube and preferentially enhances sound waves at between 3 and 4 kHz, thus increasing the hearing sensitivity of the human auditory system at these frequencies. The range of sensitivity and audibility diminishes with age.

The outer and middle ears serve also as a two-stage mechanical amplifier of sound waves. The outer ear has a much larger surface area compared to the smaller tympanic membrane. Also, the surface of the tympanic membrane is much larger than that of the stapedial footplate. So, there is a mechanical amplification: a small movement over a large area is converted to a larger movement of a smaller area. In addition, the middle ear ossicles act as a chain of levers which serve to amplify the sound waves. The outer and middle ears amplify sound waves on its passage from the outer to the inner ear by about 1000-fold or 30 dB.

In addition to the usual course of the sound wave transmission through the external auditory meatus and the middle ear ossicles described thus far, hearing may also be mediated by way of the bones of the skull. The latter has been designated as bone conduction to distinguish it from the air conduction route reserved for the former. Under ordinary conditions, sound pressures in the air cause almost no vibration in the skull bones; therefore bone conduction is less significant than air conduction in hearing. Tapping the jaw or holding vibrating devices such as a tuning fork against the skull can cause vibrations of sufficient amplitude in the skull to elicit bone-conducted sound. Intense airborne sound can also impart sufficient energy to the skull bones to initiate bone-conducted hearing. In this case, vibration of the skull is transmitted directly to the fluid of the inner ear and causes the basilar membrane to move. After it reaches the organ of Corti, the transmission of sound to the auditory cortex is the same as that for air conduction.

There are three widely accepted routes by which bone-conducted sound stimulates the cochlea: These are the compressional, inertial, and osseotympanic theories of bone conduction.

Compressional bone conduction implies that the cochlear shell is compressed slightly in response to the pressure variations caused by sound. Because the cochlear fluids are relatively incompressible, because there are volume differences between the scala vestibuli and scala tympani and because the oval window is stiffer than the round window, a pressure difference may develop across the basilar membrane resulting in its displacement and the production of a traveling wave.

The inertial wave bone conduction theory suggests that, for low-frequency vibrations, a relative motion is set up between the ossicular chain and the temporal bone. The temporal bone containing the cochlea vibrates as a whole. The middle ear ossicles, because of their inertia and flexible attachment to the temporal bone, move in opposition to the cochlea. The net result of this action is an apparent movement of the stapedial footplate in and out of the oval window, leading to cochlear stimulation in much the same manner as in air conduction. An additional inertial effect may be due to a relative motion between the perilymphatic fluids and the cochlear shell.

The osseotympanic theory refers to a mechanism by which a relative movement of the skull, with respect to the mandible, sets up pressure variations in the air present in the auditory meatus. When the bones of the skull are driven by a vibrating device, the mandible attached to the lower jaw lags behind or does not move at all. This results in relative displacements of the cartilaginous skeleton of the auditory meatus, causing sound to be generated in the auditory meatus and transmitted to the inner ear via the ossicles.

Since perception of microwave pulses is correlated with the capacity to hear high-frequency sound, as will be seen in later chapters, it rules out inertial or osseo-tympanic bone conduction as potential mechanisms for microwave auditory effect.

3.3.2   Loudness and Pitch

The perceived loudness of a sinusoidal sound wave is determined by both its ampli-tude and its frequency. Loudness varies with sound intensity, which is proportional to the square of pressure amplitude. Figure 3.15 shows the threshold of audibility and tactile sensation in terms of the weakest intensity of sound can be heard or felt as a function of frequency. At any given frequency, the loudness varies as the loga-rithm of intensity. The threshold intensity for tactile sensation is about 10^{12} times higher than that for hearing at 1 kHz. It is interesting to note that hearing is keenest in the range of 1 to 4 kHz and decreases quickly for lower and higher frequencies. On the other hand, the threshold feeling is steady. The fundamental and major overtones of the human voice are all at lower frequencies. Middle C is about 260 Hz. Sound intensity must be about 100 times greater to "just" hear 260 Hz rather than 1000 Hz.

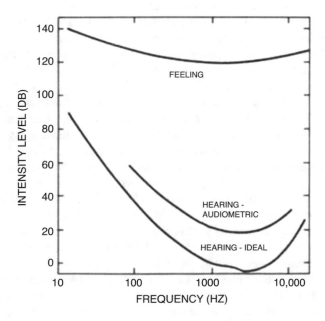

Fig. 3.15 Audibility curve and threshold of tactile sensation in humans

Although the pitch of sound is determined primarily by the sound frequency, loudness also plays a part. In general, tones below 500 Hz seem lower and tones above 4 kHz seem higher as the loudness increases. The pitch rises as the duration increases from 0.01 to 0.1 s, and the pitch of a tone cannot be perceived unless it lasts for 0.01 s or longer.

3.3.3   Sound Localization

The process of projecting a sound to its source is referred to as localization. Although the difference in time between the arrival of the sound wave in the two ears is most important in determining the direction from which a sound impinges, the differences in phase of the sound waves and the loudness on the two sides are also important. At frequencies below 1 kHz the time difference is a determining factor, and at frequencies above 1 kHz, the loudness difference appears most significant. The auditory cortex is necessary for sound localization in many experimental animals and in humans.

For sound sources in the vertical plane, located at an equal distance from the two ears, the sound waves arriving at the right and left ears are identical functions of time (< 1 ms) for all angles of elevation of the sound source. The ability to locate the sound source accurately in this case requires the following: the sound must be complex; the sound must include frequencies above 7 kHz; and the auricles or pinnae must be present. This suggests that when a complex sound with a broad spectrum impinges on the head, it is diffracted by the head and the auricles. The pinna selectively increases the high-frequency sound intensity. Its rumpled shape catches higher-frequency sounds and channels them into the ear canal. It also blocks some higher-frequency sound coming from behind, helping to identify whether the sound comes from the front or the back. For each direction, characteristic changes are superimposed on the incident sound wave which are recognized and utilized to determine the location of the sound source. The ability to hear high-frequency sound is more essential than low-frequency sound for this purpose; it also explains why sound localization becomes difficult with a high-frequency hearing loss.

If no other directional cues are present, irrespective of the actual direction, sound waves with energy predominantly around 1 kHz are localized behind the listener. Frequencies below 500 Hz and around 3 kHz appear in front of the subject. Sound waves with most of their energy centered around 8 kHz are localized overhead (Fig. 3.16). In general, sound localization abilities in the horizontal plane depend primarily on acoustic cues arising from differences in arrival time and loudness level of sound waves at the two ears [Bernstein, 2001]. Localization of unmodulated signals up to approximately 1500 Hz is known to depend on the interaural time difference arising from disparities in the detailed structure of the waveform. The major cue for localization of high-frequency signals is the interaural level difference cue.

The experience of hearing sound as originating from within the head when listening over earphones or headsets has previously been explained based on the

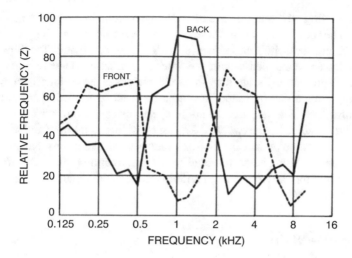

Fig. 3.16 Plane wave sound localization in the median plane by sound spectrum. Sound waves with frequencies predominantly around 1 kHz and above 8 kHz are localized behind the listener (back). Sound waves below 500 Hz and 3 kHz produce front localization. Sound waves containing mostly 8 kHz energy appear to originate from overhead

adaptive nature of sound coming through earphones, because earphones follow head motions. Furthermore, the earphone sound waves arrive at the two tympanic membranes at approximately the same instant of time. The phenomenon has been attributed to the difference in spectral characteristics between earphone listening and free field listening [Schroeder, 1975]. With earphones, standing waves are created in the auditory meatus between the tympanic membrane and the membrane of the earphone. These standing waves have time-varying spectra which are different from those caused by diffraction at the subject's head in a free-field listening situation. Thus, the subject can associate no external location with earphone listening and consequently associates the sound sources with inside the head, which is the only alternative location. Moreover, sound waves can be made to appear as arriving from any direction by appropriately filtering the frequency from a single sound signal before broadcasting it from two loudspeakers positioned in front of the listener [Schroeder, 1993].

In a process called binaural fusion, the brain compares arrival time and the intensity of sound waves information received from each ear and then merges the differences into an integrated perception of a single sound emanating from a specific region of space [Konishi, 1993]. As an example, the brain uses timing and intensity when the auditory signals are delivered separately to each ear through a headset. Instead of perceiving two distinct signals, the listener hears one sound, originating from somewhere inside or just outside the head. If the auditory signals received by the ears are equally intense and arrive simultaneously, it is perceived as a sound arising from the middle section of the head.

The midbrain and forebrain have been shown to contain neurons tuned to sound direction. The receptive fields of these neurons result from sensitivity to interaural time and sound level differences over a broad frequency range. The binaural and spectral cues are combined within the inferior colliculus [Devore and Delgutte, 2010], which projects to the auditory thalamic nucleus in the midbrain, thus enabling cortical representations of sound location. To reach a unified representation of sound source localization space, the auditory nervous system needs to combine the information provided by the two well-known cues: interaural time and sound level. Recent demonstration of the existence of auditory cortical neurons that are sensitive to sound source location helps in providing the information to form a unified representation of auditory space in human auditory cortex [Salminen, et al., 2015].

3.3.4 Masking Effect

The dictionary definition of the term "masking" includes concealing, covering, hiding, screening, and shielding. The auditory masking effect may involve all these actions or phenomena, precipitated by the presence of interfering sound. The boisterous sound has a greater masking effect if the gentle sound lies within the same frequency range, but masking also occurs when the soft sound is outside the frequency range of the loud sound. The effect may also involve the classic paradigm of embedding a tonal signal in a masking noise. Physiological correlates of masking have been identified at various levels of the auditory system [Oxenham, 2013]. Masking might be due either to the spread of the excitation produced by the masker to the place of the tone signal along the cochlea or to the suppression of the response to the signal by the masker. In cat auditory nerves, masking was found to be both excitatory and suppressive, with the relative contribution of the two mechanisms depending on the frequency separation between signal and masker [Delgutte, 1990]. The responses of the mammalian basilar membrane to the classical stimulus combination have also been reported [Recio-Spinoso and Cooper, 2013]. Specifically, the basilar membrane responses to click and noise stimuli indicated the click response component was suppressed by the noise. The simultaneous suppression of the tone and noise response components was also observed under certain stimulus conditions. Detection thresholds of the tone stimuli increased in a near-linear fashion with noise level increments. Thus, many attributes of auditory masking, observed in both neural and psychophysical experiments, are established at the level of mechanical transduction in the cochlea. The neural components of the round window response decreased because of either simultaneous or previous stimulation. The masking effect is particularly sensitive if the frequency spectrum of the masking noise overlaps that of the stimulus.

The most straightforward masking effect involves a linear auditory process, such as the constant signal-to-noise ratio at threshold over a large dynamic range, when listening to the same sound over a wide range of sound levels. Note that nonlinear neural phenomena can be produced by the interaction of noise and tones, such as frequency-specific suppression of tones by noise. However, the precise nature of the relationship between masking of sounds by noise signals and mechanical suppression in the cochlea remains to be clarified.

3.3.5 Deafness and Hearing Loss

Deafness, including partial hearing loss, is classified into two major categories: conduction deafness and nerve deafness. Any condition which interferes with the transmission of sound through the external and middle ears to the cochlea is classified as a conductive hearing loss. Common causes are wax or foreign body in the external auditory meatus, repeated blockage of the auditory tube, destruction of middle ear ossicles, perforation of the ear drum, thickening of the tympanic membrane as a result of infection, and abnormal rigidity of the attachments of the stapes. Nerve deafness means failure of the auditory nerve impulses to reach the cerebral cortex because of damage to the cochlea itself or to the central neural pathways for auditory signals. Causes of nerve deafness include chemotoxic degeneration of the auditory nerve produced by streptomysin, tumors of the auditory nerve, and damage of the hair cells induced by exposure to excessive noise or toxin. Neural hearing loss has also been attributed to viruses such as mumps, as well as to old age. Almost all older people develop some degree of neural hearing loss especially for very high-frequency sound.

Damage to hair cells would lead to hearing loss. If the outer hair cells are damaged and cease its ability to stimulate the inner hair cells, a low-frequency hearing loss for low-intensity sound is produced. For intense sound, the inner hair cells can respond if stimulated directly, so the capacity to hear louder sounds could continue. The inner hair cells are more rugged and not as sensitive as outer hair cells and are much less likely to be damaged by aging, noise, or most ototoxic drugs. If they do they tend to produce hearing loss but not deafness. Also hearing loss from loud sound often occurs first at 3 to 4 kHz due to prolonged exposure because the ear is most efficient in transmitting sounds between 3 and 4 kHz. Thus, the hearing apparatus is also most sensitive to intense stimulus produced at these frequencies, and the outer hair cells which respond to these frequencies are most at risk.

3.3.6 Hearing Acuity and Audiometry

Auditory activity (hearing acuity) is commonly measured with an audiometer. This device is also used clinically to distinguish conduction and nerve deafness. It presents the subject with pure tones which vary from 250 to 8.000 Hz at half or octave intervals. The sound intensity used can vary from zero dB to 100 dB.

The decibel (dB) scale is a relative measure of the root-mean-square (RMS) sound pressure. The standard reference sound pressure is 20 μPa (or

0.02 mPa = 0.0002 dyne/cm²) in air. This reference was adopted by the Acoustic Society of America, and it approximates the auditory threshold of the average young adult at 1000 Hz. The sound pressure level (SPL) in dB is therefore given by

$$SPL(dB) = 20\log\left(\frac{P}{P_o}\right)$$
(3.1)

where P is the RMS sound pressure, P_0 is the reference sound pressure, and log is the logarithm to base 10. It is useful to note that because sound intensity is proportional to the square of sound pressure, Eq. (3.1) may also be written as

$$SPL(dB) = 10\log\left(\frac{S}{S_o}\right)$$
(3.2)

where S and S_0 are the measured and reference sound intensities, respectively.

The reference sound pressure value used in audiometry, however, differs from the above threshold value by 15 to 20 dB. This is because the audiometric reference is the average of normal hearing for different pure tones and the measurements were made in less than ideal conditions (see Fig. 3.15).

An audiogram is a plot of a subject's auditory threshold for various frequencies relative to normal hearing. It provides an objective measurement of the degree of deafness and an assessment of the total frequency range affected. Figure 3.17 shows the audiogram of a subject with "normal" hearing. Figure 3.18 displays the audiogram of a subject who has conductive hearing loss. Approximately 50 dB of extra sound intensity had to be used in order for the subject to hear the sound at 4 kHz through air conduction. However, the hearing was even better than normal for

Fig. 3.17 Audiogram of subject with normal hearing

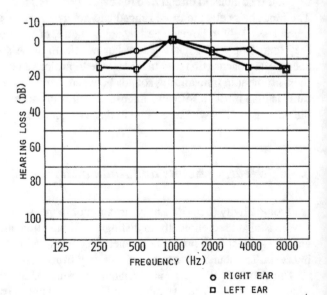

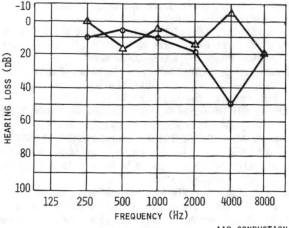

Fig. 3.18 Audiogram of a subject with conductive hearing loss for airborne sound

bone-conducted sound, which means that the cochlea and central auditory pathways were normal. The conduction of sound through the middle ear ossicular system must therefore have been impaired. If both air and bone conduction routes showed considerable loss, some degree of hearing loss or nerve deafness would have been indicated.

Note that experimental results have shown that human subjects with high-frequency hearing loss have difficulty in perceiving microwave-induced sound.

References

von Bekesy G (1957) The ear. Sci Am 197:66–78

von Bekesy G (1960) Experiments in hearing. McGraw, New York

Bernstein LR (2001) Auditory processing of interaural timing information: new insights. J Neurosci Res 66:1035–1046

Cunningham CL, Müller U (2019) Molecular structure of the hair cell mechanoelectrical transduction complex. Cold Spring Harb Perspect Med 9:a033167

Delgutte B (1990) Physiological mechanisms of psychophysical masking: observations from auditory-nerve fibers. J Acoust Soc Am 87:791–809

Devore S, Delgutte B (2010) Effects of reverberation on the directional sensitivity of auditory neurons across the tonotopic axis: influences of interaural time and level differences. J Neurosci 30:7826–7837

Djupesland G, Zwislocki JJ (1972) Sound pressure distribution in the outer ear. Scand Audiol 1:197–203

Eldredge DH (1974) Inner ear-cochlea mechanics and cochlea potential. In: Keidel WD, Neff WD (eds) Handbook of sensory physiology 5(1). Springer, New York

Engebretson AM (1970) A study of the linear and nonlinear characteristics of microphonic voltage in the Cochlea. Sc. D. dissertation, Washington U, St. Louis

Guinan JJ Jr, Peake WT (1967) Middle ear characteristics of anesthetized cats. J Acoust Soc Am 50:1237–1261

Hood LJ (2015) Auditory neuropathy/dys-synchrony disorder: diagnosis and management. Otolaryngol Clin N Am 48(6):1027–1040

Konishi M (April 1993) Listening with two ears. Sci Am 268:66–73

Moller AR (1974) Functions of the middle ear. In: Keidel WD, Neff WD (eds) Handbook of sensory physiology 5(1). Springer, New York

Oxenham AJ (2013) Mechanisms and mechanics of auditory masking. J Physiol 591(10):2375

Recio-Spinoso A, Cooper NP (2013) Masking of sounds by a background noise: cochlear mechanical correlates. J Physiol 591:2705–2721

Salminen NH, Takanen M, Santala O, Lamminsalo J, Altoè A, Pulkki V (2015) Integrated processing of spatial cues in human auditory cortex. Hear Res 327:143–152

Schroeder MR (1975) Models of hearing. Proc IEEE 63:1332–1350

Schroeder MR (1993) Listening with two ears. Music Percept 10(3):255–280

Shaw EAG (1974) The external ear. In: Keidel WD, Neff WD (eds) Handbook of sensory physiology 5(1). Springer, New York

Vernon I, Meikle M (1974) Electrophysiology of the cochlea. In: Thompson RF, Patterson MM (eds) Bioelectric recording techniques, part C. Academic Press, New York

Wever EG, Lawrence M (1954) Physiological acoustics. Princeton University Press, Princeton

Wiener FM, Ross DA (1946) The pressure distribution in the auditory canal in a progressive sound field. J Acoust Soc Am 18:401–408

General References

Cotran RS, Kumar V, Robbins SL (1994) Pathologic basis of disease, 5th edn, Philadelphia/Baltimore

Grant ILB (1972) Grant's atlas of anatomy, 6th edn. Williams & Wilkins, Baltimore

Lin JC (1978) Microwave auditory effects and applications. CC Thomas, Springfield

Nolte J (1993) The human brain, 3rd edn. Mosby, St Louis

Silverthorn DU (2016) Human physiology: an integrated approach, 7th edn. Pearson, Upper Saddle River

Chapter 4
Microwave Property of Biological Materials

The interaction of microwave radiation with biological systems is influenced by the electromagnetic properties of tissue media, specifically dielectric permittivity and magnetic permeability according to the four Maxwell's equations of the electromagnetic theory summarized in Chap. 2.

Biological tissues are heterogeneous and consist of macromolecules, organelles, cells, tissue, organs, and water. There are also various ions, polar molecules, membranes, and membrane-bound components. These are the constituents that form the structure of the biological body, including humans, and are often referred to as dielectric materials owing to their similarity to the electric, magnetic, physical, and chemical nature of such materials.

True to the popular saying: water is the substance of life. Biological materials may be classified into three major groups according to their water content. The first group is of high-water content (90% or more) that consists of fluids containing electrolytes, macromolecules, and other cellular materials and includes blood, vitreous humor, and cerebral spinal fluid (CSF). The second group is of moderate water content (80% or less) and consists of skin, muscle, brain, and internal organs. The last group is made up of tissues with low water content (less than 40%) such as fat, bone, and tendons.

The frequency-dependent characteristics of the dielectric properties of biological materials may be described by the relaxation processes associated with many dielectric materials displaying a time-dependent response to sudden excitation. This chapter describes the dielectric relaxation processes. It also presents a brief summary of the measured tissue dielectric permittivity data on dielectric constants and conductivities as functions of frequency and temperature.

© Springer Nature Switzerland AG 2021
J. C. Lin, *Auditory Effects of Microwave Radiation*,
https://doi.org/10.1007/978-3-030-64544-1_4

4.1 Frequency Dependence of Dielectric Permittivity

The electromagnetic properties of biological materials have been extensively studied [Schwan, 1957, 1963; Schwan et al., 1976; Schwan and Foster, 1980; Gabriel et al., 1996a, b]. A basic understanding has been achieved of the structures and mechanisms that determine the RF and microwave properties of biological materials.

Biological materials have magnetic permeability values that are close to free space and are independent of frequency. It is noted however that biological tissues are not free of ferromagnetic materials. They exhibit very high dielectric constants compared with many other types of homogeneous solids and liquids. This is because biological tissues are heterogeneous and consist of ions, macromolecules, cells, and other membrane-bound substances. An example of the frequency-dependent character of biological tissue materials (muscle) is given in Fig. 4.1. There are three principal regions of dispersions described as α, β, and γ dispersions, respectively. Each dispersion is defined either by a single relaxation frequency or a small group of relaxation frequencies.

The α dispersion at extremely low frequency (<10 kHz) is the result of cell membrane capacitance variations with frequency. The membrane capacitance undergoes a pronounced decrease from 10 $\mu f/cm^2$ to 1 $\mu f/cm^2$ as the frequency decreases. The β dispersion originates from the heterogeneous nature of biological tissues (proteins and macromolecules). At frequencies between 1 kHz and 10 MHz, the applied electric field causes electric charges to accumulate at boundaries separating tissue regions of different dielectric property (e.g., membranes separating intra- and extracellular fluids). Finite periods of time are required before the boundaries can reach charge neutrality, giving rise to the relaxation phenomenon.

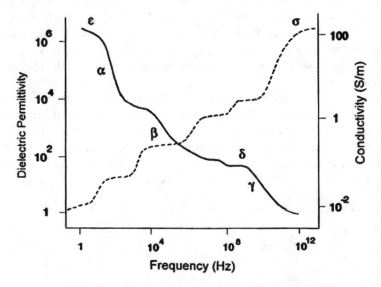

Fig. 4.1 Dielectric permittivity and electrical conductivity of muscle-like biological materials as a function of frequency from 1 Hz to 1 THz

As the frequency increases, insufficient time is available during each cycle to allow complete charging of the cell membranes. The total charges per cycle must decrease, along with the membrane capacitance, as the frequency is increased. This is manifested as a decrease in the dielectric constant.

For still higher frequencies, the membrane capacitance change stabilizes until the rotational and vibrational properties of polar molecules of water become significant. This dispersion is characterized by a single relaxation frequency slightly lower than that of pure water.

Also shown in Fig. 4.1, there is a minor relaxation termed the δ dispersion, occurring between the β and γ dispersions. This is caused by the relaxation of water molecules bound to the surface of macromolecules and rotation of amino acids as well as charged side groups of proteins [Schwan, 1977; Grant et al., 1978].

The conductivities of biological materials change in a similar manner. Cell membranes have a relatively high capacitance (about 1 $\mu f/cm^2$) at frequencies near the β dispersion. They become progressively short circuited for frequencies above the β dispersion, permitting the intracellular fluid to participate in current conduction. This causes the conductivity to increase as the frequency increases. In the γ dispersion region, the rotation of water molecules is accompanied by viscous loss which accounts for the principal mechanism for increased conductivity.

It can be seen from the dispersion behavior of muscle tissue summarized above that the dielectric and conductivity properties of biological materials in the RF and microwave frequency range are largely determined by the relaxation properties of biological membrane, protein macromolecule, and tissue water. More detailed descriptions may be found in Debye [1929], von Hipp [1954], and Daniel [1967].

For additional information about dielectric permittivity properties of biological materials, the reader is referred to Schwan [1957], Cole [1968], Hill et al. [1969], Grant et al. [1978], and Gabriel et al. [1996a, b].

4.2 Relaxation Processes

The frequency-dependent nature of the electrical properties of biological materials may be described by relaxation processes associated with many dielectric materials displaying a time-dependent response to sudden excitation. Polar molecules and cellular components rotate in response to an applied electric field. This rotation is impeded by inertia and by viscous forces. Therefore, the orientation of polar molecules does not occur instantaneously, giving rise to the time-dependent behavior (relaxation). Moreover, cells and tissues composed of structural components of different properties, when subjected to a step function electrical stimulation, require a finite length of time for charges to accumulate at the interfaces. The accumulation of charges at interfaces continues until a condition of current equilibrium is reestablished, thus the relaxation. Many types of relaxation processes can occur in tissue material (see Fig. 4.1), owing to dipoles and charges existing in the material.

When the dipole distribution is uniform, positive charges of the dipole cancel the effect of the negative charges from an adjacent dipole. However, when dipole distribution changes from point to point, complete cancellation cannot occur. At an interface, the ends of the dipoles leave an uncanceled charge on the surface, which become an equivalent "bound" charge in the material. The relaxation behavior may therefore be examined by considering the response of bound charges in an applied electric field. For the model shown in Fig. 4.2, the dynamic force balance equation is

$$m\frac{d^2 x}{dt^2} = qE - m\omega_s^2 x - mv\frac{dx}{dt} \tag{4.1}$$

where x is the charged particle displacement, E is the electric field, and q and m are the particle charge and mass, respectively. The force given by the product between mass and particle acceleration, on the left hand side of the equation, consists of the electric driving force qE, an elastic force proportional to displacement x with spring constant denoted as $m\omega_s^2$, and a retarding damping force proportional to velocity dx/dt with damping constant mv. The spring damping constants are chosen in this notation because ω_s is the characteristic frequency of the spring-mass system and v is the particle collision frequency.

If the applied electric field varies harmonically in time ($e^{j\omega t}$), Eq. 4.1 may be rearranged to give

$$x(\omega) = \frac{(q/m)}{\omega_s^2 - \omega^2 + j\omega v} E \tag{4.2}$$

Note that the equilibrium position for the charge (x = 0) represents local electrical neutrality in the medium. When the charge is displaced from its equilibrium position, a dipole field is established between the charge itself and the "hole" left behind bound in the molecular or membrane structure. A dipole moment p of charge q times the displacement x is formed. For a medium with volume-bound charge density ρ, the total dipole moment per unit volume or polarization P is

$$P = \rho p = \frac{\rho\left(\frac{q^2}{m}\right)E}{\omega_s^2 - \omega^2 + j\omega v} \tag{4.3}$$

Fig. 4.2 Behavior of bounded charges in an applied electric field

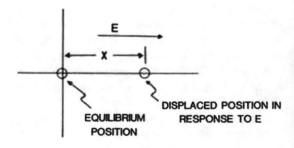

The electric flux density D may be expressed in terms of the electric field E and polarization P as $D = \varepsilon_0 E + P$. For isotropic media, the permittivity ε may be related to D by the expression $D = \varepsilon E$. These relations, together with (4.3), give an expression for the permittivity,

$$\varepsilon = \varepsilon_0 \left(1 + \frac{\omega_p^2}{\omega_s^2 - \omega^2 + j\omega v} \right) \tag{4.4}$$

where

$$\omega_p^2 = \frac{pq^2}{m\varepsilon_0}$$

and ε_0 is the free-space permittivity. Clearly ε is a complex quantity and can be denoted as

$$\varepsilon = \varepsilon' - j\varepsilon'' \tag{4.5}$$

where ε' and ε'' are the real and imaginary parts of the permittivity. It can be obtained by equating the real and imaginary parts of (4.4) and (4.5), respectively.

The velocity of bound charge motion $v = \dfrac{dx}{dt}$ is obtained from (4.2)

$$v(\omega) = \frac{\dfrac{q}{m} E}{v - j \left[\dfrac{\omega_s^2 - \omega^2}{\omega} \right]} \tag{4.6}$$

The finite velocity of charge motion in the material indicates that particle cannot respond instantaneously to a suddenly applied electric field. This time-delay phenomenon, or relaxation, gives rise to a frequency-dependent behavior of charge displacement leading to changes in permittivity with frequency.

It was mentioned earlier that relaxation is exhibited by all biological tissues and many physical materials. In what follows, the general development will focus on two classes of dielectric materials of interest to the biophysical aspects of electromagnetic interactions with biological systems.

4.2.1 Low-Loss Dielectric Materials

For low-loss dielectric materials characterized by low collision frequency, v, ε', and ε'' can be derived from (4.4) with $\omega \neq \omega_s$

$$\frac{\varepsilon'}{\varepsilon_0} = 1 + \left[\frac{\omega_p^2}{\omega_s^2 - \omega^2} \right] \tag{4.7a}$$

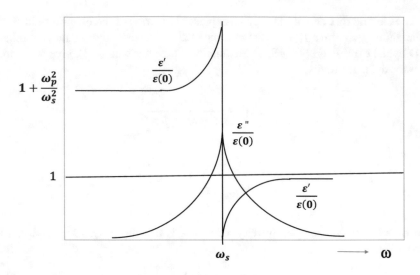

Fig. 4.3 Frequency dependence of permittivity of low-loss dielectric materials

$$\frac{\varepsilon''}{\varepsilon_0} = \frac{\omega v \omega_p^2}{\left(\omega_s^2 - \omega^2\right)^2} \tag{4.7b}$$

A graphical representation of this result is shown in Fig. 4.3. The part of the permittivity is usually high at low frequencies, increasing to extremely high values at the characteristic frequency ω_s and returning to ε_0 at higher frequencies. The imaginary part of permittivity is small at all frequencies except near ω_s. It is high because of the large particle displacement at the characteristic frequency giving rise to large collisional and thus absorption effects. For most solid dielectric materials of practical interest to microwave biophysics (e.g., Plexiglas), the frequency ω_s is in the optical spectrum or above. They are thus characterized by low loss and slowly increasing dielectric constant with frequency. Values of $\dfrac{\varepsilon'}{\varepsilon_0}$ and $\dfrac{\varepsilon''}{\varepsilon'}$ for some low-loss materials are given in Table 4.3, as a function of frequency.

4.2.2 Lossy Dielectrics at Low Frequencies

At frequencies low compared to the characteristic frequency ($\omega \ll \omega_s$), (4.4) reduces to

$$\frac{\varepsilon}{\varepsilon_0} = 1 + \frac{\dfrac{\omega_p^2}{\omega_s^2}}{1 + \dfrac{j\omega v}{\omega_s^2}} \tag{4.8a}$$

This equation may be expressed in terms of the permittivity at zero frequency

$$\varepsilon(0) = \varepsilon_0 \left(1 + \frac{\omega_p^2}{\omega_s^2} \right)$$

and the permittivity at infinite frequency $\varepsilon(\infty) = \varepsilon_0$. Note that both these limiting values of complex permittivity are real numbers. Thus, (4.8a) written in the Debye form becomes

$$\varepsilon = \varepsilon(\infty) + \frac{\varepsilon(0) - \varepsilon(\infty)}{1 + j\omega\tau} \qquad (4.8b)$$

where $\tau = v / \omega_s^2$ is the relaxation time and is inversely related to the relaxation frequency ω_s. Relaxation time τ, proportional to v, is a measure of how fast charges move in response to an applied field. A low value of v means fewer collisions and a faster response, giving rise to a shorter relaxation time τ. From (4.8b), the real and imaginary parts of ε may be written as

$$\varepsilon' = \varepsilon(\infty) + \frac{\varepsilon(0) - \varepsilon(\infty)}{1 + (\omega\tau)^2} \qquad (4.9)$$

$$\varepsilon'' = \frac{\omega\tau \left[\varepsilon(0) - \varepsilon(\infty) \right]}{1 + (\omega\tau)^2} \qquad (4.10)$$

The loss mechanism described above applies to a model in which there are only bound charges. In biological materials and many other liquids and solids, there exists an appreciable number of free charges. The loss mechanisms in these materials are described by the conductivity relating current density to applied field (see Eqs. 2.9 and 2.16). This may be visualized as free charges moving randomly because of their thermal velocities and frequently experiencing collisions with other particles making up the material. The applied electric field produces a general drift in the direction of the applied field with a nonzero average velocity. This component of velocity and the resulting current are in phase with the applied field at frequencies low compared with the collision frequency and represent an ohmic or joule loss. At any frequency, this loss and that for the bound charge add directly. If one is interested only in the macroscopic behavior, it is customary to include the conduction loss in the imaginary part of the permittivity or vice versa. If one is to be concerned with microscopic properties, it would then be necessary to keep the two mechanisms separate.

Furthermore, in biological materials, it is impossible to separate the two contributions from measurements made at a given frequency. Therefore, the presence of a finite ε'' has the effect of producing a total electrical conductivity σ, and a finite conductivity is equivalent to a total imaginary part of the permittivity as ε''. The relationship between σ and ε'' may be derived from two of Maxwell's equations, (2.10) and (2.11), or

$$\sigma = \omega\varepsilon''$$

where σ, an equivalent conductivity representing all losses, is given by

$$\sigma = \frac{\omega^2 \tau \left[\varepsilon(0) - \varepsilon(\infty) \right]}{1 + (\omega \tau)^2} \tag{4.11}$$

This equation for conductivity can be expressed in an alternate fashion such as

$$\sigma = \sigma(0) + \left[\sigma(\infty) - \sigma(0) \right] \frac{(\omega \tau)^2}{1 + (\omega \tau)^2} \tag{4.12}$$

where $\sigma(0)$ *and* $\sigma(\infty)$ are conductivity values far below and above relaxation frequency ω_s. The conductivity term $\sigma(0)$ has been added to account for the ionic and frequency-independent contribution to Eqs. (4.8b) and (4.12), which are special cases of the Kramers-Kronig relationship [Boettcher, 1952]. They show that the frequency response of the permittivity determines that of the conductivity and vice versa.

It is also convenient to define a relative dielectric constant through dividing ε' by the free space permittivity ε_0:

$$\varepsilon_r = \frac{\varepsilon'}{\varepsilon_0} \tag{4.13}$$

This notation will be used often in this book, as this notation is simple and facilitates mathematical manipulations. It should be mentioned that ε_r is usually referred to simply as the dielectric constant and it is dimensionless (See also Eq. 2.19).

The ratio $\varepsilon''/\varepsilon'$ is also a commonly used parameter for dielectric materials, and it is referred to as the loss tangent. For low-loss materials such as those given in Table 4.1, the loss tangent is much less than unity. On the other hand, biological materials are characterized by consideration amounts of losses and therefore have loss tangents close to or greater than 1.

The variation with frequency of the real and imaginary parts of the permittivity, and electrical conductivity is illustrated in Fig. 4.4. It is seen that the dielectric constant ε' falls from a higher value of $\varepsilon(0)$ to $\varepsilon(\infty)$ as the frequency increases through the dispersion region, while the conductivity rises from a small value of $\sigma(0)$ *to* $\sigma(\infty)$. The imaginary part of the permittivity ε'' peaks at $\omega \tau = 1$ and falls off for both higher and lower frequencies.

4.2.3 Biological Materials

The results summarized in Eqs. (4.9, 4.10, and 4.11) depict the dielectric properties of biological tissues in most RF and microwave regions. For a particular range of frequencies, the value of $\varepsilon(\infty)$ *and* $\varepsilon(0)$ may be taken to be the permittivity at frequencies far above and below the relaxation frequency ω_s.

Table 4.1 Permittivity of common low-loss laboratory dielectric materials at 25 °C

	$\varepsilon'/\varepsilon_0$				$\varepsilon''/\varepsilon'$			
	Frequency (Hz)							
	10^6	10^8	3×10^9	2.5×10^{10}	10^6	10^8	3×10^9	2.5×10^{10}
Fined quartz	3.78	3.78	3.78	3.78	1×10^{-4}	2×10^{-4}	6×10^{-5}	2.5×10^{-4}
Glass–soda borosilicate	4.84	4.84	4.82	4.63	3.6×10^{-3}	3×10^{-3}	5.4×10^{-3}	9×10^{-3}
Foamed polystyrene	1.03	1.03	1.03	1.03	2×10^{-4}	2×10^{-4}	1×10^{-4}	1×10^{-4}
Polystyrene	2.26	2.26	2.26	2.26	2×10^{-4}	2×10^{-4}	3.1×10^{-4}	6×10^{-4}
Teflon (polytetrafluoroethylene)	2.1	2.1	2.1	2.08	2×10^{-4}	2×10^{-4}	1.5×10^{-4}	6×10^{-4}
Beeswax	2.53	2.45	2.39		9.2×10^{-3}	9×10^{-3}	7.5×10^{-3}	

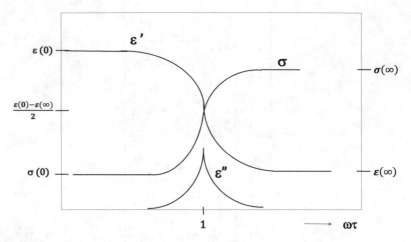

Fig. 4.4 Frequency-dependent characteristics of permittivity and conductivity of lossy dielectric materials

Since the dielectric properties of biological materials are complex and it requires a distribution of relaxation processes for proper representation throughout the RF and microwave frequencies [Cole and Cole, 1941]. In this case, the dielectric behavior may be modeled as sum of relaxation processes, with each process being a non-instantaneous exponential relaxation from one state to another, as depicted in Eq. (4.9, 4.10, and 4.11). The corresponding responses in the frequency domain are of the form, often referred to as Debye equation,

$$\varepsilon = \varepsilon_0 + \sum_{n=1}^{N} \frac{\Delta\varepsilon_n}{1 + j\omega\tau_n} \tag{4.14}$$

where $\Delta\varepsilon_n$ is the difference between the permittivity far below and far above the relaxation frequency, and τ_n is the relaxation time associated with each relaxation process.

As presented earlier, the dielectric properties of biological materials can often be characterized by using three distinct relaxation processes. The relaxation times are well separated such that $\tau_1 \gg \tau_2 \gg \tau_3$. The corresponding relaxation frequencies for each region are around 100 Hz, 50 kHz, and 25 GHz at body temperature (37 °C) for most biological materials.

4.3   Temperature Dependence of Dielectric Properties

In considering relaxation processes in Sect. 4.2, it was indicated that the electrical properties of biological tissues throughout the RF and microwave frequency are governed by structural and rotational relaxation phenomena. There are in biological materials abundant layers of tissues and membranes with different dielectric constants and conductivities. When an electric field is applied, surface charges build up at the interface between two adjacent layers, thus generating varying permittivity values. Furthermore, molecular dipoles tend to orient in an applied electric field. Since the reorientation of polar molecules does not occur instantaneously, it gives rise to rotational relaxation effect.

The movement of charges, whether to accumulate at tissue interface or to realign through rotation, is hindered by collisions with particles in the surrounding medium. The speed of movement depends therefore, on temperature, among other factors. Thus, both dielectric constants and conductivities are temperature sensitive.

A mathematical treatment of the temperature dependence of dielectric constant and conductivity, however, is not straightforward. The few studies concerned with the temperature dependence of dielectric constants and conductivities have shown that temperature dependence of the conductivity is much more pronounced than that of dielectric constant. Near a relaxation frequency, the relationship

$$\frac{d\varepsilon}{\varepsilon} = \frac{d\sigma}{\sigma} \frac{\varepsilon(0) - \varepsilon(\infty)}{\varepsilon(0) + \varepsilon(\infty)} \tag{4.15}$$

may be derived for the relative change in dielectric constant to relative change in conductivity [Schwan and Foster, 1980]. Since $\varepsilon(0)$ and $\varepsilon(\infty)$ are fairly independent of temperature and $\varepsilon(0)$ is much larger than $\varepsilon(\infty)$, the change of the dielectric constant with temperature must be smaller than that of the conductivity.

The dielectric properties of biological materials are characterized by three distinct dispersion regions. For each dispersion, the temperature coefficient reflects those of $\varepsilon(0)$, $\varepsilon(\infty)$ *and* ω_s. The dispersion may be due to frequency-dependent membrane capacitance arising from ionic gating currents through or counterions surrounding membranes. These ionic activities have temperature coefficients similar to that of the conductivity of electrolytes, i.e., about 2% per °C. The dispersion is caused by polarization effects in which the cellular membranes are charged through the electrolytes. Thus the temperature coefficient is equal to that of the conductivity of electrolytes. The dispersion originates from the rotational relaxation of water molecules. Hence, its temperature dependence is equal to that of water, which again is close to 2% per °C. Accordingly, the temperature coefficient of conductivity for biological materials has a maximum value of about 2% per °C for tissues with higher water content, such as blood, cerebral spinal fluid (CSF), and most organs.

Further discussions on the temperature dependence of dielectric constant and conductivity are given in Sect. 4.4.1 for water and in Sect. 4.4.4 for some representative data of measured tissue dielectric permittivity.

4.4   Measured and Modeled Tissue Permittivity Data

This section presents some of the measured tissue permittivity data or dielectric constants and conductivities of biological materials most relevant to the aims and objectives of this book. It will be seen that the dielectric properties of biological materials are indeed described by the relaxation processes detailed above. It is emphasized that although the permittivity values are derived from tissues of many different species, there is little variation among the measured values for a given tissue type. An exception is fatty tissue, which is characterized by low dielectric constant and conductivity and by low electrolyte content. Indeed, the water content in fatty tissues ranges from a few percent to more than 40%, depending on the animal species [Schwan, 1957]. Since the dielectric constant and conductivity of water is high, the permittivity of fatty materials may differ significantly with variations in water content.

4.4.1 Permittivity of Water

Water by far is the most abundant constituent of animals and constitutes approximately 60% of the total body mass in humans. For example, water makes up 93% of the blood, about 80% of skeletal muscle, less than 40% of fat, and approximately 70% of white matter. The fluid nature of water allows both the dissolved electrolytes and the suspended substances to diffuse to different parts of the cell or tissue. When the cell or tissue losses its water, life is endangered or extinguished.

Of the total body water, about 62% is in the intracellular space and 38% in the extracellular space. Thus, water content exerts major influences on the permittivity properties (dielectric constant and conductivity) of biological materials. The following is a brief discussion of permittivity properties of water as a function frequency and temperature.

It is well known that dielectric constant and conductivity display characteristic dispersions at microwave frequencies. A graph showing the frequency dependence of dielectric constant and conductivity of water at 37 °C is given in Fig. 4.5. A relaxation frequency is seen at 32 GHz. Both dielectric constant and conductivity are constant and invariant from DC to 1 GHz. At frequencies above 3–5 GHz, water starts to disperse. The dielectric constant falls from a value of 74 to about 28 at

Fig. 4.5 Permittivity (dielectric constant and conductivity) of free water at 37 °C from 0.01 to 1000 GHz (10 MHz to 1 THz)

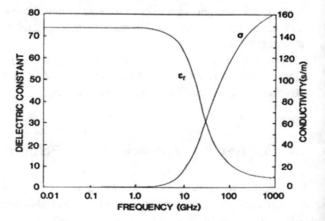

35 GHz before reaching a lower limit near 4.5. Conversely, the conductivity increases monotonically for frequencies up to beyond the microwave range. Clearly, the dielectric behavior of water is characterized by a single relaxation process between DC and microwave frequencies, which is governed by the induced rotations of individual water molecules as polar dipoles in a viscous fluid. Furthermore, they are represented by the simple Debye equations of (4.9 and 4.12), where τ is the relaxation time constant and $\sigma(0)$ is the DC conductivity, essentially zero for water.

The temperature dependence of the dielectric permittivity of water for six frequencies are given in Fig. 4.6 [Grant et al., 1978]. These graphs indicate that the dielectric constants and conductivity ($\sigma = \omega\varepsilon''$) of water at 0.58, 1.74, and 3.00 GHz decrease with increasing temperature, while those at 9.4, 23.7, and 34.9 GHz may increase, peak, and decrease with increasing temperature. These subtleties in response to temperature change at higher frequencies probably stem from the fact that water starts to disperse above 5 GHz.

Note that the relaxation frequency of water increases with temperature in an exponential fashion. In particular, the relaxation frequency at 0 °C is 8.84 GHz becomes 39.8 GHz at 60 °C. Furthermore, in the presence of polar solute molecules, the relaxation frequency shifts to lower frequencies because the dispersion of the solute molecules takes place at frequencies far below that for water. This will give rise to second or third regions of dispersion, which is usually well separated from that of water.

4.4.2   Measured Tissue Permittivity Data

Availability of accurate information and reliable knowledge about the permittivity properties of biological tissues is important for both studies on macroscopic and microscopic interactions of RF and microwave radiation with biological systems. On a macroscopic level, these properties govern the coupling and power deposition in tissue upon exposure by RF and microwave radiation. On a microscopic level, they suggest mechanisms that may underlie the absorption of RF and microwave energy by mammalian tissues. This knowledge and quantitative information are also essential for analyzing the relationships among various observed biological effects or responses from RF and microwave interactions and to help assess potential applications such as therapeutic effectiveness and diagnostic usefulness of RF and microwave radiation as well as for applications under safety and security scenarios.

Over the years, a large body of experimental data about dielectric permittivity of mammalian tissues have been accumulated through in situ, ex vivo, and in vivo laboratory studies by many investigators. A concerted effort has been devoted to consolidate the data by using modern measurement techniques to provide new data to fill gaps in knowledge with respect to tissue types and frequency ranges [Gabriel et al., 1996a, b].

The following are some representative data of measured tissue dielectric permittivity with a special emphasis on tissues in the head. It also includes results for modeled data on dielectric constants and conductivities.

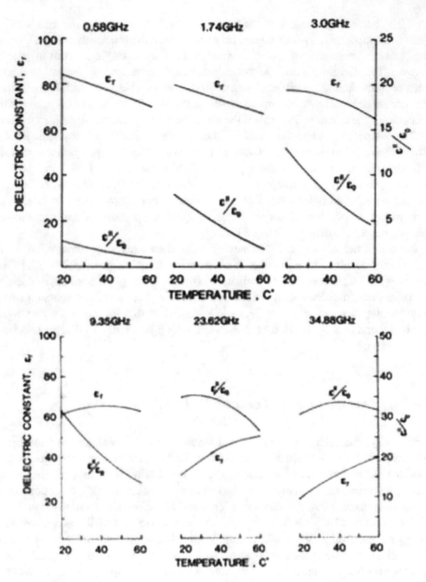

Fig. 4.6 Temperature dependence of dielectric permittivity (dielectric constant and conductivity) at six microwave frequencies

4.4.2.1 Brain Tissue Permittivity Data

Dielectric permittivity and conductivity of gray matter in the brain from reported studies at 37 °C are presented in Fig. 4.7 for frequencies from 10 Hz to 20 GHz. The dispersion behaviors are obvious and rather broad, suggesting overlap of

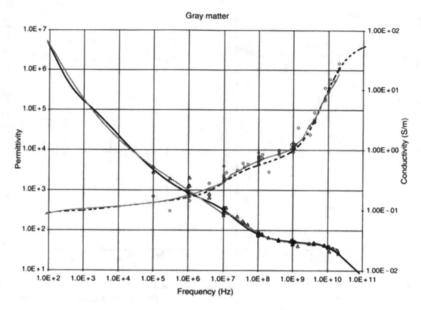

Fig. 4.7 Dielectric permittivity and conductivity of gray matter in the brain at 37 °C. Gray lines, circles, and triangles are experimental data; solid and dash black lines are model predictions. [From Gabriel et al., 1996a]

individual relaxation processes in biological materials. Changes in dielectric permittivity and conductivity are quite large and are in opposite directions. For example, dielectric permittivity, ε_r, decreases from 80 to 45 as the frequency increases from 100 MHz to 10 GHz, while conductivity, σ, increases by nearly two orders of magnitude.

A comparison of the permittivity properties for fresh human brain tissue with those for human, canine, ovine, primate, and swine brain tissues at 37 °C in the microwave frequency range of 2 to 4 GHz is provided in Fig. 4.8. The dielectric constant remains unaltered for all species listed within the test microwave frequency range [Lin, 1975]. However, in addition to the slight increase with frequency, the measured conductivity values showed a greater variability in the composite representation of data from brains of other species. Nevertheless, the similarity between the two sets of data suggests within the accuracy of these measured data that the dielectric properties of brain tissues from different mammalian species are identical. This observation is supported by mouse and rat brain tissue measurements performed at 37°C also, which again showed no significant differences in dielectric permittivity of brain tissues between these two species of rodents [Steel and Sheppard 1985]. Indeed, the validity of this conclusion prevails to date. Although the available permittivity values are derived from tissues of different mammalian species, there is little variation among the measured values for a given tissue type.

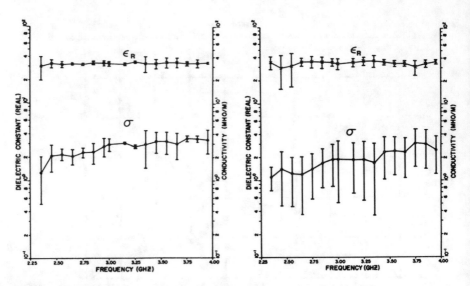

Fig. 4.8 Permittivity properties of homogenized fresh brain tissues at 37 °C in the microwave frequency range 2 to 4 GHz. (**a**) Fresh human data; (**b**) combined fresh canine, ovine, primate, swine, and human. [From Lin, 1975]

4.4.2.2 Permittivity Data of Other Tissues

The relative permittivity and values for the 25 tissue types represented in the human head are listed in Table 4.2. They include bones, CSF, fat, gray matter, white matter, muscle, teeth, etc. A Cole and Cole [1941] extrapolation technique was used to determine dielectric permittivity values for the tissues at the RF and microwave frequencies of 64, 300, and 400 MHz [Gabriel et al., 1996a]. Note that, in general, the relative permittivity and conductivity of blood, brain, eyes, CSF, muscle, and other tissues with higher water content are an order of magnitude higher than the corresponding parameters for bone, fat, tooth, and other tissue with low water content. The values of dielectric constant and conductivity for tissues with intermediate water content such as cartilage, lens, ligaments, and nerves are lower by 30 to 40% from the respective values for high-water content tissues.

4.4.3 Debye Modeling of Tissue Permittivity Data

As an example of modeling the frequency-dependent permittivity properties of various tissues using measured permittivity values by the Debye Eq. (4.14), permittivity for frequencies between 10 MHz and 3000 MHz may be modeled as the sum of two relaxation processes with two constants such that Eq. 4.14 becomes

$$\varepsilon^*(\omega) = \varepsilon_0 \left[\varepsilon_\infty + \frac{\varepsilon_{s1} - \varepsilon_\infty}{1 + j\omega\tau_1} + \frac{\varepsilon_{s2} - \varepsilon_\infty}{1 + j\omega\tau_2} \right] \qquad (4.16)$$

Table 4.2 Relative permittivity (dielectric constant) and conductivity of tissues in the human head at 37 °C

Tissue type	64 MHz		300 MHz		400 MHz		Density
	σ (S/m)	ε_r	σ (S/m)	ε_r	σ (S/m)	ε_r	ρ (kg/m³)
Blood	1.206	86.51	1.316	65.69	1.350	64.18	1058
Blood vessel	0.429	68.69	0.537	48.36	0.562	47.00	1040
Body fluid	1.503	69.13	1.517	69.02	1.529	69	1010
Bone (cancellous)	0.161	30.89	0.215	23.18	0.235	22.44	1920
Bone (cortical)	0.059	16.69	0.083	13.45	0.091	13.15	1990
Bone marrow	0.021	7.215	0.027	5.76	0.029	5.67	1040
Cartilage	0.452	62.96	0.552	46.81	0.587	45.47	1097
Cerebellum	0.718	116.5	0.972	59.82	1.030	55.99	1038
CSF	2.065	97.35	2.224	72.79	2.251	70.99	1007
Eye (aqueous humor)	1.503	69.13	1.517	69.02	1.529	69	1009
Eye (cornea)	1.000	87.45	1.150	61.43	1.193	59.28	1076
Eye (lens)	0.586	60.57	0.647	48.97	0.669	48.15	1053
Eye (retina)	0.883	75.35	0.975	58.93	1.004	57.67	1026
Eye (sclera/wall)	0.883	75.35	0.975	58.93	1.004	57.67	1026
Fat	0.035	6.511	0.039	5.635	0.041	5.579	916
Glands	0.778	73.98	0.851	62.47	0.877	61.55	1050
Gray matter	0.511	97.54	0.691	60.09	0.738	57.43	1038
Ligaments	0.474	59.52	0.537	48.00	0.560	47.29	1220
Lymph	0.778	73.98	0.851	62.47	0.877	61.55	1040
Mucous membrane	0.488	76.80	0.630	51.96	0.669	49.89	1040
Muscle	0.688	72.27	0.770	58.23	0.796	57.13	1047
Nerve (spine)	0.312	55.11	0.418	36.95	0.447	35.41	1038
Skin/dermis	0.435	92.29	0.640	49.90	0.688	46.78	1125
Tooth	0.059	16.69	0.083	13.45	0.091	13.15	2160
White matter	0.291	67.91	0.413	43.82	0.445	42.07	1038

ρ (kg/m³) is mass density corresponding to each tissue type

where ε_∞ is the high-frequency permittivity and the static (or low frequency) permittivity is given by

$$\varepsilon_s = \varepsilon_{s1} + \varepsilon_{s2} - \varepsilon_\infty \tag{4.17}$$

For the results shown here, the measured properties of biological tissues (muscle, fat, bone, blood, intestine, cartilage, lung, kidney, pancreas, spleen, lung, heart, brain/nerve, skin, and eye) were obtained from the literature. Optimized values for $\varepsilon_{s1}, \varepsilon_{s2}, \varepsilon_\infty, \tau_1$, and τ_2 in Eq. (4.16) were obtained by nonlinear least squares fitting to the measured data for fat and muscle (Table 4.2), with τ_1 and τ_2 being the average of the optimized values for fat and muscle, respectively. All other tissues have properties falling between these two types of tissues. The measured and modeled tissue properties in the RF and microwave frequency range of 10 MHz to 3000 MHz, using two Debye relaxation constants for muscle and fat, are shown in Fig. 4.9.

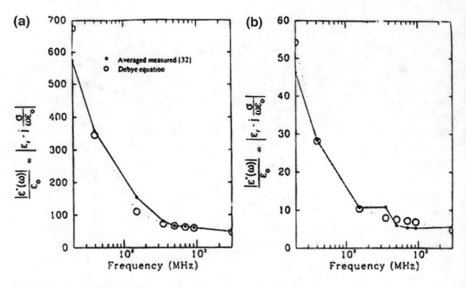

Fig. 4.9 Measured and modeled tissue properties using two Debye relaxation constants for (**a**) muscle and (**b**) fat. [From Lin and Gandhi, 1996]

4.4.4   Temperature Dependence of Measured Tissue Dielectric Permittivity

The temperature dependence of the dielectric properties of biological tissues in the microwave frequency range has only been studied more vigorously since about 2000, although there were some discussions on the topic especially regarding water and electrolytes around 1980. It is interesting to observe this in the context of scientific investigations of therapeutic heating of deep-seated issues like muscle by microwave radiation that began in the mid-1940s. Further clinical studies of microwave diathermy led to its acceptance as a therapeutic instrument by the American Medical Association in 1947. Microwave diathermy's therapeutically effectiveness in traumatic conditions such as sprains and stresses, especially when combined with massage, is clearly indicated and remains as an integral part of the treatment modalities in physical and sports medicine.

Furthermore, microwave hyperthermia as a clinical treatment for localized malignant disease began in the late 1970s. The objective of this controlled temperature elevation is the treatment of tumors, directly by microwave-induced irreversible biological damage or indirectly by enhancing the therapeutic effectiveness of other treatment regimes such as nonionizing radiation or chemotherapy. There has been a steady stream of publications over the last 40 years on all aspects of hyperthermia therapy, showing hyperthermia is gaining wider use in clinical practice in treating a variety of malignant tumors.

Table 4.3   Debye relaxation time constants $\tau_1 = 46.2 \times 10^{-9}$ s and $\tau_2 = 0.91 \times 10^{-10}$ s for listed tissues

Tissue type	ε_∞	ε_{s1}	ε_{s2}
Muscle	40.0	3948	59.09
Bone/cartilage	3.4	312.8	7.11
Blood	35.0	3563	66.43
Intestine	39.0	4724	66.09
Liver	36.3	2864	57.12
Kidney	35.0	3332	67.12
Pancreas/spleen	10.0	3793	73.91
Lung (1/3 value)	10.0	1224	13.06
Heart	38.5	4309	54.58
Brain/nerve	32.5	2064	56.86
Skin	23.0	3399	55.59
Eye	40.0	2191	56.99

Microwave and RF catheter ablation for cardiac arrhythmias was first announced in 1987 for tachyarrhythmia [Huang and Wilber, 2000; Lin, 2003]. Shortly thereafter, this minimally invasive microwave technique was adopted to help ablate malignant liver nodules. Microwave catheter ablation along with the use RF energy has become an important therapy in the management of related diseases.

As mentioned, for quite some time, the scientific attention has mostly been directed toward the frequency dependence of dielectric permittivity properties. In recent years, perhaps given the momentum by the successful applications of minimally invasive ablation techniques in the RF and microwave frequency range, it has done much to stimulate interest in temperature dependence of dielectric permittivity of biological tissues, since about 2000. Even so, only limited data on temperature-dependent dielectric permittivity is available (Table 4.3).

Table 4.4 summarizes published papers on measured temperature-dependent microwave dielectric permittivity, relative dielectric constant, and conductivity of freshly excised animal tissues ranging in temperatures from 5 to 90 °C and their coefficients of temperature dependence. The tissue type tested includes blood, kidney, liver, and muscle from several different species. These data corroborate the decrease of dielectric constants and increase of conductivity as a function of frequency in the microwave spectrum region. Most of the data were obtained for liver in the 25 to 80 °C temperature range, clearly displaying the contemporary interest in liver ablation. While the permittivity values are derived from tissues of many different species, there is little variation among the measured values for a given tissue type, except for the well-known frequency dependent drop in dielectric constant and rise in conductivity.

The data in Table 4.4 shows a slight average decrease (−0.20% per °C) of dielectric constant with increasing temperature for frequencies from 400 to 2450 MHz. In contrast, the conductivity increases by 1.01% per °C on average. These results are consistent with the general notion that dielectric constants are independent of temperature, and the change of the dielectric constant with temperature is smaller than

Table 4.4 Measured temperature-dependent dielectric permittivity of animal tissues at microwave frequencies

Frequency (MHz)	Tissue type	Temp range (°C)	ε	ε Coeff. (%/°C)	σ (S/m)	σ Coeff. (%/°C)	References
400	Human blood	25–45	–	−0.10	0.7	1.13	Jaspard and Nadi [2002]
468	Porcine muscle	36–60	65.03	−0.30	1.11	0.92	Fu et al. [2014]
468	Porcine kidney	36–60	58.36	−0.42	1.04	1.30	Fu et al. [2014]
468	Porcine liver	36–60	51.43	−0.35	0.80	1.03	Fu et al. [2014]
915	Bovine liver	50–80	48.10	−0.13	1.03	1.82	Chin and Sherar [2001]
915	Bovine liver	10–90	52.56	−0.04	0.88	1.14	Stauffer et al. [2003]
915	Bovine/porcine liver	37–60	49.50	−0.20	0.99	1.33	Lazebnik et al. [2006]
915	Animal liver	5–50	48.0	−0.22	0.94	1.29	Brace [2008]
1000	Human blood	25–45	–	−0.11	0.70	0.98	Jaspard and Nadi [2002]
2450	Bovine/porcine liver	37–60	47.60	−0.17	1.77	0.20	Lazebnik et al. [2006]
2450	Animal liver	5–50	45.50	−0.18	1.62	−0.20	Brace [2008]
2450	Bovine liver	15–80	43.52	−0.15	1.74	−0.13	Lopresto et al. [2012]

that of the conductivity. Unexpectedly, two reports indicated slight decreases of 0.13 to 0.20% per °C in conductivity at 2450 MHz.

It is interesting to note the temperature coefficients of dielectric constant and conductivity (−0.10 and 1.13%/°C) at 400 MHz and (−0.11 and 0.98%/°C) at 1000 MHz, respectively, for blood. As anticipated, the dielectric constants show very minor changes, but the conductivity showed an increase about 1% per °C. The δ, γ dispersions over which 400 and 1000 MHz reside are prone to membrane capacitance arising from ionic gating currents through or counterions surrounding membranes (Fig. 4.10). These ionic activities have temperature coefficients that are expected to be like that of the conductivity of electrolytes, i.e., about 2% per °C. The dispersions are also caused by polarization effects and the rotational relaxation of water molecules in which the cellular membranes are charged through the electrolytes. Hence, the temperature coefficient should be equal or close to that of the conductivity of electrolytes. Hence, its temperature dependence is equal to that of water, which again is close to 2% per °C. Thus, the temperature coefficient of conductivity for blood materials would have a maximum value of about 2% per °C, as for most tissues with higher water content.

A graph of the temperature dependence of relative permittivity and conductivity of animal liver tissue at 2450 MHz is presented in Fig. 4.11 [Rossmanna and Haemmerich, 2014]. While the basic data derived from Brace [2008] and Lopresto

Fig. 4.10 Dielectric constant and conductivity of blood at 37 °C for the microwave frequency region

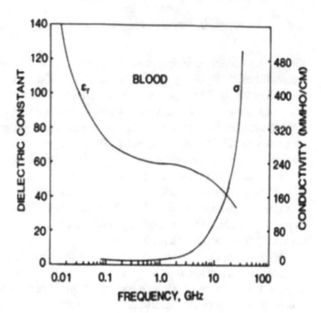

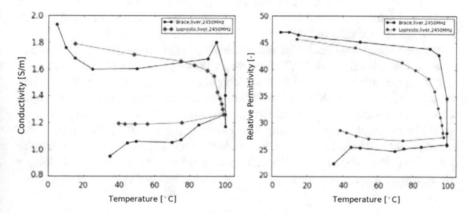

Fig. 4.11 Temperature dependence of relative permittivity and conductivity of animal liver tissue at 2450 MHz. [From Rossmanna and Haemmerich, 2014]

et al. [2012] were limited, they demonstrate modest changes with temperatures from below 25 °C to near 100 °C. The severe irreversible decrease of dielectric permittivity and electrical conductivity at 2450 MHz is expected as the temperatures approached a high of 100 °C. Protein denaturation and other cellular and molecular damages obviously have already been taking effect at 65 °C or below, well ahead of the boiling point of water at 100 °C. So, what happens to tissue damages at 100 °C is a non-sequitur. Note that aside from temperature the extent of tissue changes depends also on exposure duration.

4.4.5   Dielectric Permittivity at Low Temperatures

Most of these reported temperature-dependent dielectric permittivity changes involved temperatures at physiological regime or greater (36 to 80 °C), except for a few that was below 25 °C. The reported temperature change coefficients of dielectric permittivity and electrical conductivity in Table 4.4 are small, all within 2%/°C or less.

However, the dielectric permittivity of biological materials can vary over a wide range below 25 °C and especially when they are in frozen states [Lin, 1988]. Furthermore, the presence of cryoprotective agents strongly influence the dielectric properties of tissue materials. Some dielectric constant and conductivity properties of kidney tissues with and without the presence of cryoprotective agents (5% DMSO) are given in Table 4.5 for temperatures of −35 to +25 °C. Note that the changes in dielectric constant and conductivity can be as great as 20- to 30-fold between warm and frozen states.

For example, the dielectric constant for kidney tissue at a frequency of 2450 MHz decreases from greater than 50 to about 2.2 as the temperature decreases from +25 to −35 °C, while conductivity decreases from about 2.3 to 0.07 S/m. Note that the loss tangent, a quantity proportional to the ratio of conductivity to dielectric constant (see Chap. 2), not only increases but also shows a peak at a temperature of about 0 °C for kidney tissue. The dielectric constant for kidney tissue perfused with 5% DMSO decreases from 52 to about 4.6 as the temperature decreases from +25 to −20 °C, while conductivity decreases from about 2.58 to 0.18 S/m.

The power transmission coefficient and depth of penetration (see Chaps. 2 and 5) at the air-tissue interface for 2450 MHz along with the loss tangent listed in Table 4.5 give some measures of microwave power deposition or energy absorption. Note that the differences in dielectric constant and conductivity yield a penetration depth, which is the distance through which the power density decreases by a value of e^{-2}, for warm kidney about four time shorter than kidney tissue at −20 °C. While the transmitted power is substantial (42% of the incident power) at +25 °C, it is about a factor of two lower than at −20 °C. A loss tangent of about one-third indicates that kidney tissues behave more as a dielectric than a conducting material at these temperatures.

Table 4.5 Temperature effect on dielectric permittivity and power deposition characteristics of warm and frozen kidney tissue at 2450 MHz

Temperature (°C)	Dielectric constant	Conductivity (S/m)	Loss-tangent	Transmission coefficient	Penetration depth (cm)
Normal tissue					
−35	2.2	0.07	0.23	0.96	11.33
−20	4.3	0.17	0.29	0.87	6.55
0	49.0	2.00	0.30	0.43	1.88
20	56.0	2.10	0.28	0.41	1.91
25	50.0	2.30	0.34	0.42	1.66
With 5% DMSO (Dimethlsulfoxide)					
−20	4.6	0.18	0.29	0.86	6.39
25	51.9	2.58	0.36	0.41	1.51

References

Böttcher CJF (1952) Theory of electric polarization. Elsevier, Amsterdam

Brace CL (2008) Temperature-dependent dielectric properties of liver tissue measured during thermal ablation: toward an improved numerical model. Conf Proc IEEE Eng Med Biol Soc:230–233

Chin L, Sherar M (2001) Changes in dielectric properties of ex vivo bovine liver at 915 MHz during heating. Phys Med Biol 46(1):197–211

Cole KS (1968) Membranes, ions and impulses. University of California Press, Berkeley

Cole KS, Cole RH (1941) Dispersion and absorption in dielectrics: alternating current characteristics. J Phys 9:341–351

Daniel VV (1967) Dielectric relaxation. Academic, New York

Debye P (1929) Polar molecules. Reinhold, New York

Fu F, Xin SX, Chen W (2014) Temperature- and frequency-dependent dielectric properties of biological tissues within the temperature and frequency ranges typically used for magnetic resonance imaging-guided focused ultrasound surgery. Int J Hyperth 30(1):56–65

Gabriel S, Lau RW, Gabriel C (1996a) The dielectric properties of biological tissues: II. Measurements in the frequency range 10 Hz to 20 GHz. Phys Med Biol 41:2251–2269

Gabriel S, Lau RW, Gabriel C (1996b) The dielectric properties of biological tissues: II Parametric models for the dielectric spectrum of tissues. Phys Med Biol 41:2271–2293

Grant EH, Sheppard RJ, South GP (1978) Dielectric behavior of biological molecules in solution. Clarendon, Oxford

Hill NE, Vaughan WE, Price AH, Davies M (1969) Dielectric properties and molecular behavior. D. Van Nostrand, Princeton

Huang SKS, Wilber DJ (eds) (2000) Radiofrequency catheter ablation of cardiac arrhythmias: basic concepts and clinical applications, 2nd edn. Futura, Armonk/New York

Jaspard F, Nadi M (2002) Dielectric properties of blood: an investigation of temperature dependence. Physiol Meas 23(3):547–554

Lazebnik M, Converse MC, Booske JH, Hagness SC (2006) Ultrawideband temperature-dependent dielectric properties of animal liver tissue in the microwave frequency range. Phys Med Biol 51(7):1941–1955

Lin JC (1975) Microwave properties of fresh mammalian brain tissues at body temperature. IEEE Trans Biomed Engg 22:74–76

Lin JC (1988) Electromagnetic heating techniques for organ rewarming. In: Pegg D, Karow A (eds) Biophysics of organ cryopreservation. Plenum Press, pp 315–335

Lin JC (2003) Chapter 36: Minimally invasive medical microwave ablation technology. In: Hwang NHC, Woo SLY (eds) New frontiers in biomedical engineering. Kluwer/Plenum, New York, pp 545–562

Lin JC, Gandhi OP (1996) Handbook of biological effects of electromagnetic fields. CRC Press, pp 337–402

Lopresto V, Pinto R, Lovisolo GA, Cavagnaro M (2012) Changes in the dielectric properties of ex vivo bovine liver during microwave thermal ablation at 2.45 GHz. Phys Med Biol 57(8):2309–2327

Rossmanna C, Haemmerich D (2014) Review of temperature dependence of thermal properties, dielectric properties, and perfusion of biological tissues at hyperthermic and ablation temperatures. Crit Rev Biomed Eng 42(6):467–492

Schwan HP (1957) Electrical properties of tissues and cell suspensions. Advances in biological and medical physics. Academic, New York, pp 147–209

Schwan HP (1963) Electric characteristics of tissues. Biophys J 1:198–208

Schwan HP (1977) Field interaction with biological matter. Ann N Y Acad Sci 303:198–213

Schwan HP, Foster KR (1980) RF-field interactions with biological systems: electrical properties and biophysical mechanisms. Proc IEEE 68(1):104–113

Schwan HP, Sheppard RJ, Grant EH (1976) Complex permittivity of water. J Chem Phys
    64:2257–2258
Stauffer PR, Rossetto F, Prakash M, Neuman DG, Lee T (2003) Phantom and animal tissues for
    modelling the electrical properties of human liver. Int J Hyperthermia 19(1):89–101
Steel MC, Sheppard RJ (1985) Dielectric properties of mammalian brain tissue between 1 and 18
    GHz. Phys Med Biol 30:621–630
von Hipp AR (1954) Dielectric materials and applications. MIT, Cambridge, MA

Chapter 5
Dosimetry and Microwave Absorption

Regardless of the mechanism of interaction, incident RF and microwaves must be coupled into the system, and energy must be transferred, deposited, and absorbed in the biological body for the system to respond in some manner. Thus, to gain a better knowledge of biological responses, the exposure to RF and microwave radiation that is effective in exerting its influence must be quantified and correlated with any observed response or phenomenon. This issue goes beyond the general question of how similar human response is to that of prototypical laboratory animals. The factors and parameters obtained from animal experiments may not always be directly applicable to humans and especially regarding exposure and dosimetry. The same exposure and quantities may cause markedly different induced fields and absorption inside the body and thus potentially different biological effects in various mammalian species. Aside from potential differences in biological endpoints, one must also consider the difficult problem of extrapolating the quantitative dosimetric results from laboratory animals to humans.

Dosimetry, in the context of this book, may be defined as the quantification of RF and microwave radiation's distribution and absorption in biological materials or animal bodies under exposure [Lin, 2007]. It is noteworthy that unlike ionizing radiation such as X- or Gamma-ray, the exposure may produce highly complicated distributions of RF and microwave energy within the subject, regardless of the uniformity of external exposures. Indeed, the same applied RF and microwave radiation can produce significantly different absorptions in different bodies or subjects. This chapter discusses the coupling, dosimetry, and distribution of RF and microwaves in biological bodies, especially humans. These topics are complex functions of not only the exposure sources and scenario but also the shape, size, composition, and structures of the exposed subjects, as well as orientation and position of the subject with respect to the source, among others.

© Springer Nature Switzerland AG 2021
J. C. Lin, *Auditory Effects of Microwave Radiation*,
https://doi.org/10.1007/978-3-030-64544-1_5

5.1   Dosimetric Quantities and Units

To establish scientifically any biological response to the exposure of a physical agent such as RF and microwave radiation, the agent and its exposure that are exerting the influence must be characterized, quantified, and correlated with the observed biological effect.

The commonly employed metrics or dosimetric quantities include incident electric and magnetic fields, induced fields, incident power density, transmitted (or induced) power density, specific absorption rate (SAR), and specific absorption (SA) in biological bodies or tissue materials. The metric SAR (in W/kg) referred to in Sect. (2.4) is a derived quantity and is given by the time derivative of the incremental energy absorbed by or dissipated in an incremental mass contained in a volume of a given density [Chou et al., 1996; Lin, 2007; NCRP, 1981]. This definition allows SAR to be used as a metric for RF and microwaves in both the near and far field. Specific absorption (in J/kg) is the total amount of energy deposited or absorbed and is given by the integral of SAR over a finite interval of time. Information on SA and SAR is of interest because it can serve as an index for extrapolation of experimental results from cell to cell, cell to organ, organ to animal, animal to animal, and animal to human exposure. It is also useful in analyzing relationships among various observed biological effects in different experimental models and subjects.

The induced field is of primary interest because it relates the RF and microwave radiation to specific responses of the body, facilitates understanding of biological phenomena, and is independent of mechanisms of interaction. Once the induced field is known, quantities such as SAR can be derived by a simple conversion formula. For example, from an induced electric field E in V/m,

$$\text{SAR} = \frac{\sigma E^2}{\rho} \tag{5.1}$$

where σ is the bulk electrical conductivity and ρ is the mass density (kg/m^3) of tissue. At present, the smallest isotropic implantable electric field probe, available with sufficient sensitivity for practical use, is about 1 mm in diameter and expensive. Consequently, a common practice in experimental dosimetry relies on the use of temperature elevation produced under a short-duration (<30 s), high-intensity exposure condition. (The short duration is insufficient for significant convective or conductive heat contribution to tissue temperature rises, and the intensity is sufficient to produce a measurable temperature elevation.) This condition is sometimes referred to as thermal containment. In this case, the time rate of initial rises in temperature (slope) can be related to SAR through an alternative or secondary method, so that

$$\text{SAR} = c \frac{\Delta T}{\Delta t} \tag{5.2}$$

where c is the specific heat capacity of tissue (J/kg-°C), ΔT is the temperature increment (°C), and Δt is the duration (s) over which ΔT is measured. It is important to distinguish the use of SAR and its alternative derivation from temperature-based measurements. The quantity of SAR is merely a metric for power deposition or energy absorption, and it should not be construed to imply any mechanism of inter-action, thermal, or otherwise. However, it is a quantity that pertains to a macro-scopic phenomenon by virtue of the use of bulk electrical conductivity and mass density in its derivation in Eq. (5.1) and the use of specific heat capacity of tissue in Eq. (5.2).

It is important to note the use of bulk electrical conductivity, specific heat capac-ity, and mass density (kg/m^3) of tissue in the derivation of SAR from electric field strength and temperature elevation. Their use in the definition means that a volume of tissue mass must be selected over which SAR is determined [Lin, 2007]. In com-mon usage, 1 or 10 g of tissue in the form of a cubic volume is specified. Obviously, the numerical value of SAR would be the same regardless of what mass or volume is chosen if the induced field and power deposition are uniform. A variance arises when the absorption is non-uniform or tissues with differing properties are within the same volume of averaging mass. In principle, a 10-g averaging mass could underestimate SARs of non-uniform fields by up to a factor of 10 compared with a 1 g averaging mass [Cavarnaro and Lin, 2019]. It is emphasized that recent advances suggest that spatial resolutions comparable to a 0.01 g or less are routinely obtain-able using available computational algorithms and resources to provide higher spa-tial precision in SAR determination.

A complex biological system, such as a human body, consists of multiple layers of tissue, curved surfaces with different dielectric properties, organs of different dimensions, cylindrical limbs, and a spheroidal head and, of course, the whole body that would all affect the absorbed energy distribution. Our discussions begin with quantification of RF and microwave distribution and absorption at planar tissue boundaries and progressing to include anatomical human body models.

5.2   Boundary of Planar Surfaces

At boundaries separating regions of different biological materials (including air), microwave energy is reflected or transmitted (Fig. 5.1). For a plane wave impinging normally from a medium, the reflection coefficient (R) and transmission coefficient (T) given in Eqs. (2.55 and 2.56) provide a measure of microwave energy coupling; R and T are related through $T = 1 + R$. The fraction of incident power reflected by

Fig. 5.1 A plane wave impinging normally on a planar tissue medium

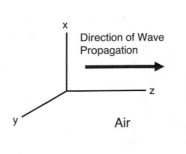
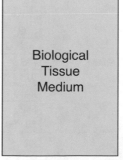

the discontinuity is R^2, and the transmitted fraction is $T^2 = (1 - R^2)$. Table 5.1 gives the calculated reflection coefficient from air-tissue and tissue-tissue interfaces. Clearly, about one half of the incident power is reflected at these boundaries. Further, the reflection coefficient for tissue-tissue interfaces generally is smaller than air-tissue interfaces. The percentage of reflected power for tissue-tissue interfaces ranges from a low of 5 for muscle blood to a high of 50 for bone-biological fluid interfaces. This suggests that the closer are the dielectric properties on both sides of the interface, the smaller is the power reflection [Lin and Bernardi, 2007].

As the data from Table 5.1 suggests the power transmitted, $T^2 = (1 - R^2)$ at air-tissue interfaces is variable but can be quite substantial at RF and microwave frequencies. Figure 5.2 shows that the power transmission coefficient is highly

Table 5.1 Magnitude of reflection coefficient (in percent) between air and tissue and between biological tissues at 37 °C

	Frequency (MHz)	Tissues					
		Air	Fat (bone)	Lung	Muscle (skin)	Blood	Saline
Air	433	0	46	76	82	81	83
	915	0	43	73	78	79	80
	2450	0	41	71	76	77	79
	5800	0	39	70	75	76	78
	10,000	0	37	70	74	76	78
Fat (bone)	433		0	46	56	56	60
	915		0	43	52	54	57
	2450		0	42	50	53	57
	5800		0	42	50	53	56
	10,000		0	45	52	54	58
Lung	433			0	14	13	19
	915			0	12	14	18
	2450			0	10	15	19
	5800			0	10	14	19
	10,000			0	10	13	18
Muscle (skin)	433				0	4	6
	915				0	4	7
	2450				0	5	10
	5800				0	4	9
	10,000				0	3	9
Blood	433					0	6
	915					0	4
	2450					0	5
	5800					0	5
	10,000					0	6
Saline	433						0
	915						0
	2450						0
	5800						0
	10,000						0

Fig. 5.2 Power transmission coefficients at different tissues interfaces as functions of frequency

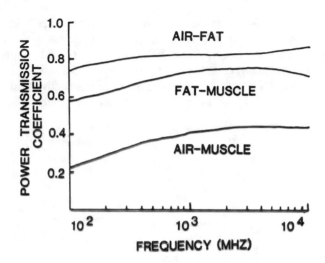

Table 5.2 Depth of penetration of microwaves in biological tissues as a function of frequency

Frequency	Tissues				
(MHz)	Saline	Blood	Muscle (Skin)	Lung	Fat (bone)
Depth of penetration (cm)					
433	2.8	3.7	3.0	4.7	16.3
915	2.5	3.0	2.5	4.5	12.8
2450	1.3	1.9	1.7	2.3	7.9
5800	0.7	0.7	0.8	0.7	4.7
10,000	0.2	0.3	0.3	0.3	2.5

frequency dependent, more so at lower microwave frequencies. Clearly, the coupling of RF and microwaves from air into planar tissue is greater for low-water content tissue compared to that of high-water content tissue. It is greater for higher microwave frequencies than for lower ones and ranges from about 15 to 80%.

As the transmitted field propagates in tissue medium, microwave energy is extracted from the field and absorbed by the medium, resulting in a progressive reduction of power density of the field as it advances in the tissue. This reduction is quantified by the penetration depth (see Eq. 2.52), which is the distance through which the power density decreases by a factor of e^{-2}. Table 5.2 presents the calculated depth of penetration in selected tissues. It is seen that the penetration depth is inversely frequency dependent, and the higher the frequency, the lower the penetration depth. The depth takes on different values for different tissues. In particular, the penetration depth for fat and bone is nearly five times greater than for higher-water-content tissues.

Microwave radiation of general or unspecified polarization may be decomposed into its orthogonal linearly polarized components whose electric or magnetic field parallels the interface (see Eqs. 2.65 and 2.66). These components can be treated separately and combined afterward. Figures 5.3 and 5.4 illustrate the magnitude and phase of the reflection coefficients of representative tissue interfaces at a temperature of 37 °C for 2450 MHz microwave exposures. The figures clearly show the

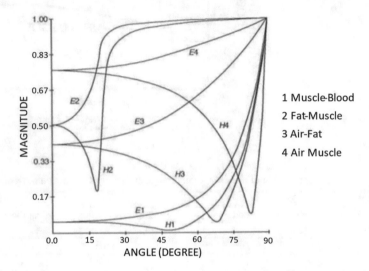

Fig. 5.3 Magnitude of reflection coefficients for E and H polarized microwaves at 2450 MHz with a plane wavefront

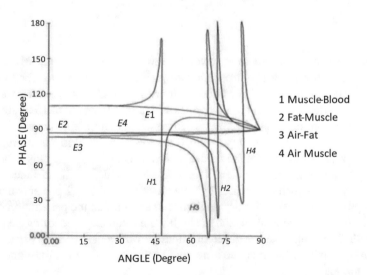

Fig. 5.4 Phase of reflection coefficients for E and H polarized 2450 MHz microwaves with a plane wavefront

difference between E and H polarization. E polarization, also called perpendicular polarization, and H polarization, often referred to as parallel polarization, are defined in Chap. 2 of this book. For E polarization, there is only a slight variation in magnitude and phase of the reflection coefficient with incidence angle. For H polarization, however, there is a pronounced dependence on incidence angle. The reflection coefficient reaches a minimum magnitude and has a phase angle of 90° at Brewster's angle (Eq. 2.70). Thus, the H polarized wave is totally transmitted into the muscle medium at Brewster's angle.

5.3   Multiple Tissue Layers

In a layered tissue structure with different tissue permittivity, the reflection and transmission characteristics become more complicated. Multiple reflections can occur between the skin and subcutaneous tissue boundaries, with a resulting modification of the reflection and transmission coefficients (Eqs. 2.55 and 2.56). In general, the transmitted wave combines with the reflected to form standing waves in each layer. The peaks of the standing waves can result in greater deposition of RF and microwave energy into the tissue layer. The standing-wave phenomenon becomes especially pronounced if the thickness of each layer is greater than the penetration depth for that tissue and is approximately one-half wavelength or longer at the RF and microwave frequency. This dependence of standing-wave oscillation peaks on layer thickness is a manifestation of layer resonance, which can enhance power transmission and energy absorption.

For the layered tissue model depicted in Fig. 5.5, the electric field strength in the fat layer is given by

$$E_f = F_l \, E_0 \left[e^{-(\alpha_2 + j\beta_2)\,z} + \Gamma_{32}\, e^{(\alpha_2 + j\beta_2)\,z} \right] \tag{5.3}$$

and the electric field in the underlying muscular tissue is given by

$$E_m = F_t \, E_0 \, e^{-(\alpha_3 + j\beta_3)z} \tag{5.4}$$

where α_2, β_2, and α_3, β_3, are the attenuation and propagation coefficients in fat and muscle, respectively. The layer function F_l and the transmission function F_t are given by

$$F_l = \frac{T_{12}}{e^{(\alpha_2 + j\beta_2)\,l} + \Gamma_{21}\,\Gamma_{32}\,e^{-(\alpha_2 + j\beta_2)\,l}} \tag{5.5}$$

$$F_t = \frac{T_{12}\,T_{23}}{e^{(\alpha_2 + j\beta_2)\,l} + \Gamma_{21}\,\Gamma_{32}\,e^{-(\alpha_2 + j\beta_2)\,l}} \tag{5.6}$$

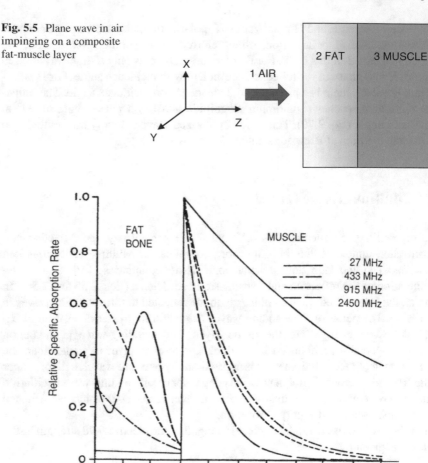

Fig. 5.5 Plane wave in air impinging on a composite fat-muscle layer

Fig. 5.6 Relative specific absorption rate in planar fat-muscle layers

where T_{12} and T_{23} are the transmission coefficients at the air-fat and fat-muscle boundaries, respectively. Γ_{21} and Γ_{32} denote the reflection coefficients at these boundaries, respectively, and l is the thickness of the fat layer. The power deposition in a given layer can be obtained from equations given in Sect. (2.4).

Computed results of SAR distribution in fat-muscle layers for four different frequencies are given in Fig. 5.6. The values are normalized to the SAR in muscle at the fat-muscle boundary. Note the absorbed energy is much lower in fat than in muscle. The standing-wave maximum becomes bigger in fat, and the penetration into muscle becomes less as the frequency increases. For fat tissues less than 3 cm thick, the corresponding values can be simply obtained by deleting portions of the curves left of the actual fat thickness displayed. The electromagnetic energy absorbed in models composed of planar layers of skin, fat, and muscle may be analyzed in a similar manner, except the distribution of absorbed energy becomes more complex. (See [Lin 2000] for an example of the dependence of peak SAR on thickness of the fat layer in a planar tissue model of skin-fat-muscle layers at different frequencies. The skin is assumed to be 2 mm thick. The incident power density is 10 W/m^2.)

5.4 Influence of Orientation and Polarization on Elongated Body

For non-planar, elongated bodies such as a human body or a prolate spheroidal model, where the height-to-width ratio is larger than one, the coupling of RF and microwave energy is influenced by the orientation of the electric field vector (polarization) with respect to the body.

The three principal polarizations of the impinging plane wave to be distinguished are E-polarization in which the electric vector is parallel to the long axis of the body, H polarization in which the magnetic vector is parallel to the long axis of the body, and K-polarization in which neither the electric nor magnetic vector is parallel to the long axis of the body. For K-polarization, the plane wave impinges along the long axis of the body. The frequency at which the highest (resonant) absorption occurs is a function of both polarization and the exposed subject. In general, the shorter the subject, the higher the resonance frequency and vice versa. Moreover, E-polarization couples microwave energy most efficiently to the body in a plane wave field for frequencies up to and slightly above the resonance region, where the body dimension and wavelength are approximately equal.

For RF frequencies well below resonance, such that the ratio of long body dimension (L) to free space wavelength (λ_o) is less than 0.2, the average SAR is characterized by a f^2 dependence. SAR goes through the resonance region in which $0.2 < L/\lambda_o < 1.0$. Specifically, the SAR rapidly increases to a maximum near $L/\lambda_o = 0.4$ and then falls off as $1/f$. At frequencies (f) for which $L/\lambda_o > 1.0$, the whole-body absorption decreases slightly but approaches asymptotically to about one-half of the incident power is transmitted into biological body (approximately equal to 1- power reflection coefficient). The resonances are not nearly as well defined for H-polarization as for E-polarization. The average SAR for H-polarization gradually reaches a plateau throughout the RF spectrum [Lin and Gandhi, 1996].

5.5 Spheriodal Head Models

Although depth of penetration and reflection and transmission characteristics in planar tissue models provide considerable insights into coupling and distribution of RF and microwave energy, biological structures generally are more complex in form and exhibit substantial curvature that can affect microwave energy transmission, reflection, and distribution. These complexities place limitations on analytical calculations of reflected and transmitted microwave energy for bodies of various shape with complex permittivity. Nevertheless, analytical calculations can produce insights into their fundamental interactions. This section presents a summary of results for homogeneous and multilayered spheroidal models that may be applied to approximate especially mammalian head structures.

For a plane wave linearly polarized in the x direction and propagating along the positive z direction, the incident and induced electromagnetic fields may be expanded in terms of vector spherical wave functions. The expansion coefficients are then found from the boundary conditions at the surface of the sphere. The general formulation and derivation details are readily available [Stratton, 1941]. Note the idealized model does not account for the effect of the neck and the rest of the body.

Several investigators have presented results of computer calculations of microwave energy absorption using spherical head models [Shapiro et al., 1971; Kritikos and Schwan, 1972; Lin et al., 1973; Ho and Guy 1975]. An example of SAR distributions through the center of the brain sphere along the coordinate axes calculated from analytical formulations for homogeneous spherical model of the human adult (9-cm radius) brain exposed to 915 MHz plane waves is given in Fig. 5.7. Note the location of a peak SAR near the center of the brain sphere, although the center peak for an adult brain is about 20% down from peak absorption at the leading brain surface in this case [Lin and Bernardi, 2007].

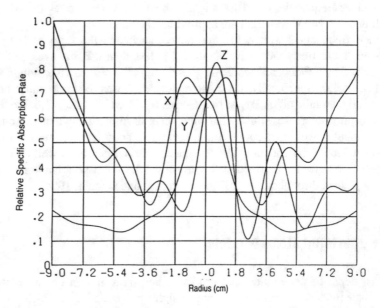

Fig. 5.7 SAR distribution in a homogeneous spherical brain model (18 cm diameter) exposed to 915 MHz plane wave. The direction of microwave propagation is along the z axis

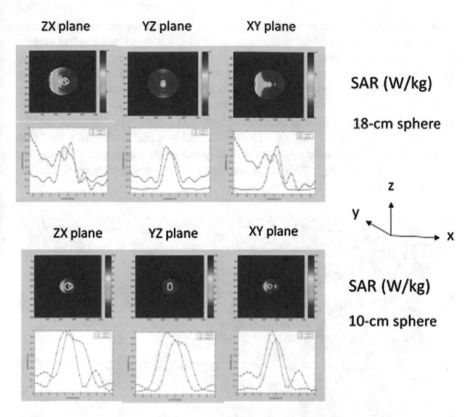

Fig. 5.8 FDTD computed SAR distributions through the center of the sphere over the three principal cross sections for homogeneous spherical model of the human adult or child brains exposed to 915-MHz plane waves. The plane wave microwave incidents from the negative x direction

Similar SAR characteristics are shown in two-dimensional (2D) displays of microwave absorption in the canonical brain models (Fig. 5.8). These examples of FDTD-computed SAR distributions, through the center of the sphere over the three principal cross sections, are for homogeneous spherical model of the human adult (18-cm diameter) or child (10-cm diameter) brains exposed to 915 MHz plane waves [Lin and Wang, 2005]. The line graphs give the SAR distributions through the center along the axial directions. The plane wave microwave exposure impinges from the negative x direction. The results for the adult head at 915 MHz are in good agreement with the analytic calculations. As noted above for an adult-size brain sphere, the peak SAR moves to an anterior location toward the incident microwaves. However, a small peak SAR appears near the center of an adult brain. Both exposed human child and adult brain spheres exhibit clear peak absorption occurring near the center and inside of the head. Note that the location of peak SAR for the smaller child-size brain sphere appears prominently near the center for 915 MHz microwaves. This result reveals the fact that for microwave exposure the same plane wave source can produce very different SAR distributions inside different sized human bodies.

A 2D display of microwave absorption in a 9-cm radius homogeneous brain sphere for the 400 MHz plane wave is given in Fig. 5.9. Note that the left side (in 2D graphs) and right side (in line graphs) of the model show equal magnitudes in the transverse planes because of symmetry of the brain model and uniform transverse fields of the incident microwave. Also, the SAR distributions have different peaks deep inside the brain sphere, and the local peaks may be several times greater than that due to exponential or penetration losses in planar homogeneous models. The enhancement is the result of refraction of incident plane wave into the brain sphere by the curved tissue surface and a geometrical resonance phenomenon (standing wave). Although SAR in the shadow region is lower compared with the front surface, it can still be substantial.

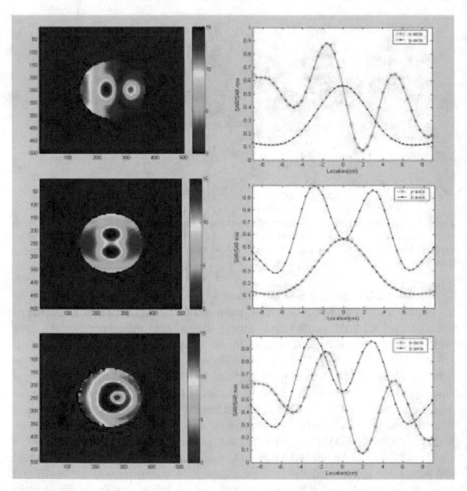

Fig. 5.9 SAR distributions inside a 9 cm radius adult brain sphere exposed to 400 MHz plane wave. The three rows represent XY, ZX, and YZ plane patterns, respectively. The corresponding line distributions are along (red) the direction of propagation and transverse (blue) to the direction of propagation shown on the right. The direction of wave propagation is along the X axis

Clearly, microwaves can overcome exponential losses and produce enhanced coupling at greater depth in bodies with curved surfaces. Typically, if the largest dimension of the body is comparable to the wavelength of the impinging microwave, energy deposition and distribution will be influenced by the surface curvature of the whole body or body part and the tissue composition. In particular, the ratio of geometric dimension and wavelength or the factor, ka (k is the wave number and a is radius), affects the amount and characteristics of microwave energy coupling.

In general, standing-wave SAR patterns with multiple peaks and valleys are observed. The peak absorption may be several times greater than the average, and the enhanced absorptions near the center of these brain models may be significantly greater than that expected from the planar tissue models. The increased absorptions are due to a combination of high dielectric constant and curvature of the model, which produces a strong focus of energy toward the interior of the sphere that more than compensates for the transmission and propagation losses through the tissue.

An example of this geometric resonance phenomenon is given in Fig. 5.10, where the relative absorption coefficients for a 14 cm diameter homogeneous sphere of the brain (and muscle equivalent materials) as a function of frequency is shown. The absorbed energy varies widely with sphere size and frequency. In principle, the absorption increases rapidly with increasing radius and is then followed by some

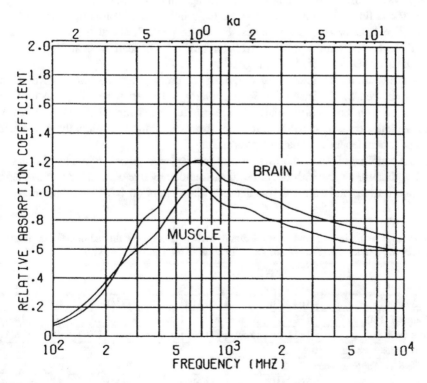

Fig. 5.10 Frequency dependence of absorption in a 14 cm diameter spherical model of the human brain and muscle equivalent materials from 100 MHz to 10 GHz

wavelength-dependent geometrical resonant behavior. However, with increasing frequency, microwave energy is absorbed in a decreasingly smaller volume as a result of shortened penetration depth. The peaks of these resonant oscillations are related to the maxima, or hot spots, in the SAR distributions inside the head model. For (ka = $2\pi a/\lambda_0$) < 0.4, where a is the sphere radius and λ_0 is the wavelength in vacuum or free space, hot spots or SAR peaks do not occur inside the sphere. However, for some combinations of exposure frequency and radius (0.4 < ka < 2.0), hot spots will occur. For larger brain spheres, the maximum absorption appears at the anterior portion (first exposed surface) of the brain sphere, and the penetration depth at the surface becomes a dominating factor for exposures at frequencies in this range. The planar model discussed previously may be applied to obtain a theoretical estimation of absorbed energy distribution in this case.

In addition to wavelength and body size, the type of microwave sources and exposure configuration can also influence SAR distribution and peak absorption. A comparison of computed maximum SAR in two spherical head models for plane wave and birdcage sources for different frequencies are given in Table 5.3. In this case, the SAR values are normalized to 20 W of total absorbed power. The peak SARs vary widely for the two head models and for different frequencies, as expected. They are considerably higher for child-size models across all frequencies listed for plane-wave sources. Note that for MRI birdcage antennas, the variations are modest both for child and adult head models, especially when comparing within their respective field or frequency ranges. However, peak SARs at the lower 1.5 and 3.0 T (Tesla) or 64 and 128 MHz levels tend to be lower than those at the higher 7.0 and 9.4 T or 300 and 400 MHz ranges.

A study of the interaction of circularly or cross-polarized plane waves [Lin, 1976] showed SAR distribution for circularly polarized waves is more uniform compared with the linearly polarized case. In fact, the absorbed energy distribution in the planes transverse to the direction of propagation is rotationally symmetric, that is, it is independent of angular variation.

The coupling of plane microwaves into multilayered models of the head structure was studied for a spherical core of brain surrounded by five concentric shells of skin, fat, skull, dura, and cerebral spinal fluid. The results on SAR distribution of the

Table 5.3 Comparison of computed maximum SAR in two spherical head models for different sources and frequencies (Normalized values to 20 W of total absorbed power)

Parameter	Frequency (MHz)							
	Plane wave							
Frequency	64 MHz		128 MHz		300 MHz		400 MHz	
Diameter (cm)	10	18	10	18	10	18	10	18
SAR (W/kg)	111.49	15.59	134.83	18.77	79.82	19.71	94.91	26.05
MRI birdcage coil antenna								
B Field and RF Frequency	1.5 T		3.0 T		7.0 T		9.4 T	
	64 MHz		128 MHz		300 MHz		400 MHz	
SAR (W/kg)	56.49	7.94	57.93	8.33	60.11	12.66	66.38	12.85

six-layer model of a 10 cm radius adult head for 915 and 2450 MHz are increased SAR in the skin. Absorptions in fat and skull are found to be the lowest among the tissue layers [Lin and Bernardi, 2007]. Furthermore, the peak and average SARs may be several times greater than for homogeneous models. The enhancement is apparently due to resonant coupling of microwaves into the brain sphere by the outer tissue layers. Although SAR in the shadow region is lower compared to the front surface, it is still considerable. In addition, SARs at the top and bottom or left and right sides of the model have equal magnitudes for the uniform incident plane wave.

5.6   Absorption in Anatomical Models

While spheroidal models can provide insightful results that may resemble those from more realistic head models, the SAR values and their distributions obtained from spheres or simplified head models do not provide the quantitative detail, numerical accuracy, and degree of realism that can only come from anatomical models.

Many millimeter (mm), and in some cases, sub-mm resolution models of the human body have been developed as better or more realistic representations of the complex geometry and structural organization of the human body. Some of the models have been used to compute RF and microwave energy absorption and their distribution in the human body. The models include constructions using small-volume-cubic cells or cell meshes and anatomically based models generated from x-ray computerized tomography (CT) and magnetic resonance imaging (MRI) data. The most frequently used high-resolution, three-dimensional (3D) human models come from the "Visible Human Project" of the National Library of Medicine [Ackerman, 1998].

5.6.1   The Visible Human Anatomical Model

The visible human (VH) is a digital 3D image library representing an adult human male and female. The data set for both the male and female include photographic images obtained through cryosectioning of human cadavers and digital images acquired through CT and MRI of cadavers. The male data set, the first to be con-structed (released in 1994), consists of 1871 digital axial images obtained at 1.0 mm intervals, with a pixel resolution of 1 mm, while the female data set contains 5189 digital axial images (released in 1995), attained with a finer spatial grid of 0.33 mm.

While these digital data sets represent a unique tool to explore human anatomy, their direct use for computational RF or microwave dosimetry is limited by the fact that the image data set cannot be directly used as an input for a numerical electromagnetic, computational algorithm or tool, but must be converted to a so-called "segmented" version by segmentation of the original image sets, which is complex and time-consuming. A segmented model is a model where every pixel,

usually called in such models "voxel," does not contain information about the color
(like in digital images), but rather a label that is uniquely assigned to a given tissue.
In such a way, it is possible to know which tissue fills each of the model voxels and
hence assign the correct complex permittivity values to be used in numerical computations and simulations.

The male VH images were segmented at the Air Force Research Laboratory,
Brooks Air Force Base, Texas, USA [Mason et al., 2000]. The final segmented
model, (ftp://starview.brooks.af.mil/EMF/dosimetry_models/), made freely available to the scientific community, consists of $586 \times 340 \times 1878$ voxels with a resolution of $1 \times 1 \times 1$ mm^3. The model was segmented into about 40 different tissue types
with frequency-specific tissue permittivity, such as those given in Chap. 4, which
are used to represent each tissue type in the heterogeneous model of the human
body. The VH model has been widely used to study both whole-body and localized
human exposure to RF and microwaves generated by various types of sources. They
have been included in several commercially available electromagnetic simulation
tools with capabilities for dosimetric evaluation.

The original VH model has a height of 1.88 m and a mass of 103 kg (Fig. 5.11).
The VH model has often been scaled in its axial cross sections (e.g., with a 0.91
scaling factor in the shoulder-to-shoulder direction and 0.83 in the front-to-back
direction) to arrive at a mass of about 80 kg to mimic a more typical mass for adult
male [Piuzzi et al., 2011]. Still others have engaged lighter models obtained by
reducing the digital cell dimensions on the axial plane.

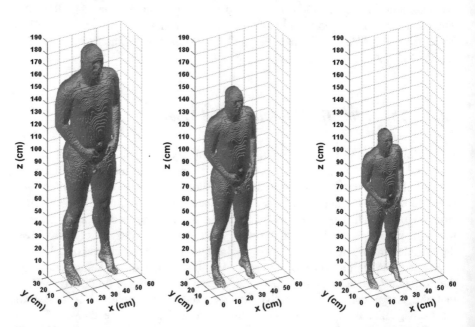

Fig. 5.11 Original anatomical VH model of an adult male (left) and lighter models obtained by
reducing the digital cell dimensions

In fact, the larger than average height and mass of the VH phantom model has prompted development of several anatomically realistic models with close to average dimensions. Examples include an upright phantom model developed using MRI data and scaled to the height of 1.76 m and mass of 73 kg, near the measurements of the ICRP [1994] reference man with voxels having approximately a 2 mm side and consisting of 38 tissue types [Dimbylow, 2002], a fine resolution male model (1.72 m, 65 kg) with 2 mm voxels and segmented into 51 tissue types [Nagaoka et al., 2004], and a 3.6 mm adult male model with a height of 1.77 m and mass of 78 kg, composed of 31 different tissue types [Lacroux et al., 2008].

Anatomical body models with postures other than upright have been reported in support of studies pertaining to different exposure scenarios. These include postures with outstretched arms [Findlay and Dimbylow, 2005] and seating stances [Allen et al., 2005; Findlay and Dimbylow, 2006].

5.6.2   Family of Anatomical Computer Models

Several computer models with mm resolution have been developed to further improve the anatomical realism and fidelity of human models for various application scenarios. These are mostly based on MRI files of healthy adults and children and have been segmented into as many as 80 distinct tissue types. Examples are the Japanese voxel model [Nagaoka et al., 2004; Nagaoka and Watanabe, 2009], Virtual Family data set [Christ, et al., 2010; FDA, 2017, https://www.fda.gov/about-fda/cdrh-offices/virtual-family], and Chinese anatomic models of adult and infants [Li et al., 2015; Wu et al., 2011]. Developments have also included variable postures and two-body infant and adult models with different skin-to-skin contact positions and regions [Dimbylow, 2007; Li and Wu, 2015; Wang et al., 2006].

5.7   Computing SAR in Anatomical Models

This section discusses the use of anatomically realistic human body models to study both whole-body and partial-body (head and shoulder) exposure to RF and microwave radiation from various types of sources. It involves solving Maxwell's equations in the differential form (Chap. 2) for the computation of induced fields. The section begins with a brief description of the computational algorithm based on the numerical method of finite difference, time domain (FDTD) technique [Kunz and Luebbers, 1993; Yee, 1966]. For further information on mathematical, algorithmic, and computational details of SAR and microwave absorption in biological bodies, interested reader is referred to Lin [2018], among others given in the references. Also, more details on the use of FDTD algorithms for calculation of the dependence of induced sound pressure on the characteristics of microwave pulses are presented in Chap. 9. The FDTD-based analyses are applied to determine the dependence of microwave-induced sound pressure on features of the head anatomy and structure.

5.7.1 The FDTD Algorithm

The FDTD technique solves Maxwell's equations by directly modeling the propagation of microwaves into a volume of space containing the biological body. In repeatedly implementing a finite difference representation of Eqs. (2.5 and 2.6) at each cell of the corresponding space lattice (Fig. 5.12), the incident wave is tracked as it first propagates to the body and then interacts with it through surface current excitation, transmission, and diffraction. This wave-tracking process is completed when the steady-state behavior is observed at each lattice cell. Considerable simplification is achieved by analyzing the interaction of the wavefront with a part of the body surface at a time, rather than attempting a simultaneous solution of the entire problem.

Specifically, by introducing a magnetic current ($\mathbf{J}_m = \sigma_m \mathbf{H}$) term, the time-dependent Maxwell's equations from Chap. 2 may be written as,

$$\nabla \times \mathbf{E} = \mathbf{J}_m - \frac{\partial \mathbf{B}}{\partial t} \tag{2.5}$$

$$\nabla \times \mathbf{H} = \mathbf{J}_e + \frac{\partial \mathbf{D}}{\partial t} \tag{2.6}$$

are implemented for a lattice of subvolumes (or Yee cells) that may be cubical or parallelepiped with different dimensions, Δx, Δy, and Δz in the x, y, or z directions, respectively. The components of $\mathbf{E}$ and $\mathbf{H}$ are positioned about each of the cells as shown in Fig. 5.12 and are calculated alternately with half-time steps where the time step $\Delta t = \delta/2c$. Here δ is the smallest of the dimensions used for any of the cells and c is the maximum phase velocity of the fields in the modeled space. Since some of the modeled volume is air, in this case, c corresponds to the velocity of electromagnetic waves in air.

Fig. 5.12 Unit cell of Yee lattice showing positions for various electric and magnetic field components

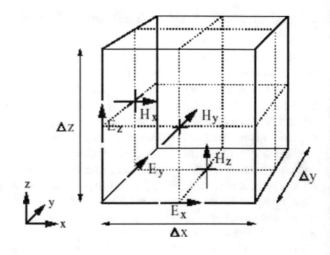

Thus, in rectangular coordinates, Maxwell's curl equations can be expressed in components such that

$$\mu \frac{\partial H_x}{\partial t} = \frac{\partial E_y}{\partial z} - \frac{\partial E_z}{\partial y} - \sigma_m H_x \tag{5.7}$$

$$\mu \frac{\partial H_y}{\partial t} = \frac{\partial E_z}{\partial x} - \frac{\partial E_x}{\partial z} - \sigma_m H_y \tag{5.8}$$

$$\mu \frac{\partial H_z}{\partial t} = \frac{\partial E_x}{\partial y} - \frac{\partial E_y}{\partial x} - \sigma_m H_z \tag{5.9}$$

and

$$\varepsilon \frac{\partial E_x}{\partial t} = \frac{\partial H_z}{\partial y} - \frac{\partial H_y}{\partial z} - \sigma_e E_x \tag{5.10}$$

$$\varepsilon \frac{\partial E_y}{\partial t} = \frac{\partial H_x}{\partial z} - \frac{\partial H_z}{\partial x} - \sigma_e E_y \tag{5.11}$$

$$\varepsilon \frac{\partial E_z}{\partial t} = \frac{\partial H_y}{\partial x} - \frac{\partial H_x}{\partial y} - \sigma_e E_z \tag{5.12}$$

where E_x, E_y, E_z are electric field components in the direction of x, y or z, and H_x, H_y, H_z are magnetic field components in the direction of x, y, or z. Following the Yee's notation, a point in space is given by $(x, y, z) = (i\Delta x, j\Delta y, k\Delta z)$. The 3D FDTD simulation space consists of cubic elements with sides Δx, Δy, and Δz. The E_x, E_y, E_z, H_x, H_y, and H_z defined in each cell are shown in Fig. 5.12. It can be seen that the E-field components are encircled by three H-field components and vice versa. According to Taylor's series expansion, where Δx, Δy, Δz are the spatial steps in x, y, z direction, respectively, and Δt is the time step, the FDTD implementation of the Maxwell's equations are given by (Note that in the absence of conduction and magnetic currents, both σ_m and σ_e may be set to zero.)

$$E_x^{n+1}\left(i+1/2,j,k\right) = \frac{2\varepsilon - \sigma_e \Delta t}{2\varepsilon + \sigma_e \Delta t} E_x^n\left(i+1/2,j,k\right) + \frac{2\Delta t}{2\varepsilon + \sigma_e \Delta t}$$

$$\left\{ \begin{array}{l} \frac{1}{\Delta y}\left[H_z^{n+1/2}\left(i+1/2,j+1/2,k\right) - H_z^{n+1/2}\left(i+1/2,j-1/2,k\right)\right] \\[2mm] -\frac{1}{\Delta z}\left[H_y^{n+1/2}\left(i+1/2,j,k+1/2\right) - H_y^{n+1/2}\left(i+1/2,j,k-1/2\right)\right] \end{array} \right\} \tag{5.13}$$

$$E_y^{n+1}\left(i,j+1/2,k\right)=\frac{2\varepsilon-\sigma_e\Delta t}{2\varepsilon+\sigma_e\Delta t}E_y^n\left(i+1/2,j,k\right)+\frac{2\Delta t}{2\varepsilon+\sigma_e\Delta t}$$

$$\left\{\begin{array}{l}\dfrac{1}{\Delta z}\left[H_x^{n+1/2}\left(i,j+1/2,k+1/2\right)-H_x^{n+1/2}\left(i,j+1/2,k-1/2\right)\right]\\[3mm]-\dfrac{1}{\Delta x}\left[H_z^{n+1/2}\left(i+1/2,j+1/2,k\right)-H_z^{n+1/2}\left(i-1/2,j+1/2,k\right)\right]\end{array}\right\} \qquad (5.14)$$

$$E_z^{n+1}\left(i,j,k+1/2\right)=\frac{2\varepsilon-\sigma_e\Delta t}{2\varepsilon+\sigma_e\Delta t}E_z^n\left(i,j,k+1/2\right)+\frac{2\Delta t}{2\varepsilon+\sigma_e\Delta t}$$

$$\left\{\begin{array}{l}\dfrac{1}{\Delta x}\left[H_y^{n+1/2}\left(i+1/2,j,k+1/2\right)-H_y^{n+1/2}\left(i-1/2,j,k+1/2\right)\right]\\[3mm]-\dfrac{1}{\Delta y}\left[H_x^{n+1/2}\left(i,j+1/2,k+1/2\right)-H_x^{n+1/2}\left(i,j-1/2,k+1/2\right)\right]\end{array}\right\} \qquad (5.15)$$

and

$$H_x^{n+1/2}\left(i,j+1/2,k+1/2\right)=\frac{2\mu-\sigma_m\Delta t}{2\mu+\sigma_m\Delta t}H_x^{n-1/2}\left(i,j+1/2,k+1/2\right)-\frac{2\Delta t}{2\mu+\sigma_m\Delta t}$$

$$\left\{\begin{array}{l}\dfrac{1}{\Delta y}\left[E_z^n\left(i,j+1,k+1/2\right)-E_z^n\left(i,j,k+1/2\right)\right]\\[3mm]-\dfrac{1}{\Delta z}\left[E_y^n\left(i,j+1/2,k+1\right)-E_y^n\left(i,j+1/2,k\right)\right]\end{array}\right\} \qquad (5.16)$$

$$H_y^{n+1/2}\left(i+1/2,j,k+1/2\right)=\frac{2\mu-\sigma_m\Delta t}{2\mu+\sigma_m\Delta t}H_y^{n-1/2}\left(i+1/2,j,k+1/2\right)-\frac{2\Delta t}{2\mu+\sigma_m\Delta t}$$

$$\left\{\begin{array}{l}\dfrac{1}{\Delta z}\left[E_x^n\left(i+1/2,j,k+1\right)-E_x^n\left(i+1/2,j,k\right)\right]\\[3mm]-\dfrac{1}{\Delta x}\left[E_z^n\left(i+1,j,k+1/2\right)-E_z^n\left(i,j,k+1/2\right)\right]\end{array}\right\} \qquad (5.17)$$

$$H_z^{n+1/2}\left(i+1/2,j+1/2,k\right)=\frac{2\mu-\sigma_m\Delta t}{2\mu+\sigma_m\Delta t}H_z^{n-1/2}\left(i+1/2,j+1/2,k\right)-\frac{2\Delta t}{2\mu+\sigma_m\Delta t}$$

$$\left\{\begin{array}{l}\dfrac{1}{\Delta x}\left[E_y^n\left(i+1,j+1/2,k\right)-E_y^n\left(i,j+1/2,k\right)\right]\\[3mm]-\dfrac{1}{\Delta y}\left[E_x^n\left(i+1/2,j+1,k\right)-E_x^n\left(i+1/2,j,k\right)\right]\end{array}\right\} \qquad (5.18)$$

In the FDTD method, it is necessary to represent not only the microwave absorbing target, such as the human body or a part thereof, but also the sources, including their shapes, excitations, etc., if these sources are in the near-zone region. The far-zone sources, on the other hand, are described by means of incident plane-wave fields prescribed for a "source" plane, typically six to ten cells away from the exposed body. The source-body interaction volume is subdivided into Yee cells of the type shown in Fig. 5.12. The interaction space consisting of several hundred thousand to a few million cells is truncated by means of absorbing boundaries. The prescribed incident fields are tracked in time for all cells of the interaction space. The solution is considered complete when the fields have died either off or for sinusoidal excitation, when a sinusoidal steady-state behavior for $\mathbf{E}$ and $\mathbf{H}$ is observed for the interaction space.

The biological body of interest is mapped into the lattice space by first choosing the lattice increment and then assigning values of dielectric permittivity and conductivity to each cell. The boundary conditions at material interfaces are inherently generated by Maxwell's equations. Thus, once a computer program is developed, the basic routines need not be changed for different model geometries. In fact, heterogeneities and fine structural details can be modeled with a maximum resolution of one-unit cell.

Time-stepping for the FDTD method is accomplished by an explicit finite difference procedure [Kunz and Luebbers, 1993; Yee, 1966]. For a cubic-cell-lattice space, this procedure involves positioning the electric and magnetic field components about a unit cell of the lattice and then evaluating the components at alternate half-time steps. In this manner, centered difference expression can be used for both the space and time increments without the need to solve simultaneous equations to compute the fields at the latest time step.

The variation or distribution of SAR (i, j, k) is calculated from the peak amplitude of the three electric field components evaluated in the center of each FDTD cell from

$$SAR(i,j,k) = \frac{\sigma(i,j,k)\left[E_x^2(i,j,k) + E_y^2(i,j,k) + E_z^2(i,j,k)\right]}{2\rho(i,j,k)} \tag{5.19}$$

where $\sigma(i, j, k)$ in S/m and $\rho(i, j, k)$ in kg/m^3 represent the conductivity and density of the tissue in cell (i, j, k).

The following sections provide examples of results obtained from applying the FDTD algorithm for computation of SAR distribution and microwave absorption in anatomical human body and head models. Note that quantitative details of SAR and absorption may vary with specific anatomic body and head models.

5.7.2   SAR in Anatomical Human Head in MRI or Near-Zone Exposures

Examples of FDTD computation of local SAR from microwave absorption in a human anatomical head model under near-zone exposure are given here, specifically in the near field produced by coil antennas employed in MRI procedures. The birdcage and transverse electromagnetic (TEM) coils are the most engaged coil antennas in MRI technology (Fig. 5.13). Both coil antennas can provide the homogeneous RF magnetic field required for the MRI procedures. However, at higher fields, significant mutual coupling between the birdcage coil and the subject may occur. The TEM coils can produce relatively more homogeneous RF magnetic fields over a large spatial volume. However, there are technological differences between the two cylindrical coil antennas. In both antenna systems, external metallic shields provide essential functions. For a birdcage coil, the shield is disconnected from the inner conductor elements. It serves to contain the fields inside the coil through RF reflection to prevent excessive RF radiation loss. The TEM coil's inner conductors are not connected to their adjacent elements but instead are connected directly to the shield to serve as a return path for currents from the inner conductors.

For the results shown below, the cylindrical TEM coil antenna used is of conventional design [Jin, 1993]. The dimensions are 30 cm in diameter and 16 cm in length. The antenna consisted of 16 rungs with a 1 cm rung width, equipped with a cylindrical shield with a diameter and length of 38 cm and 24 cm, respectively. Current sources are placed at each of four break points of each rung element. Each current source had a sinusoidal waveform with a 22.5° phase shift between adjacent rungs, producing a circularly polarized field from the antenna.

The anatomical human head model with a 3 mm resolution was derived from segmented images obtained from the VH project. The VH human head model used has approximate dimensions of 137x182x180 mm and consisted of 25 tissue types, including gray matter, white matter, cerebral spinal fluid (CSF), muscle, fat, bone,

Fig. 5.13 MRI Birdcage (left) and TEM volume (right) coil antennas with 16 rungs (http://mriquestions.com/)

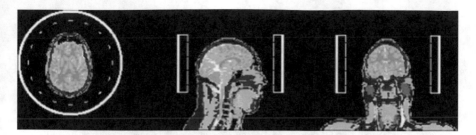

Fig. 5.14 The head model embedded in the center of the cylindrical TEM coil antenna

tooth, etc. A 4-Cole-Cole extrapolation technique was used to determine values for the dielectric properties of the tissues. The relative permittivity and conductivity values at 64, 300, and 400 MHz are listed in Table 4.2.

The center of the head model coincides with the center of the cylindrical TEM coil antenna (Fig. 5.14) and is embedded in the computational domain, which consisted of a 150x150x150 grid of 3x3x3 mm voxels. A Berenger's perfectly matched layer (PML) with eight layers was implemented as the electromagnetic absorption boundary in the FDTD algorithm [Berenger, 1996]. SAR distributions in the anatomic head model for TEM coil operating frequencies of 64 MHz (1.5 T), 128 MHz (3.0 T), 200 MHz (4.7 t), 300 MHz (7.0 T), 340 MHz (8.0 T), and 400 MHz (9.4 T) are shown in Fig. 5.15 for 1 g and 10 g averaging mass. Note that results shown in the figures are normalized values for a whole-head averaged SAR of 3 W/kg.

Since the head is heterogeneous with complex permittivity properties and asymmetric, its frequency-dependent interactions with RF fields from the TEM coil antennas lead to variable, nonuniform, and asymmetric local SAR distributions, especially at higher frequencies. These observations are clearly visible in the axial (two) slices and sagittal and coronal slices through the human head model (Fig. 5.15).

At lower frequencies, the computed SARs show fairly uniform distributions but exhibit local peak absorptions in the antero-peripheral regions such as the area of the eye and nose. SAR decreases with distance toward the center and becomes nearly zero at the center of the head. With increasing RF frequency, the SAR results not only show local absorption peaks in the peripheral regions but also show higher values in several regions deep in the brain. Specifically, local peak SARs are found in the right-temporal and central-occipital regions. The appearance of multiple SAR peaks inside the brain clearly demonstrates a pattern of standing-wave-like resonant RF absorption inside the head from 300 MHz (7.0 T) to 400 MHz (9.4 T). A region of high SAR in the upper portion of the head drifts in the anterior direction and migrates toward the center of the head, which coincides with the center of the TEM coil. It is most visible on the coronal plane at higher frequencies, but not so apparent below the center of the head model – a consequence of the upper portion of the head with its spheroidal geometry and the asymmetries in the neck and shoulders inside the lower part of the TEM coil antenna.

The peak SAR in the head occurs at different locations for different frequencies or magnetic field strengths. Thus, the peak SARs are found in peripheral regions

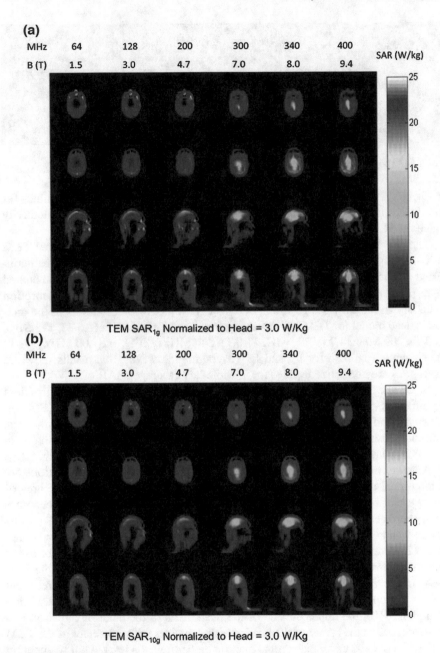

Fig. 5.15 SAR distributions in the anatomic head model for TEM coils of different frequencies or field strengths for 1 g (**a**) and 10 g (**b**) averaging mass, normalized to an average SAR of 3 W/kg) in the head

Table 5.4 Variation of peak SARs as functions of averaging mass size and frequency or field strength

RF Frequency	Magnetic field	Peak SAR$_{1g}$		Peak SAR$_{10g}$	
MHz	Tesla (T)	W/kg	Tissue type	W/kg	Tissue type
64	1.5	19.65	Muscle	11.65	Fat
128	3.0	18.71	Muscle	12.35	Ligament
200	4.7	19.59	Muscle	13.34	Muscle
300	7.0	16.10	White matter	12.57	Grey matter
340	8.0	24.30	White matter	17.01	Grey matter
400	9.4	23.72	White matter	15.49	Grey matter

such as fat and muscle tissues at the lower frequencies, whereas they are associated with more centrally located gray or white matter at higher frequencies. Furthermore, they vary considerably for SAR averaging schemes based either on 1 g or 10 g of tissue mass (Table 5.4). Indeed, the peak 1 g SARs are much higher than the corresponding 10 g ones, indicating that 1 g averaging mass is a more accurate or more precise representation of local tissue SAR.

5.7.3 SAR in Anatomical Body Models Exposed to Plane Waves

At distances far from the microwave source, the radius of curvature of the propagating spherical wave becomes so large that the wavefront would essentially appear as a plane to an exposed subject. For a plane wave, the electric and magnetic fields are uniform in planes normal to the direction of wave propagation and the power density varies only in the direction of propagation. Plane waves are an important class of exposure scenarios since their behaviors are well quantified (Sect. 2.6).

In this case, both electric and magnetic fields of the propagating microwave are orthogonal in space and lie in the plane of the wavefront and are related through the impedance or dielectric permittivity of the medium. In other words, in the far or radiation zone, the electric and magnetic fields have only transverse components to the direction of propagation. Thus, in the far zone of a microwave source, the exposure field is an outgoing plane wave, the induced SAR is independent of source configuration.

Since introduction of the VH human anatomical digital phantoms or models, many studies have been published on anatomical models of the human body with fine spatial resolutions. These body models have been used to calculate RF and microwave absorption as a function of frequency for plane waves [Dimbylow 1997, 2002, 2007; Mason et al., 2000; Lin, 2018]. However, aside from source configurations, the structure, geometry, and composition of the biological body will have influence on the induced field, power deposition, and SAR distribution inside the body.

Table 5.5 Whole-body average SAR_{WB} for plane wave exposure of adult human-body model to an incident power density of 1 W/m²

Frequency (MHz)	SAR_{WB} Voxel size	
	3-mm³ cell	5-mm³ cell
200	0.48	0.51
400	0.64	0.60
600	0.67	0.66
800	0.64	0.63
1000	0.63	0.61
1200	0.61	0.60
1400	0.60	0.59
1600	0.58	0.59
1800	0.56	0.60
2000	0.55	0.60

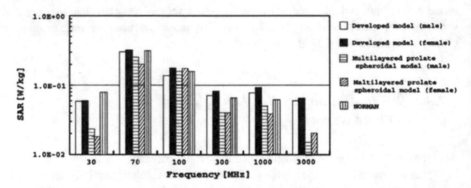

Fig. 5.16 Comparison of computed whole-body average SAR_{WB} for E-polarized plane wave exposure of different adult human-body models as functions of frequency at an incident power density of 10 W/m²

The data summarized in Table 5.5 indicate the whole-body average SAR_{WB} has a weak dependence on model voxel resolution [Mason et al., 2000]. Only minor difference in SAR_{WB} is found for frequencies between 200 MHz and 2 GHz in fine resolution human models exposed to an incident power density of 1 W/m². Furthermore, the variation of SAR_{WB} is small by changing the model resolution from 3 mm³ to 5 mm³.

The frequency dependence of FDTD computed whole-body average SAR_{WB} for E-polarized plane wave exposure of two different male (Norman and Japanese) and a female human anatomical body models at 10 W/m² [Nagaoka et al., 2004] are shown in Fig. 5.16. The similarity of these SAR_{WB} data for the two different adult male-body models in the frequency range of 30 MHz to 3000 MHz suggests consistency and agreement of reported FDTD results. Figure 5.16 also gives for comparison the whole-body average SAR_{WB} calculated for prolate spheroidal models of female and male bodies with the long axes being 161 cm and 176 cm, respectively.

It is interesting to note that for whole-body average SAR$_{WB}$, the anatomical details are not nearly as influential as exposure frequency and Yee cell size. The selection of appropriate computational algorithms for SAR calculation is an important and significant consideration [Laakso et al., 2010; Nagaoka et al., 2004; Uusitupa et al. 2010; Wang et al. 2006].

Examples of computed SAR distributions for an anatomical model of a human body exposed to plane wave sources of microwaves are shown in Figs. 5.17. The anatomical body model (Duke from the Virtual Family data set) was employed for the simulation. Model Duke is comprised of 110 × 58 × 360 cell elements with 78 different types of tissues. The dielectric permittivity properties of the 78 different tissues were taken from ITIS database (http://www.itis.ethz.ch/itis-for-health/tissue-properties/database/), with their origin from Gabriel et al. [1996a, b]. The FDTD algorithm used a 1 mm spatial resolution.

Figure 5.17 presents SAR distributions in the central coronal section of the Duke body model for plane-wave exposures at three microwave frequencies: 700 MHz, 1800 MHz, and 2700 MHz [Cavarnaro and Lin, 2019]. Shown are computed values, normalized to 1 W/kg (at 0 dB). These results indicate highly non-uniform SAR distribution inside the body at all three frequencies, although SAR distribution at 700 MHz is considerably more uniform than at 2700 MHz, where higher SAR values are seen in more superficial tissues including the head region. As expected, the

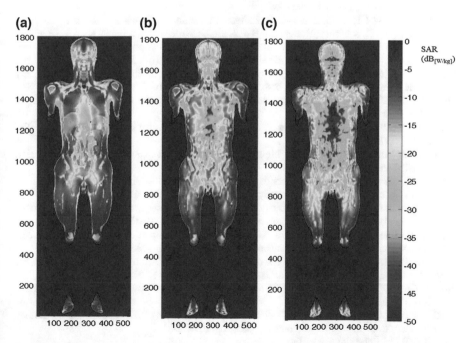

Fig. 5.17 Plane wave SAR distribution in the central coronal section of the Duke human body model at three microwave frequencies. (**a**) 700 MHz; (**b**) 1800 MHz; (**c**) 2700 MHz normalized to 1 W/kg (0 dB)

higher the microwave frequency, the shallower the penetration depth of microwaves. Note that the higher-frequency SARs in the center of this coronal section of the body and head are negligible in comparison to the skin and subcutaneous tissues. Thus, there is relatively a strong concentration of microwave power deposition close to the body and head surface at higher microwave frequencies. However, SARs are much more uniformly distributed among all tissues in the arms and legs, mainly because of their narrower cross sections.

References

Ackerman MJ (1998) The visible human project. Proc IEEE 86:504

Allen SJ, Adair ER, Mylacraine KS, Hurt W, Ziriax J (2005) Empirical and theoretical dosimetry in support of whole body radio frequency (RF) exposure in seated human volunteers at 220 MHz. Bioelectromagnetics 26:440–447

Berenger J (1996) Perfectly matched layer for the FDTD solution of wave-structure interaction problems. IEEE Trans Antennas Propag 44(1):110–117

Cavarnaro M, Lin JC (2019) Importance of exposure duration and metrics on correlation between RF energy absorption and temperature increase in a human model. IEEE Trans Biomed Eng 66(8):2253–2258

Chou CK, Bassen H, Osepchuk J, Balzano Q, Peterson R, Meltz M, Cleveland R, Lin JC, Heynick L (1996) Radio frequency electromagnetic exposure: tutorial review on experimental dosimetry. Bioelectromagnetics 17:195–208

Christ A, Kainz W, Hahn EG, Honegger K, Zefferer M, Neufeld E, Rascher W, Janka R, Bautz W, Chen J, Kiefer B, Schmitt P, Hollenbach H-P, Shen J, Oberle M, Szczerba D, Kam A, Guag JW, Kuster N (2010) The Virtual Family—development of surface-based anatomical models of two adults and two children for dosimetric simulations. Phys Med Biol 55:23–38

Dimbylow PJ (1997) FDTD calculations of the whole-body averaged SAR in an anatomically realistic voxel model of the human body from 1 MHz to 1 GHz. Phys Med Biol 42:479–490

Dimbylow PJ (2002) Fine resolution calculations of SAR in the human body for frequencies up to 3 GHz. Phys Med Biol 47:2835

Dimbylow P (2007) SAR in the mother and foetus for RF plane wave irradiation. Phys Med Biol 52:3791–3802

FDA (2017) Virtual Family data set, https://www.fda.gov/about-fda/cdrh-offices/virtual-family. U.S. Food and Drug Administration

Findlay RP, Dimbylow P (2005) Effects of posture on FDTD calculations of specific absorption rate in a voxel model of the human body. Phys Med Biol 50:3825–3835

Findlay RP, Dimbylow P (2006) FDTD calculations of specific energy absorption rate in a seated voxel model of the human body from 10 MHz to 3 GHz. Phys Med Biol 51:2339–2352

Gabriel S, Lau RW, Gabriel C (1996a) The dielectric properties of biological tissues: II. Measurements in the frequency range 10 Hz to 20 GHz. Phys Med Biol 41:2251–2269

Gabriel S, Lau RW, Gabriel C (1996b) The dielectric properties of biological tissues: II parametric models for the dielectric spectrum of tissues. Phys Med Biol 41:2271–2293

Ho HS, Guy AW (1975) Development of dosimetry for RF and microwave radiation – II. Health Physics 29(2):317–324

ICRP (International Commission on Radiation Protection) (1994) ICRP Publication 66: human respiratory tract model for radiological protection. ICRP 24:1–3

Jin J (1993) The finite element method in electromagnetics. Wiley, New York

Kritikos HN, Schwan HP (1972) Hot spots generated in conducting spheres by electromagnetic waves and biological implications. In IEEE Transactions on Biomedical Engineering, BME-19(1):53–58

Kunz KS, Luebbers RJ (1993) The finite-difference time-domain method for electromagnetics. CRC Press, Boca Raton

Laakso I, Uusitupa T, Ilvonen S (2010) Comparison of SAR calculation algorithms for the finite-difference time-domain method. Phys Med Biol 55:N421–N431

Lacroux F, Conil E, Carrasco AC, Gati A, Wong MF, Wiart J (2008) Specific absorption rate assessment near a base-station antenna (2,140 MHz): some key points. Ann Telecommun 63:55–64

Li C, Wu T (2015) Dosimetry of infant exposure to power-frequency magnetic fields: Variation of 99th percentile induced electric field value by posture and skin-to-skin contact. Bioelectromagnetics 36:204–218

Li C, Chen Z, Yang L, Lv B, Liu J, Varsier N, Hadjem A, Wiart J, Xie Y, Ma L, Wu T (2015) Generation of infant anatomical models for evaluating the electromagnetic fields exposure. Bioelectromagnetics 36:10–26

Lin JC (1976) Interaction of two cross-polarized electromagnetic waves with mammalian cranial structures. IEEE Trans Biomed Eng 23:371–375

Lin JC (2000) Mechanisms of Field coupling into biological systems at ELF and RF frequencies. In: Advances in electromagnetic fields in living systems, vol 3. Kluwer/Plenum, New York, pp 1–38

Lin JC (2007) Dosimetric comparison between different possible quantities for limiting exposure in the RF band: rationale for the basic one and implications for guidelines. Health Phys 92(6):547–453

Lin JC (2018) Computational methods for predicting electromagnetic fields and temperature increase in biological bodies, Chapter 9. In: Bioengineering and biophysical aspects of electromagnetic fields, 4th edn, pp 299–397

Lin JC, Bernardi P (2007) Chapter 10: Computer methods for predicting field intensity and temperature change. In: Barnes F, Greenebaum B (eds) Handbook of biological effects of electromagnetic fields. CRC Press, Boca Raton, pp 293–380

Lin JC, Gandhi OP (1996) Computer methods for predicting field intensity. In: Polk C, Postow E (eds) Handbook of biological effects of electromagnetic fields. CRC Press, pp 337–402

Lin JC, Wang ZW (2005) SAR and temperature distributions in canonical head models exposed to near- and far-field electromagnetic radiation at different frequencies. Electromagn Biol Med 24(3):405–421

Lin JC, Guy AW, Kraft GH (1973) Microwave selective brain heating. J Microwave Power 8(3):276–286

Mason AP, Hurt WD, Walters TJ, D'Andrea JA, Gajšek P, Ryan KL, Nelson DA, Smith KI, Ziriax JM (2000) Effects of frequency, permittivity, and voxel size on predicted specific absorption rate values in biological tissue during electromagnetic-field exposure. IEEE Trans Microwave Theory Tech 48:2050

Nagaoka T, Watanabe S (2009) Voxel-based variable posture models of human anatomy. Proc IEEE 97:2015–2025

Nagaoka T, Watanabe S, Sakurai K, Kunieda E, Watanabe T (2004) Development of realistic high-resolution whole-body voxel models of Japanese adult male and female of average height and weight and application of models to radio-frequency electromagnetic-field dosimetry. Phys Med Biol 49:1–15

NCRP (1981) Radio frequency electromagnetic fields: properties, quantities, units, biophysical interactions, and measurements, Rpt 67, NCRP, Bethesda

Piuzzi E, Bernardi P, Cavagnaro M, Pisa S, Lin JC (2011) Analysis of adult and child exposure to uniform plane waves at mobile communication systems frequencies (900 MHz – 3 GHz). IEEE Trans Electromagn Compat 53:38–47

Shapiro R, Lutomirski RF, Yura HT (1971) Induced fields and heating within a cranial structure irradiated by an electromagnetic plane wave. IEEE Trans Microwave Theory Tech 19(2):187–196

Stratton JA (1941) Electromagnetics theory. McGraw Hill, New York

Uusitupa T, Laakso I, Ilvonen S, Nikoskinen K (2010) SAR variation study from 300 to 5000 MHz for 15 voxel models including different postures. Phys Med Biol 55:1157–1176

Wang J, Fujiwara O, Kodera S, Watanabe S (2006) FDTD calculation of whole-body average SAR in adult and child models for frequencies from 30 MHz to 3 GHz. Phys Med Biol 51:4119–4127

Wu T, Tan L, Shao Q, Zhang C, Zhao C, Li Y, Conil E, Hadjem A, Wiart J, Lu L, Wang N, Xie Y, Zhang S (2011) Chinese adult anatomical models and the application in evaluation of wideband RF EMF exposure. Phys Med Biol 56:2075–2089

Yee KS (1966) Numerical solutions of initial boundary value problems involving Maxwell's equations in isotropic media. IEEE Trans Ant Propag 14:302–307

Chapter 6
The Microwave Auditory Effect

In Chap. 1, the perception of pulse-modulated microwave radiation via the human auditory system was introduced. Indeed, the microwave auditory effect has been widely recognized as one of the most interesting and significant biological phenomena from microwave exposure. The hearing of pulsed microwaves is a unique exception to the airborne or bone-conducted sound energy, normally encountered in human auditory perception. The hearing apparatus commonly responds to airborne or bone-conducted acoustic or sound pressure waves in the audible frequency range (up to 20 kHz). However, the hearing of microwave pulses involves electromagnetic waves whose frequency ranges from hundreds of MHz to several GHz. Since electromagnetic waves (e.g., light) are commonly seen or visible, but not heard or audible, the report of auditory perception of microwave pulses was at once astonishing and intriguing. Moreover, it stood in sharp contrast to the responses associated with continuous-wave microwave radiation [Lin, 1990, 2004; Lin and Wang, 2010].

This chapter describes the research studies on humans and animals leading to documenting scientifically that absorption of a single microwave pulse impinging on the head may be perceived as an acoustic zip, click, or knocking sound. Also, depending on the incident microwave power, a train of microwave pulses to the head may be sensed as an audible buzz, chirp, or tone by humans. The chapter describes neurophysiological, psychophysical, and behavioral observations from laboratory studies involving humans and animals as experimental subjects. The objective is to present what is scientifically known about the microwave auditory effect. In reviewing the accumulated experimental data and scientific knowledge, it will focus on important results that are of lasting scientific significance.

Studies have shown that the microwave auditory phenomenon does not arise from an interaction of microwave pulses directly with the central auditory system. Instead, the microwave pulse, upon absorption by soft tissues in the head, launches an acoustic pressure wave that travels by bone conduction to the inner ear. There, it activates the cochlear receptor cells via the same process involved for normal airborne or bone-conducted sound hearing.

© Springer Nature Switzerland AG 2021
J. C. Lin, *Auditory Effects of Microwave Radiation*,
https://doi.org/10.1007/978-3-030-64544-1_6

6.1 A Historical Perspective

The earliest reports of the auditory perception of microwave pulses were provided anecdotally by airmen and radar operators during World War II or shortly thereafter. In particular, the Airborne Instruments Laboratory [1956] described in an advertisement observations made in 1947 on human auditory detection of microwave signals at radar installations. Others had passed on their experiences with microwave auditory effect to their physicians, journalists, or magazine editors. The witnesses described an audible sound, often as a zip, click, or buzz that occurred apparently at the pulse repetition frequency of radars when standing in the radiation beam of radar antennas. A concerted effort was initiated nearly a decade later by interviewing people who had experienced the auditory sensation. The information collected from people who reported experiencing the sensation was evaluated and collated with a variety of RF and microwave radar transmitters. The information also helped to provide clues to the effect's characteristics and suggestions for experimental testing of persons who had reported the sensation. Those results were summarized in a technical note [Frey, 1961].

The report found in field experiments that radar transmitter frequencies from 200 MHz to 3000 MHz had elicited auditory responses from tested subjects, who were over 100 ft. away from the radomes enclosing the transmitter's antenna. The experimental subjects reported hearing a buzzing or knocking sound for pulse repetition frequencies that varied from 200 Hz to 400 Hz. Subjects blindfolded with tight-fitting blackened goggles reported perception which coincided perfectly with pulsed microwave exposure. When earplugs were used to attenuate the ambient noise level by 80 dB, the subjects indicated a reduction in ambient noise level and an apparent increase in the level of microwave-induced sound. Moreover, in a paired test, it was found that persons shielded from the impinging microwave radiation ceased to report perception. Subjects who were not shielded continued to report hearing microwave-induced sonic signals. Also, the sensation occurred instantaneously and appeared to originate from within or near the back of the head; the orientation of the subject in the microwave field was not an important factor.

Another finding was that subjects who were asked to compare the perceived sound with conventional acoustic sound invariably choose their parallels from the higher frequencies and eliminated all frequencies below 5 kHz. (This appears to be the same as the later report on human's ability to match sounds caused by RF pulses in the MHz range to acoustic frequencies near 4.8 kHz [Frey and Eichert, 1985]). Anyway, the subject report showed that the human auditory system can respond to pulse-modulated RF and microwave radiation and is referred to this auditory phenomenon as "RF sound." Although the report conceded that in classic physiology the auditory and visual systems are distinguished by the fact that the two systems respond to different types of energy, acoustic and electromagnetic, respectively. However, the report also contended that it had obtained data which suggested that "this fact may not be correct."

The earlier observations were extended to a wider range of microwave frequencies (200 MHz to 8.9 GHz, but with no auditory response at 8.9 GHz) during the following 2 years [Frey, 1962, 1963]. These two papers contained much of the same field tests results conveyed in 1961. It also reported that people with a notch in their audiogram around 5 kHz may have difficulty to perceive microwave-induced sound. However, the stated objective of the 1962 paper was to bring a new phenomenon to the attention of physiologists and likewise that the 1963 paper was to acquaint clinicians with the phenomena. These papers mentioned that with appropriate modulation, the perception of various sounds can be induced by microwave pulses in human subjects at inches and up to thousands of feet from the radar transmitter. It suggested that further experimental work with these phenomena may yield information on auditory system functioning and, more generally, information on nervous system function. It discussed some preliminary inquiries into evidences for the various possible neurophysiological sites of microwave or electromagnetic sensors and had specifically ruled out locations peripheral to the cochlea, such as the middle ear. Indeed, a region over the temporal brain lobe was favored as a sensitive area for detecting microwave auditory effect (see the next chapter). This assertion was repeated in a later publication [Frey, 1967]. It was mentioned that their data suggested the microwave auditory effect was neuronal in nature and possibly involving the auditory (eighth cranial) nerve, central auditory nervous system, and the auditory cortex.

There were two other reports on this subject from that same year [Constant, 1967; Ingalls, 1967]. One of them reported sensation of 1–2μs wide microwave pulses at 6500 MHz [Constant, 1967]. Interestingly, Ingalls, [1967] was from the same university as the earlier papers [Frey, 1961, 1962, 1963]. The reported rationale for that study was that, at the time, there was an account of someone hearing a radar at an installation in Turkey and having located a similar radar in the United States. The findings in this mostly descriptive paper mirrored those described in the other papers. It concluded by saying that, albeit with very little material support, the microwave auditory effect takes place by direct stimulation of the central nervous system, perhaps in the brain, bypassing the ear and much of the associated auditory system.

In any case, the reports were met with skepticism, and the hearing sensation from radars was dismissed by most scientists and engineers in the United States as an artifact. Perhaps because the reports tended to be speculative and the studies were conducted as field test on radar installations with limited control measures. The field studies also raised questions that could not be answered at the time. However, beginning in the early1970s, the situation started to change.

The psychophysical technique of magnitude estimation was used in a controlled laboratory experiment to study the perception of sound induced by pulse-modulated microwave radiation inside a microwave anechoic chamber [Frey and Messenger, 1973]. The microwave anechoic chamber constructed of microwave energy absorbers minimized microwave reflections. A pulse microwave power source was able to generate amplitude modulated carrier signals at 1245 MHz. A pulse repetition rate was selected so that it produced a buzzing sound. Trained human subjects with clinically normal hearing were tested individually within the microwave anechoic chamber. The test subjects sat on a wooden stool with back to a horn antenna. The subject indicated with a hand switch to signal microwave pulse-induced sound perception. The data obtained indicated that the perception of microwave-induced sound was primarily a function of the peak power density and secondarily dependent on pulse width. The report also suggested that a band of optimal pulse widths seems to exist for the microwave auditory sensation.

Note that while a definitive statement was not offered in this paper concerning potential mechanism of induction or interpretation of auditory perception of pulsed microwaves, it mentioned that the perception did not involve any transduction of electromagnetic to acoustic energy. It further suggested that the perception differed from the electrophonic effect and could not be accounted for by an explanation involving radiation pressure against the body surface. However, the data reported by the subjects in this psychophysical experimentation was received as consistent and reliable. It served to affirm previous field reports of the microwave auditory effect and helped to facilitate documentation of the phenomenon of microwave auditory effect. The perceived reliability of the data had encouraged other investigators to embark on further psychophysical experiments in attempts to confirm the observations. In this sense, the Frey and Messenger [1973] report served as a water shed moment, which led the microwave auditory effect to be accepted as a scientific fact.

One investigation commenced using a 2450 MHz laboratory microwave source capable of providing up to 10 kW peak power pulses with pulse width varying from 0.5μs to 32μs [Guy et al., 1973, 1975a]. Pulsed microwave energy was launched through an S-band aperture horn antenna. Microwave absorbing materials were placed around the vicinity of the subject to reduce microwave reflections. The horn antenna, absorbing material, and test subject were situated in a shielded room completely isolated from the microwave power-generating equipment and experimenter in order to eliminate disturbing noises and unwanted subject-experimenter interactions (Fig. 6.1). The incident power density at the location of the exposed surface of the subject's head

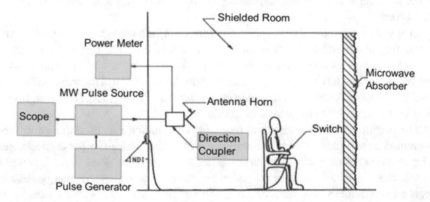

Fig. 6.1 Experimental arrangement for studying microwave pulse-induced auditory effect in human subjects

Fig. 6.2 Experimental subject with the back of the head toward a microwave horn antenna

was measured with the subject being absent. The subject sat with the back of the head directly in front of the horn antenna, as shown in (Fig. 6.2), and used a light switch to signal the experimenter when an auditory sensation was perceived.

This investigation found that each individual pulse could be heard as a distinct zip or click originating from somewhere within or near the back of the head. Short pulse trains could be heard as chirps with the tone corresponding to the pulse repetition frequency. When the pulse generator was keyed manually, transmitted digital codes could be accurately interpreted by the subject. It is significant to note that the energy required for auditory perception by a subject with normal hearing was approximately a third to a quarter of that required for a subject with sensori-neural conduction impairment.

There was another laboratory experiment that followed up on the microwave-induced auditory sensation in humans [Cain and Rissmann, 1978]. In this case, each subject placed his or her head under a horn antenna which was driven by 15μs or 20μs pulses of 3000 MHz radiation. The study essential applied the same psycho-physical protocols mentioned above. Five subjects reported hearing a click which was reported to originate from inside the head. Although there were some variations in hearing ability, none of the subjects had a pronounced hearing loss greater than 25 dB. This study also hinted at the possibility that microwave-induced sound in humans may contain a significant portion of its energy above 8 kHz.

The outcome of these two controlled laboratory studies confirmed the earlier field reports, and in particular, the Frey and Messenger [1973] paper demonstrated that humans can indeed consistently hear pulsed microwave radiation. Specifically, humans can perceive an auditory sensation when the head is exposed to 200–6500 MHz pulse-modulated microwave energy with a peak power density in the range of 1–10 kW/m^2 and pulse widths from 1μs to 100μs. The microwave-induced sound appears as an audible zip, click, popping, or knocking sound and buzz, hiss, chirp, or tune depending on such factors as pulse width and pulse repetition frequency of the impinging microwave radiation, and it is perceived as originating from within and near the back of the head. When the pulses are delivered manually, transmitted digital codes could be reliably interpreted by humans.

While the microwave auditory effect or microwave hearing was becoming an accepted fact, the precise neurophysiological site or anatomical location of trans-duction and mechanism of microwave auditory effect remained obscure for some time and had developed into a major topic of scientific research; the substance of which will be discussed in later sections and chapters. What follows next will be a discussion on details of the psychophysical studies that have assisted in confirming the phenomena of mammalian auditory perception of pulse microwave radiation.

6.2   Psychophysical Studies in Humans

Between 1961 and 1973, investigations of the microwave auditory effect had pri-marily been conducted as field tests in radar installations with limited control mea-sures. While the results were interesting and extraordinary, the reports were met with skepticism and draw little attention beyond curiosity from the scientific com-munity. But still the acquired information helped to provide clues to the effect's characteristics and suggestions for advanced study in more appropriately equipped laboratory and better controlled experimental investigation. The following are detailed descriptions of three psychophysical experiments conducted with human subjects in controlled laboratory experiments that have contributed data toward the microwave auditory effect being accepted as a scientific fact [Frey and Messenger, 1973; Guy et al., 1973, 1975a, b; Rissmann and Cain, 1975; Cain and Rissman, 1978]. Another interesting experiment on the loudness of perception for human sub-jects exposed to pulse-modulated 800 MHz microwave radiation in 18 males and females is given in Sect. 6.5.

6.2.1   Microwave Pulses at 1245 MHz

A series of psychophysical experiments with human subjects was conducted in a controlled laboratory experiment to study the perception of sound induced by pulse-modulated microwave radiation inside a microwave anechoic chamber [Frey and Messenger, 1973]. The inside wall of the microwave anechoic chamber was covered with microwave radiation absorber to minimize microwave reflections. A pulse microwave power source generated amplitude modulated carrier signals at 1245 MHz. Microwave pulses were fed through an air line, coaxial cable, and coax-to-waveguide adaptor to a standard-gain horn antenna in the anechoic chamber, which radiated vertically polarized microwaves.

Exposure parameters were measured with a half-wave dipole antenna on a wooden pole, which was positioned where the center of the subject's head would be during data collection. The dipole was connected by coaxial cable to an attenuator outside the chamber, which was in turn connected to a thermistor mount. In this study, the peak power density varied between 900 and 6300 W/m^2 and pulse width ranged from 10μs to 70μs.

Trained human subjects with clinically normal hearing were tested individually within the microwave anechoic chamber. The subject sat on a wooden stool with back to the horn antenna and chin against an acrylic rest mounted on a vertical wooden pole. Signaling of microwave pulse-induced sound perception by the subject was done with a multi-number hand switch; keying in a number would record the perceived loudness. The psychophysical technique of magnitude estimation [Stevens 1961] was used in this experiment. Each subject was informed that the first pulsed microwave sound heard in each trial would be a reference that would be assigned the number 100 and that the second sound heard would differ in loudness from trial to trial. A 50-Hz pulse repetition rate was selected to produce a buzzing sound. The reference level was chosen to approximate the middle loudness range.

The "standard sound" or reference microwave exposure was presented for 2 s. A silent period of approximately 5 s followed, and then a microwave pulse of some combinations of peak power density, average power density, and pulse width was presented for 2 s. The subject was asked to assign to each microwave-induced sound a number proportional to the apparent loudness. The test conditions – the order of presentation of microwave parameters – were randomized by using a table of random numbers. There were three randomized repetitions of the test conditions.

The threshold peak power density required for perception is around 2458 W/m^2 for this experiment. The results showed that while holding the average power density constant by changing the pulse width, the perceived loudness clearly exhibited a dependence on peak power density. However, holding the peak power density constant by increasing either the pulse width or average power density, the data showed the perceived loudness of pulse-modulated 1245 MHz exposure for the same subjects was a complex function of both, which is consistent with theoretical predictions [Lin, 1977a, b; 1980]. We will return to this topic later in this Chapter.

6.2.2   Microwave Pulses at 2450 MHz

In a study designed for the purpose of more precisely measuring the microwave exposure parameters, two investigators served primary as the human subjects in this 2450-MHz study [Guy et al., 1973, 1975a,b]. Microwave radiation was generated by a pulse signal source (Applied Microwave Laboratory, PH 40 K). The source was capable of providing up to 10 kW peak power pulses with the pulse width varying from 0.5μs to 32μs and was used to feed an S-band aperture horn antenna (32 cm × 26 cm) by means of an RG8 coaxial cable. The incident power to the horn and reflected power were monitored by means of a bolometer and power meter combination connected to a bidirectional coupler inserted between the coaxial cable and the horn. The pulse width and pulse repetition frequency were controlled by an external pulse generator and monitored on an oscilloscope (see Fig. 6.1). The subject sat with the back of the head directly in front of the horn, 15–30 cm from the aperture with ears at approximately the same height as the center of the horn

(Fig. 6.2). Placement of the subject's head in the near zone field of the horn was necessary to obtain the full dynamic range of pulse widths and power levels necessary for evoking an auditory sensation. The average power density at the location of the exposed surface of the subject's head inside the shielded room was measured with a Narda 8100 power monitor at high pulse repetition frequencies and at low peak power levels as a function of incident power to the horn without the subject.

The values for higher power and lower pulse repetition frequency were obtained by linear extrapolation from the monitored incident power to the horn. Microwave absorbing materials were placed around the vicinity of the subject position to eliminate reflections. As a blinding protocol for the experiment, the horn antenna and test subject were located inside a shielded room completely isolated from the microwave power equipment to mitigate against any subject – experimenter interactions. Cable connectors to the horn and bidirectional couplers were made through bulkhead connections on the wall of the shielded room. The subject used a light switch to signal the experimenter when an auditory sensation was perceived.

Prior to the tests, standard audiograms of the subjects were taken. One subject had normal hearing, while a pronounced notch at 3500 Hz was noted for both ears of the other human subject. Similar results were obtained for both air and bone conduction. The ambient noise level in the shielded room, exposure chamber was measured as 45 dB, re 20 mPa, with a sound level meter. Each subject was exposed to a range of microwave pulse widths varying from 1μs to 32μs. The pulses were presented as a train of three pulses 100 ms apart every second, to maintain an average power density well below 10 W/m². The psychophysical method used was closely related to the method of limits or minimal change [Sheridan, 1971]. The subjects responded to each microwave stimulus separately by depressing a light switch. In order to guard against any ordering effect, both ascending and descending series were employed.

Each individual pulse could be heard as a distinct zip or click originating from somewhere within or near the back of the head. Short pulse trains could be heard as chirps with the tone corresponding to the pulse repetition frequency. When the pulse generator was keyed manually, transmitted digital codes could be accurately interpreted by the subject. It is significant to note that the energy required for auditory perception by the subject with normal hearing was approximately a third to a quarter of that required for the subject with sensorineural conduction impairment. The effort to determine threshold for auditory sensation in humans and other observations will be discussed in the following section.

6.2.3 Microwave Pulses at 3000 MHz

Opting for a slightly higher microwave frequency by which earlier field study had been conducted, Rissmann and Cain [1975] investigated in a laboratory setting microwave-induced auditory sensation in humans. Before the experimental session, standard audiograms were obtained for each of the eight volunteer subjects.

Although there were some variations in hearing ability, none of the subjects had a pronounced hearing loss greater than 25 dB. Each subject positioned his or her head under a horn antenna that was driven by a 15µs pulse of 3000 MHz microwave radiation. All subjects wore plastic foam ear muffs to attenuate room noise. The background noise level at the head position was measured to be 45 dB using a sound level meter. Five subjects with normal audiograms reported hearing a click which originate from inside the head. Three subjects had difficulty in perceiving a micro-wave-induced sound when the maximum power output of the microwave source was used with a pulse width of 15µs, but they had no difficulty in hearing a click when the pulse width was increased to 20µs. This observation noted that although the audiograms revealed no significant differences in hearing ability among the sub-jects, they only provide values up to a maximum frequency of 8 kHz. It was sug-gested that microwave-induced sound in humans could possibly contain a significant portion of its energy above 8 kHz, and these three subjects had an inordinate amount of hearing loss at higher frequencies. This had in fact been documented to be the case in a later publication [Cain and Rissmann, 1978]. These results also corrobo-rated earlier theoretical predictions [Lin, 1977a, b; 1980].

In summary, human perception of pulse-modulated microwave energy transmit-ted through air has been demonstrated in controlled laboratory studies at RF and microwave frequencies ranging from 1000 MHz to 6500 MHz, with pulse width varying from 1µs to 1000µs. The sensation occurred instantaneously and was per-ceived as originating from within and near the back of the head. Furthermore, head orientation in the microwave field did not influence the loudness of perceived sound. While an ideal, noise-free laboratory environment is not a requirement for percep-tion, the incident microwave energy required in some case decreased by more than 6 dB when earplugs were used. As mentioned, a single microwave pulse can be perceived as an acoustic zip, click, or knocking sound, and exposure of the head to a train of microwave pulses may be sensed as an audible buzz, chirp, or tune, depending on the pulse repetition frequency. When the pulse generator was keyed manually, transmitted digital codes could be accurately interpreted by the subject.

6.3   Threshold Power Density for Human Perception

Since the first report that pulse-modulated microwave radiation induces an auditory sensation in humans, several investigators have attempted to assess the thresholds for sensation as a function of microwave parameters. The discussions to this point have led to the conclusion that the microwave auditory effect is a genuine phenom-enon and a scientific fact. The precise anatomical location, neurophysiological site of transduction, and mechanism(s) of interaction for the effect remained obscure for some time. The microwave-induced sound, whether it appears as a zip or click, buzz or chirp, is clear that it would depend on such exposure factors as pulse width, peak power, and pulse repetition frequency of the impinging microwave radiation. The average power did not significantly impact the microwave auditory effect.

6.3.1   Field Tests of Adult Humans with Normal Hearing

Although not explicitly stated, it is reasonable to assume plane-wave exposures in the far zone of radar antennas for the early field experiments. The peak incident power density of 6μs wide 1310 MHz microwaves pulsed at 244 Hz was converted from average power density measured with a loop antenna, a power bridge, and microwave wattmeter [Frey, 1961]. The ambient noise level was about 70 dB. Earplugs that attenuated tones between 125 Hz and 8000 Hz by 25–30 dB were also used. The peak power density perception threshold for eight subjects was approximately 2.67 kW/m². For the 2982 MHz experiment, the ambient noise level was about 80 dB. The same earplugs as above were used. The power density was determined in this case with horn antenna and Narda power meter. The peak incident power density threshold for seven subjects was approximately 50 kW/m² for 2982 MHz microwaves (1μs pulses at 400 Hz).

The determination of threshold power density for perception was extended to 425 MHz in another reported field experiment involving various microwave parameters [Frey, 1962]. The measured ambient sound level was 70–90 dB. The report also provided field investigations made at 216 MHz and 1310 MHz. All field measurements were summarized and tabulated in a later publication [Frey, 1963], including an average threshold at 425 MHz at 2.54 kW/m² for pulse widths of 125μs to 1000μs (Table 6.1). Note that with earplugs in place, the ambient noise levels likely are in the range of 45–50 dB for these plane wave power density thresholds.

6.3.2   Laboratory Study of Human Adults with Normal Hearing

The threshold peak power density for human adults was investigated under controlled laboratory experimental conditions at frequencies of 1245, 2450, and 3000 MHz with pulse widths of 1–70μs. In contrast to the plane-wave field

Table 6.1 Threshold power density for human auditory perception of pulsed microwave radiation obtained in far zone of antenna fields

Ambient noise level 70 to 90 dB (with Earplugs 45 to 50 dB)				
Frequency	Pulse width	Peak power density	Pulse rate	Duty factor
MHz	μs	kW/m²	Hz	
216	–	6.70	–	0.006
425	1000	2.54	27	0.028
425	500	2.29	27	0.014
425	250	2.71	27	0.0007
425	125	2.63	27	0.00038
425	–	2.54 (Ave)[a]	27	–
1310	6	2.67	244	0.0015
2982	1	52.50	400	0.0004

[a]Average threshold value calculated for all peak power densities at 425 MHz

experiments, these studies involved near-zone exposure of subjects. The results of these efforts to establish the threshold of microwave-induced auditory sensation are given below. Some interesting qualitative information on threshold of human perception has been reported at 800 MHz [Tyazhelov et al., 1979]; however, quantitative values are difficult to ascertain.

Also, the thresholds for auditory perception of pulsed RF energy absorption in the human head have been studied on six subjects with magnetic resonance imaging (MRI) head surface coils [Roschmann, 1991]. RF exposure frequencies ranged from 2.4 MHz to 170 MHz, and pulse widths varied from 3µs to 100µs (see Chap. 9 for further detail). Apart from the different frequency range explored, the use of MRI coils also provides a different mode of RF power deposition which is predominantly of an inductive nature via the interaction of RF magnetic field with lossy dielectric tissues of the head. The auditory effect RF energy thresholds were observed at 16±4 mJ per pulse. The auditory threshold of RF pulse widths greater than 200µs occurred at an average peak power level as low as 20 W for surface coils. The study noted that the results are in excellent agreement with data from horn antenna measurements at 2450 MHz [Guy et al., 1973, 1975a], using the assumption of an effective head absorption area of 400 cm² for conversion of reported peak power density data into absorbed peak power levels in the head.

In the Frey and Messenger [1973] study, adult subjects with clinically normal hearing were tested individually in a microwave anechoic chamber with an estimated sound pressure level of 45 dB. The method of magnitude estimation was used. At the rate of 244 Hz, a buzzing sound was perceived by the subjects. In Table 6.2, the number for each case represents the median of all subjects and all repetitions. It was found that the threshold peak power densities required for perception are functions of both pulse width and peak power. The threshold values ranged from 0.9 to 6.3 kW/m² which varied with pulse widths from 10µs to 70µs with average power and energy per pulse kept constant.

The experimental protocol and conditions associated with the study by Guy et al. [1973; 1975a,b] were described previously. The near-zone 2450 MHz exposure was conducted in a microwave anechoic exposure chamber with a 45 dB measured ambient sound pressure level. The psychophysical method used was the method of limits or minimal change. Measured parameter values are reported as averages of

Table 6.2 Laboratory measured thresholds at 1245 MHz

Pulse width	Peak power density	Average power density	Energy per pulse
µs	kW/m²	W/m²	J/m²
10	6.3	3.2	63
20	3.15	3.2	63
30	2.1	3.2	63
40	6.3	11.26	450
50	1.25	3.2	63
60	1.05	3.2	63
70	0.9	3.2	63

Table 6.3 Laboratory measured thresholds at 2450 MHz

Pulse width	Peak power density	Average power density	Energy per pulse
µs	kW/m²	W/m²	mJ/m²
1	400	1.2	400
2	200	1.2	400
4	100	1.2	400
5	80	1.2	400
10	40	1.2	400
15	23.3	1.1	350
20	21.5	1.2	430
32	12.5	1.2	400

Table 6.4 Laboratory measured thresholds at 3000 MHz

Pulse width	Threshold	Subject					
		#1	#2	#3	#4	#5	Ave for all 5
10µs	Peak power (kW/m²)	18	2.25	6	20	20	13.25
	Pulse energy (mJ/m²)	180	23	60	200	200	132.6
15µs	Peak power (kW/m²)	3	3	3	6	10	5
	Pulse energy (mJ/m²)	45	45	45	90	150	75

three ascending-descending series. The threshold values for pulse width, peak incident power density, and energy density per pulse are given in Table 6.3 [Guy et al., 1975a,b]. The data suggest that the threshold for a 2450 MHz microwave-induced sound was related to both pulse width and peak power for average power densities of 1.2 W/m² and energy densities of 400 mJ/m² per pulse. The threshold varied as functions of both pulse width and peak power density for 1–32µs width pulses.

The threshold of microwave-induced auditory sensation in humans at the higher-frequency band in field tests in radar installations was assessed in a laboratory with a background noise level not more than 45 dB [Rissmann and Cain, 1975]. Five subjects with normal audiograms reported hearing a click in response to 10µs and 15µs wide 3000 MHz microwave pulses delivered in the near zone of a horn antenna. The results obtained for threshold incident peak power density and energy per pulse are listed in Table 6.4 individually for each of the five subjects [Cain and Rissman, 1978]. Although there are similarities, the thresholds differ among subjects and are different for 10µs and 15µs pulses. The average threshold incident peak power densities for five subjects are 13.25 and 5 kW/m², respectively. Note that the subjects showed some variations in hearing acuity, even though they were deemed having normal hearing by audiograms. Some had appreciable amount of hearing loss for frequencies higher than 8 kHz.

Table 6.5 presents a summary of the thresholds determined under controlled laboratory conditions for peak microwave power density of auditory perception in human subjects with normal hearing. Note that while there are greater variations in measured threshold values over the entire range of 1–70µs of pulse widths involved,

Table 6.5 Thresholds of microwave-induced auditory sensation in adult humans with normal hearing determined in controlled laboratory studies

Frequency	Pulse width	Peak power density	Ambient noise level	Study
(MHz)	(μs)	(kW/m²)	(dB)	Author (year)
1245	10–70	0.9–6.3	45[a]	Frey & Messenger [1973]
2450	1–32	12.5–400	45	Guy et al., [1975a]
3000	10–15	2.25–20	45[b]	Cain & Rissman [1978]
Pulse width between 10 and 32μs				
1245	10–30	2.1–6.3	45[a]	Frey & Messenger [1973]
2450	10–32	12.5–40	45	Guy et al., [1975a]
3000	10–15	2.25–20	45[b]	Cain & Rissman [1978]

[a]Typical SPL for microwave anechoic chambers lined with absorbing materials
[b]With plastic foam earmuffs

the subset of data for 10–32μs fall within a narrower range. Considering that the ambient noise levels in all three experiments were essentially the same, it may be reasonable to conclude that the threshold power densities of 2.25–40 kW/m² or an average quantity of 13.86 (~14) kW/m² is a realistic threshold peak power density for induction of the microwave auditory effect.

6.4   Loudness of Human Perception

In general, a sensory response such as perceived loudness of sound is inversely proportional to threshold of sensation. The loudness of perception as a function of pulse width for human subjects exposed to pulse-modulated 800 MHz microwave radiation was determined in 18 male and female subjects with normal high-frequency auditory acuity [Tyazhelov et al., 1979]. A 500 W source was coupled through a 3 meter length of coaxial cable and fed the open end waveguide (15 × 27 cm), which was mounted on a foamed plastic rest that permitted transmission of microwaves to the parietal area of a subject's head. The pulses were 5–150μs wide, and the repetition frequency varied from 50 Hz to 20 kHz at constant power. The loudness for sensation at 8000 Hz is shown in Fig. 6.3 [Lin, 1980, 1981] (plotted from data provided in [Tyazhelov et al. 1979]). As the width of pulses of constant peak power density were gradually increased from 5μs to 150μs, a complex oscillatory loudness function was observed. The loudness increased as pulse width increased from 5μs to 50μs, then diminished with further increase of pulse width from 70μs to 100μs, and then increased again with longer pulse widths. It is significant to note that in this meticulously conducted experiment, the head of the subjects is at various levels above the water line with the body submerged in a large steel drum containing seawater. By pressing a button, the subject informs the researcher sound detection when a microwave pulse or an acoustic stimulus is presented, or a shift in loudness has occurred.

Fig. 6.3 Perceived loudness of pulse-induced sound level as a function of microwave pulse width in human subjects

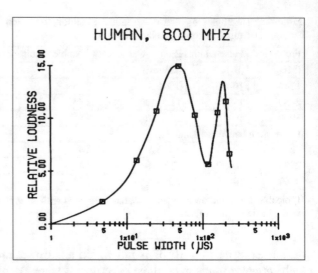

These experimental results corroborated earlier theoretical predictions based on a spherical model of the head [1976, 1977a, b]. The character of the loudness curve is consistent with the predictions (see Chap. 8 for further discussions). They are also in agreement with results obtained at different microwave frequencies by other investigators [Frey and Messenger, 1973; Guy et al., 1973, 1975a].

6.5   Behavioral Study in Animals

The previous section demonstrated that under certain conditions humans can perceive pulse-modulated microwaves as sound. Because the auditory perception involved a discrimination response to differential characteristics of impinging pulsed microwaves, a common issue in studies involving human subjects, the possibility of subjective responses, is averted by using animals. Furthermore, confirmatory evidence in lower animals can substantially enhance the observation of microwave-induced auditory sensation as a genuine biological responsive.

6.5.1   Discriminative Control of Appetitive Behavior

That microwave pulses are acoustically perceptible and can serve as a discriminatory auditory cue in behavioral experiments is reported in a discriminative control of appetitive behavior by pulse-modulated microwave energy in rats [Johnson et al., 1976].

The aim of this investigation was to substitute pulse-modulated microwave for the previously well-discriminated tone cue (acoustic click). The subjects were six female white rats (300–350 g Wistar-derived strain). The animals were partially deprived of food until their body mass fell to 80% of that before deprivation. They

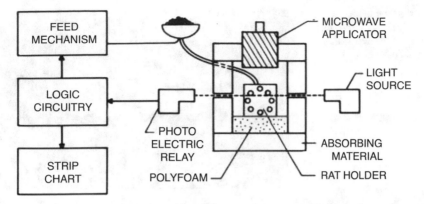

Fig. 6.4   Schematic diagram of experimental setup for testing head-raising-for food response of rats

Fig. 6.5   Rat in a conical acrylic rod-body restrainer with head extended for operant device

were then placed in a body-movement restrainer and trained to perform a head-raising response for food pellets. During daily 90 min sessions, individual rats were presented alternating 5 min stimulus-on/stimulus-off periods during which food was made available as a reward for responding only during stimulus-on periods. The initial stimulus was a 7.5 kHz acoustic click produced by a high-frequency speaker driven by a 1 volt, 3μs wide rectangular pulse at the rate of 10 Hz.

The general experimental setup for the behavioral test is shown in Fig. 6.4. The rat holder shown in Fig. 6.5 was designed to provide necessary restriction of body movement to control for microwave exposure during experimentation, while permitting sufficient movement of the animal's head and neck for the collection of behavioral data. The holder was constructed of acrylic to reduce the amount of distortion of the incident microwave field [Lin and Wu, 1976; Lin et al., 1977a]. The space rod construction provided adequate ventilation for control of the animal's surface temperature and permitted easy placement of the animal. After the first few sessions, the rats learned to position themselves in the holder by running into the cone and extending their heads through the opening. The holder with the rat was then placed in a receiver as shown in Fig. 6.6. The receiver positioned the rat in such a way that the rat could move its head in a short vertical arc.

Fig. 6.6 Rat in acrylic
body restrainer placed on a
baseplate receiver with
operant device

The small head movement, allowing its nose to interrupt the light beam, consti-
tuted the operant behavior. The interrupted light beam caused a switch to close
which led to the delivery of food. An external feeder caused a small, 45 mg food
pellet to be delivered via a polyethylene tube to a receptacle which was constructed
of the same material as the holder and located directly below the rat's head. The rat
was able to pick up and eat the food pellet with only a slight downward movement
of its head. Standard relays, counters, and recorders were used to program the stimuli and record the responses. A closed-circuit television system was used to observe
the animal's behavior during each test session. This system provided a consistent
means of investigating behavior adaptable to the special requirements of microwave
radiation in the exposure chamber [Lin et al., 1974; 1977a, b; Johnson et al., 1976].

After these animals learned to show evidence of their responses so that 85–90%
of a given session's total responses were made during the appropriate stimulus-on
periods, individual animals were then exposed to 30 s of pulse-modulated 918 MHz
microwaves at the same pulse width and pulse repetition rate as the acoustic stimuli
at average incident power densities up to 50 W/m². These animals began to respond
immediately (Fig. 6.7). In subsequent sessions when microwave, not the acoustic
click, was present during the stimulus-on periods, all animals demonstrated a continued ability to respond at the 85–90% level [Johnson et al., 1976]. The rat
responded equally well during presentation of acoustic and microwave stimulation.
This result clearly suggested an auditory component in the microwave control of
this behavior.

6.5.2   Pulsed Microwave as a Cue in Avoidance Conditioning

The detection and use of pulsed microwaves as a cue in avoidance conditioning in
animals (including cats and rats) have been reported over the years [Frey, 1965,
1966, 1971; Frey et al., 1975]. Sprague Dawley rats (male, 150 g, 125 days old)
were tested in a microwave anechoic chamber which contained two acrylic barrier
boxes [Frey and Feld 1972, 1975] to determine whether rats would perceive

Fig. 6.7 Cumulative record showing an animal's performance. Top: In response to 30 s microwave probe, rat begins to respond as if acoustic cue had been presented. Bottom: Rat responds equally well during presentation of acoustic and microwave stimulation

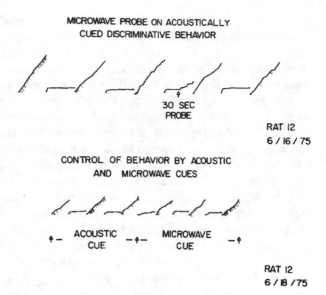

MICROWAVE PROBE ON ACOUSTICALLY
CUED DISCRIMINATIVE BEHAVIOR

30 SEC
PROBE

RAT 12
6 / 16 / 75

CONTROL OF BEHAVIOR BY ACOUSTIC
AND MICROWAVE CUES

ACOUSTIC   MICROWAVE
CUE   CUE

RAT 12
6 / 18 / 75

pulse-modulated microwave energy and respond to it behaviorally. Each box consisted of two compartments. The compartment of the right of one was shielded and the one on the left was shielded from the impinging microwaves using microwave absorbers to minimize microwave exposure of the respective sides of the boxes and to exclude any possible effect due to side preference.

The location of the subject was monitored using a switch affixed to the bottom of each compartment of the barrier box. Rectangular microwave pulses (30μs wide, 1245 MHz) were derived from a pulse source (Applied Microwave Laboratory, Model PG 5 K) at the rate of 100 Hz and were fed to the horizontally polarized standard-gain horn antenna through air lines, coaxial cables, and a waveguide adapter. The incident power density at 5 cm above the floor of each half of the boxes, when the animal was absent, was measured using a half-wave dipole and a thermister and power meter combination. The average power densities in the unshielded compartment were less than $10 \, W/m^2$. The shielded side had a value of 7% or less of the unshielded side.

After acclimation to the barrier boxes, place-avoidance conditioning was initiated with pulse-modulated microwaves as the discriminative stimuli. During each 90 min session, cumulative measurements of residence time in shielded and in unshielded compartments were taken to reflect the course and status of conditioning. Rats were assigned to either an experimental or a control group. Control sessions were run with all equipment turned on but without power output. It was reported that the number of crossings was reduced substantially in the experimental group over the entire experimental sessions. The animals did not exhibit a preference between the compartments in the absence of microwaves (control group). Rats exposed to 1.33 or 3.0 kW/m^2 peak power density exhibited an avoidance of the unshielded compartment or spent most of their time in the shielded side. The observance of avoidance behavior in the absence of explicit location cues led the investigators to conclude that the rats could perceive pulse-modulated microwave radiation.

Another behavioral investigation of microwave-induced auditory sensation was conducted by testing rats in a shuttle-box experiment, in which one compartment was exposed to 33 kW/m² of 2880 MHz microwave pulses at 100 Hz with a pulse width of 3μs and the other was shielded. Cumulative measurements of residence time in shielded and in unshielded compartments were taken to reflect the status of conditioning. Rats were found to spend significantly more time in the shielded side [Hjeresen et al., 1979]. When a high-frequency (37.5 kHz tone) acoustic stimulus was exchanged for the microwave pulses, rats exhibited a preference for the acousti- cally "quiet" side. In addition, the amount of side-to-side traversing activity was greater in rats exposed either to microwave pulses or acoustic stimuli in both sides of the box than in unexposed control animals. It suggests that rats seemed to find the pulsed microwave to be aversive and are motivated to actively avoid it.

The simultaneous presentation of a broad band "pink" acoustic noise (20–40 kHz) and pulsed microwaves produce no statistically significant differences in side pref- erence between experimental and control groups. In contrast, in all cases the num- ber of traverses made by microwave exposed rats was significantly greater than those of unexposed controlled rats. These results indicate that pulse microwave stimulus and the acoustic tone stimuli can result in similar behavior patterns and support the notion that rats acoustically detected the microwave pulses and general- ized the microwave-induced sound and conventional acoustic cues.

It is apparent from the above studies that instantaneous detection of pulse modu- lated microwaves can be mediated by the auditory system. The auditory detection occurs only if the microwave radiation is pulse modulated. Rectangular pulses seemed more effective than other pulse shapes. In neither human nor animal detec- tion was it obvious if the action is on the neural receptor cell, central nervous sys- tem, or some accessory structures or tissues. Pulse-modulated microwaves could be having a direct effect on the primary auditory nerve. It is also possible that some tissues in the head is induced to vibrate in the presence of pulsed microwaves and that this detection is mediated by the normal bone conduction hearing route. Evidences for and against various interaction mechanisms will be presented in a later chapter.

6.6   Neurophysiological Study in Animals

Behavioral investigations have suggested that microwave pulses interact with audi- tory system and are perceived by laboratory rats in the same manner as conventional acoustic pulses. It should be noted, however, that behavioral studies rely on infer- ence rather than direct measurement of the anatomical or physiological entities involved in microwave pulse interaction with the auditory system. They should therefore be complemented by direct observations in identifying the anatomical or physiological substrates. Such observations, which would contribute to definition of the characteristics, mechanisms, and site of transduction of this phenomenon, have been made through direct neurophysiological investigations.

      `   On many sites such as the cerebral cortex, thalamus, cranial nerve, and cochlea associated with the mammalian auditory sensory pathway, small electrodes may be attached or inserted to record the electrical potentials arising in response to acoustic stimulation. If the electrical potentials elicited by microwave pulses exhibited characteristics akin to those elicited by conventional acoustic pulses, this would rigorously support the behavioral findings that pulsed microwaves are acoustically perceptible. Direct and quantitative experimental findings that are related to these characteristics will further the understanding of pulsed microwave interactions with the auditory system. They may confirm or refute hypotheses about direct neural excitation. If microwave-evoked potentials are recorded from the ear and brain loci, this would lend support to the contention that microwave auditory effect is mediated at the periphery as is the sensation of a conventional acoustic stimulation.

Microwave-induced auditory response has been described in reports from several laboratories in terms of recordability from the peripheral and central nervous systems of laboratory animals and similarity between microwave-evoked electrical potentials and those produced by conventional acoustic stimuli. Indeed, there are several different types of electrical activity which may be recorded from the inner ear and the brain during pulse stimulation. These interesting studies designed to help establish the site of interaction and the mechanism involved in the pulsed microwave-induced auditory sensation are discussed in this section, together with experimental observations on the electrophysiological events that occur along the auditory sensory pathways in response to pulse-modulated microwave exposure and purposeful manipulations, including tissue and/or neural nuclei ablation, along the auditory pathway. The discussion begins with the simplest method without any need of surgical procedures – brainstem evoked electrical potential recording via a surface or scalp electrode affixed to the vertex on top of the head.

6.6.1   Brainstem Evoked Response

When acoustic pulses or clicks are presented through loudspeakers to a human or animal subject, a characteristic sequence of evoked electrical potentials or, in short, brainstem evoked response (BER) can be recorded via surface or scalp electrodes [Jewett, 1970, Jewett and Williston, 1971, Picton et al., 1974]. The averaged responses show a series of early components that occur during the first 8 ms following onset of a stimulus that represents activation of the cochlea and the auditory brainstem nuclei. Figure 6.8 shows the first 10 ms of a normal brainstem auditory evoked response recorded using electrodes placed over the vertex and the mastoid. At least seven distinct wave components are seen representing activation of the brainstem nuclei of a human subject. Intermediate components occur between 8 and 50 ms after the stimulus and arise both from cerebral tissues and extracranial

Fig. 6.8 Brainstem auditory evoked response from a normal subject for the first 10 ms. 1000 averages. Scales: 1.5μV vertical; 1 ms horizontal

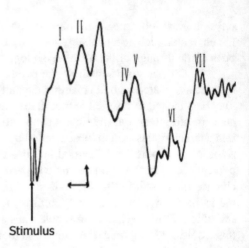

Stimulus

potentials associated with muscular reflexes. Late components that occur 50–300 ms after the stimulus are recorded most prominently over frontocentral region of the scalp and most likely represent widespread activation of frontal cortex. These scalp-recorded, average-evoked potentials in response to an acoustic pulse are highly repeatable from one subject to the next and reflect brain activity from cochlea to the auditory cortex. These complex electrical potentials are therefore of importance in the objective evaluation of hearing.

Several investigators have reported evoked brainstem auditory responses recorded from the vertex of the head of laboratory animals using surface or scalp electrodes [Chou and Galambos, 1979; Chou and Guy, 1979; Chou et al., 1985; Lin et al., 1978; 1979; 1982]. With an electrode affixed to the top of the head and a reference electrode to the side of the head under the ear (typically the mastoid), the evoked brainstem responses represent the volume conducted electrophysiological events that occur in the auditory brainstem nuclei after onset of an acoustic stimuli. The experimental procedure is straightforward. However, the level of recorded electrical signals is typically small; therefore coherent averaging of 100 events or more is generally required as a signal processing scheme. Brainstem electrical potentials evoked by microwave and acoustic pulses from the vertex of a cat are shown in Fig. 6.9. Note the comparable response characteristics. However, microwave pulse-evoked responses are always seen immediately after delivery of the pulse, without the familiar acoustic wave propagation delay associated with sound, as expected. These results show that the same pathway through the central nervous system is activated by both microwave and acoustic pulses. Clearly, microwave energy does not need to be deposited in the cochlea to initiate the microwave auditory effect.

Note that in this series of experiments involving laboratory cats [Lin, et al., 1978, 1979, 1982], it employed, for the first time, a small (15 mm in diameter, Elmed, Addison, IL) direct-contact microwave antenna (see Sect. 6.6.7 for experimental details). The use of a small direct-contact applicator antenna was innovative at the time. It has several advantages over conventional microwave exposure techniques and experimental protocols, such as horns in an anechoic chamber or open

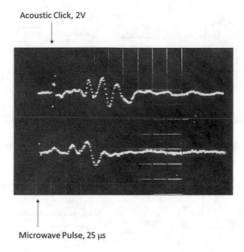

Fig. 6.9 Brainstem electrical potentials from the vertex of a cat's head evoked by acoustic clicks (100 averages) and microwave pulses (200 averages)

waveguides close to the subject. It did not require any specialized devices or equipment such as carbon-loaded, high-resistance electrodes and leads or microwave filters designed specifically for microwave exposure studies [Guy, et al., 1972; Lin, 1978].

6.6.2   Primary Auditory Cortex

In a series of studies of auditory responses in the primary auditory cortex of laboratory animals irradiated by pulsed microwaves, 2.0–3.4 kg cats were anesthetized with sodium pentobarbital (50 mg/kg) following premedication with Acepromazine and were administered atropine sulfate (0.2 mg) after induction of anesthesia [Guy et al., 1973, Taylor and Ashleman, 1974]. The cats were supported on a heating pad controlled by a rectal temperature monitor and were placed in a head holder constructed of low-loss dielectric slabs (Fig. 6.10). Each cat was fitted with a piezoelectric crystal transducer for the presentation of acoustic stimuli via bone conduction. A plastic ring (18 mm in diameter and 2 mm thick) was fitted to the dorsal surface of the frontal bone just anterior to the coronal suture and was held rigidly in place by nylon screws and dental acrylic cement. The interior of the ring was threaded to facilitate installation and to allow easy removal of the crystal transducer during microwave exposure. This prevented possible artifacts from excitation of the transducer by the microwave field or from energy concentration at the point of contact [Guy et al., 1975a,b].

Skin and soft tissue were excised to expose the temporal bone and lateral portion of the parietal bone. Portions of these bony elements were removed to expose the ectosylvian gyrus. A microwave-transparent carbon electrode was then placed,

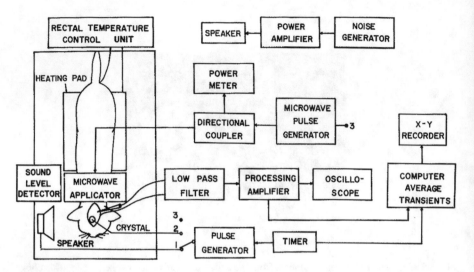

Fig. 6.10 Block diagram of laboratory system for testing the microwave-induced auditory effect in cats

under direct visualization, on the surface of the anterior ectosylvian gyrus. The evoked responses were led from the active electrodes through high resistance carbon leads to a microwave filter and then to an amplifier and an oscilloscope. Some of the signals were further processed with a signal averaging computer before printout.

Following surgical observation of the auditory cortex, the animal was allowed to stabilize until there was a consistent response waveform and latency as evoked by a piezoelectric transducer driven with 10μs wide square pulses at a rate of 1 Hz. The transducer was then removed from the mounting ring and the microwave stimuli applied at the same rate but at an increased pulse width of 32μs. The microwave stimuli consisted of rectangular pulses of 2450 MHz energy produced by a signal generator (Applied Microwave Laboratory model PH40K) and was fed through a coaxial cable to a directional coupler and a vertically polarized horn antenna. The antenna was positioned posterolaterally to the cat's head at a distance of 10 cm and an angle of 30° from the sagittal plane. The incident power levels were measured by a thermistor mount and power meter combination. Figure 6.11 shows typical evoked response signals recorded from the auditory cortex following pulsed microwave and conventional acoustic stimulation. Note the remarkable similarity between these responses.

Also, during the surgical procedures, most of the lateral and ventral surface of the bulla was revealed by reflection and removal of the overlying soft tissue. The surgical maneuver provided a convenient way to explore the role of the cochlea in the auditory response to pulsed microwaves. For this purpose, the lateral wall of the bulla was perforated with a drill, and the hole was then enlarged with a small rongeur until both round windows could be clearly visualized. When clear-cut responses were established, the cochlea was disabled by careful perforation of the round window with a microdissecting knife and aspiration of perilymph. Aspiration of the

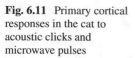
Fig. 6.11 Primary cortical
responses in the cat to
acoustic clicks and
microwave pulses

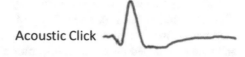

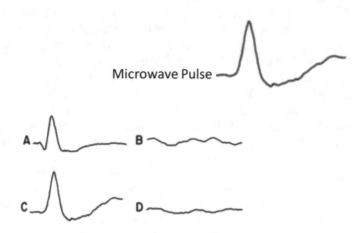

Fig. 6.12 Disablement of both cochlea in the animal resulted in total loss of the cortical responses
in the cat to acoustic click (**b**) and pulsed microwave (**d**) stimulation. (**a** and **c**) are the corresponding response signals before bilateral destruction of the cochlea.

contralateral cochlea led to marked reduction of the amplitude of the evoked potentials. Disablement of the remaining cochlea in these animals resulted in total loss of
the response signal, as shown in Fig. 6.12. These results show that the same pathway through the central nervous system is activated by both microwave and acoustic
pulses. Furthermore, the cochlea has an essential role in the auditory perception of
pulsed microwaves.

6.6.3   Central Auditory Nuclei

Direct recordings of pulse microwave-evoked brainstem nuclei potentials have been
reported for cats using microelectrodes. For example, electrophysiological data
have been recorded by using a glass microelectrode filled with Ringer's solution and
with a tip diameter of 80–100μm [Guy et al., 1972, 1973, 1975a,b, Taylor and
Ashleman, 1974]. Compound action potentials have been recorded from the medial
geniculate body of cats exposed to 918 and 2450 MHz microwave pulses. Because
the dielectric properties of Ringer's solution and brain tissues are similar, the glass
pipettes filled with Ringer's solution were essentially transparent to microwaves
when used for recording bioelectric signals from the depth of the brain.

Cats were anesthetized intravenously with alpha-chloralose (55 mg/kg) in Ringer's solution (20 ml), and 0.2 mg of atropine sulfate was administered intra-muscularly after induction of anesthesia. Cats were paralyzed with Flaxedil (20 mg) and then maintained on artificial respiration. The body temperature of the cats was held constant at 38 °C by a heating pad connected to a rectal temperature control unit. A pair of wooden ear bars was used to hold the cat in a stereotaxic instrument. To minimize the distortion of the fields around the cat's head, all metal pieces for fixing the inferior orbit and the upper jaw were replaced by wooden pieces.

Following unveiling of the dorsal surface of the skull by conventional methods of skin incision and reflection of the underlying muscle, a burr hole was made in the parietal bone. Before insertion of the electrode, each cat was fitted with a piezoelectric crystal transducer for providing acoustic stimuli by bone conduction (see previous section). The electrode was directed toward the medial geniculate body by the standard stereotaxic method [Snider and Niemer, 1961]. The electrode and associated ground connection were coupled via high resistance 1000 ohm-cm carbon-loaded plastic conductors (which are transparent to microwaves in air), through a low-pass microwave filter to a high input impedance physiological sig-nal processing and recording apparatus (see Fig. 6.10). The responses evoked by acoustic clicks from a loudspeaker were continuously monitored as the electrode was advanced vertically into the cerebrum. Proper placement of the electrode tip was assumed when the evoked responses displayed the expected latency period. The electrode placement was verified in some of the animals by histological exam-ination of the brains.

Acoustic clicks were presented to the animal by activating either the loudspeaker placed 17 cm to the right of the center line of the cat's head for air conduction or the piezoelectric transducer (for bone conduction) with square pulses 1–30µs in duration at 1 Hz from a pulse generator. Microwave pulses at 918 or 2450 MHz of the same pulse characteristics were provided by horn or aperture antennas located 8 cm away from the occipital pole of the cat and driven by an AML PH 40 K microwave signal source. In the absence of the cat, a power monitor was used to measure the average incident power density to the location where the cat's head was placed, and the bidirectional coupler and power meter were used to measure incident power to the antennas.

Figure 6.13 presents typical evoked responses recorded from the medial genicu-late body due to acoustic and 2450 MHz microwave pulse stimulation. The similari-ties between the evoked responses are evident. The recordings were based on 40 averages taken with a signal averaging computer. Other investigators also reported similar activities from the medial geniculate nucleus [Chou et al., 1976; Lin et al., 1978].

Note that damage to the cochlea led to total loss of the medial geniculate body's responses to both acoustic and microwave stimuli (Fig. 6.14). The procedure was the same as described in Sect. (6.6.2); when clear responses were established, the cochlea was disabled by perforation of the round windows.

The late slow wave in the general somatosensory thalamic region (VPL) was the same for both conventional acoustic click and pulsed microwave stimulation (Fig. 6.15). That such pulses were eliciting similar responses in regions of the brain other than

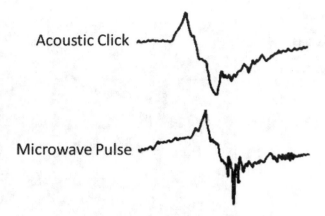

Fig. 6.13 Acoustic click and microwave pulse-evoked potentials recorded from the medial geniculate body of cat

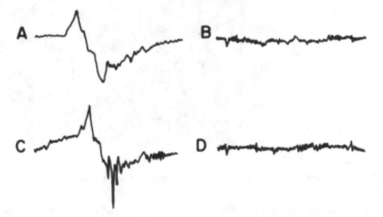

Fig. 6.14 Total loss of the medial geniculate responses in the cat to acoustic click (**c**) and pulsed microwave (**d**) stimulation following perforation of both cochlea in the animal. (**a** and **c**) are the corresponding response signals before bilateral destruction of the cochlea.

auditory areas indicated that the microwave-evoked response was not merely an artifact generated in either the animal preparation or the recording equipment.

Furthermore, using these same procedures, evoked responses from the medial geniculate body of the cat were reported for pulses at microwave frequencies between 8670 MHz and 9160 MHz [Guy et al., 1973, 1975a,b]. The required microwave energy to elicit the responses was significantly higher than those required for the lower frequencies. In this case, the X-band horn had to be placed within a few cm from the exposed brain surface of the animal (through the 1.0 cm diameter electrode access hole in the skull). No response could be elicited for an animal in which the electrode access port through the skull was limited to a diameter slightly larger than the electrode. However, when the hole in the skull was enlarged and baring the soft brain tissues, a response was observed. Apparently, deposition of microwave

Fig. 6.15 Evoked cross-modal brain responses recorded in the general somatosensory thalamic region (medial geniculate and VPL) for both acoustic click and pulsed microwave stimulation

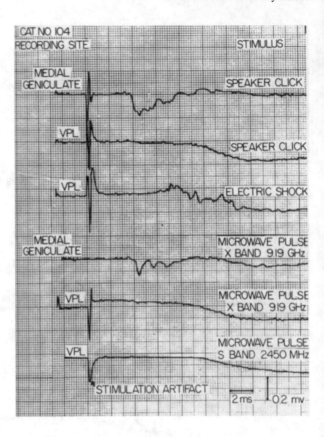

power in soft tissues in the head was necessary for microwave-induced auditory sensation.

In addition to the medial geniculate body, the experimental procedures like those just described including glass microelectrodes filled with Ringer's solution were used in a study to investigate microwave pulse-induced auditory response from the inferior colliculus of cats [Rissmann and Cain, 1975, Cain and Rissmann,1978]. Recordings of evoked electrical activities were obtained from the inferior colliculus of cats exposed to 10μs wide 3000 MHz microwave pulses. It was found also that the evoked potentials in response to acoustic and pulsed microwave stimuli disappeared in these animals following replacement of the antenna with a dummy load and following death.

Furthermore, using a small direct-contact microwave antenna, microwave-evoked auditory responses have been recorded from the medial geniculate, inferior colliculus, lateral lemniscus, superior olivary nucleus, and the vertex of cats [Lin et al., 1978, 1979, 1982]. Figure 6.16 gives a set of recorded responses both for acoustic click and 2450 MHz microwave pulse. Essentially identical tracings are obtained for both stimuli at each of the electrode sites. Note that microwave-evoked responses are always seen immediately following delivery of stimulus, without the familiar transmission delay associated conventional acoustic waves that must travel

Fig. 6.16 Microelectrode recordings of acoustic and pulse microwave-evoked brainstem nuclei potentials in cats. Medial geniculate (MG); inferior colliculus (IC); lateral lemniscus (LL); and superior olive (SO)

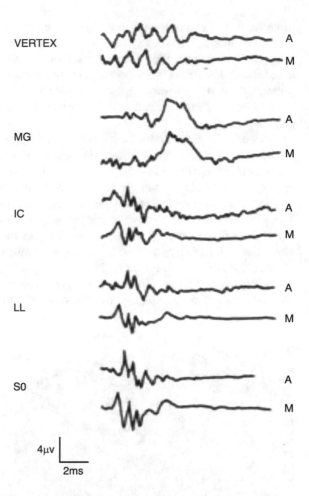

at a much slower speed in air. These recordings indicated that brainstem potentials can be evoked by microwave pulses and that they have characteristics closely resemble those evoked by conventional acoustic stimuli. Experimental details are given in Sect. 6.6.7.

An earlier study implanted coaxial metal electrodes in the brainstem of cats with the help of a stereotaxic instrument [Frey, 1967]. The electrode was affixed to the skull by nylon screws and dental acrylic plastic while the cat was under Fluothane anesthesia. After a recovery period, the cat was placed in a polystyrene head holder which was located inside a microwave absorber lined wooden exposure chamber. The electrode previously implanted was connected via coaxial cable to a preamplifier, oscilloscope, transient signal averager, and a recorder. Pulses of 10 ms wide acoustic and 1200–1535 MHz microwaves were applied at 5 min intervals. Because of the similarity of the acoustic and microwave-evoked activities, and because the responses were seen immediately before but not immediately after death, it was concluded that the signals were neural rather than an artifact of the experimental protocol. It was also suggested that the effect might be the result of direct stimulation of the auditory nervous system at a site central to conventional sound perception, reinforcing again Frey's earlier assertions. Nevertheless, the suggestion was based also on failure to observe any apparent cochlear microphonics associated with pulse-modulated microwaves in cats and guinea pigs [Frey, 1967, 1971] even with incident power densities far above that needed to induce the microwave auditory effect in cats.

Considerable caution must be taken in accepting these results, however, because of the exposure protocol and recording electrode used. In this experiment, the evoked potentials were sensed by a metal coaxial electrode. Despite the coaxial electrode was developed with the intent to avoid microwave energy induced on the electrode, and despite the reports that during extensive testing the electrode had shown no indication of microwave current pickup [Frey et al., 1968], coaxial electrode of similar construction has been shown to increase the peak microwave absorption in the brain tissue surrounding it by as much as two orders of magnitude [Guy et al., 1972] (see Fig. 6.17). Therefore, by using this type of electrode, the possibility of brain tissue stimulation by microwave current directly induced on the electrode cannot be completely ruled out. Much stronger evidence for pulsed

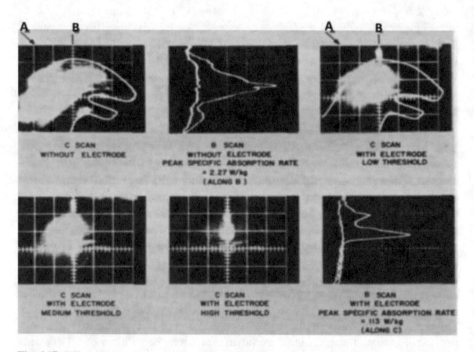

Fig. 6.17 Effect of a metal coaxial electrode on microwave absorption pattern in the brain of a cat exposed to near-zone CW 918 MHz radiation. The bright spots on the C-scan thermograms show patterns of microwave energy absorption. The high-threshold C-scans clearly show the increased microwave absorption in the region where the tip of the metal electrode is located. The incident power density is 25 W/m² and is directed along A with the electric field oriented in the plane of the paper

microwave-evoked electrophysiological activities in the brainstem auditory nuclei using high-resistance, carbon-loaded plastic conductors and/or glass microelectrode filled with Ringer's solution have been reported since then, as described above.

6.6.4   The Eighth Cranial Auditory Nerve

To further explore the microwave auditory response in animals, adult cats (2.0–3.4 kg) were anesthetized with sodium pentobarbital (50 mg/kg) following premedication with Acepromazine [Guy, et al., 1972, 1973, 1975a,b; Taylor and Ashleman, 1974]. The cats were placed in the head holder described previously. After reflection of the auricle and removal of the underlying muscles to expose the temporal bone, a hole was drilled to remove most of the squamous and a portion of the parietal bone. Through this opening, brain tissue was removed to expose the tentorium cerebelli. Using a drill and rongeur, an opening approximately 1.5 cm in diameter was made in the tentorium. The dissection was then continued with the aid of a dissecting microscope. Cerebellar tissue was removed to reveal the eighth cranial (auditory) nerve as it emerged from the internal auditory meatus. A dissecting microscope and a micromanipulator were used to insert a Ringer's solution filled 100 μm diameter tip glass microelectrode within the nerve. The microwave exposure apparatus and recording instrumentation were like that shown in Fig. 6.10. During recording, the auditory nerve and surrounding tissue were covered with warm mineral oil. Acoustic click- and microwave pulse-evoked signals in the eighth cranial auditory nerve are shown in Fig. 6.18. The response to loudspeaker shows a classic propagation delay between stimulus and response. Clearly, microwave-induced activity is like that evoked by a conventional acoustic click from a piezo-electric transducer that launched the acoustic sound signal via bone conduction to the cochlea. Unilateral ablation of the cranial nerve led to total loss of these evoked potentials to both acoustic and microwave stimuli, as expected.

Note that electrophysiological recordings from single auditory neurons in the cat also demonstrated a response to microwave pulses that was like the response to acoustic stimuli. Neuronal responses to 915 MHz microwave pulses have been studied in cats by recording extracellular action potentials of individual neurons, or "units," in the eighth cranial nerve and in the cochlear nucleus with glass microelectrodes. Post-stimulus time histograms of the auditory eighth nerve fibers and cochlear nucleus units showed time-locked microwave responses depend monotonically on pulse amplitude but nonmonotonically on pulse width [Lebovitz and Seaman, 1977; Seaman and Lebovitz, 1989]. The results of singe-unit studies not only support a microwave interaction site at or peripheral to the cochlea but are also consistent with the thermoelastic expansion theory, as shown later in this book. This study also showed auditory units with lower characteristic frequencies (a few kHz) appear to be more responsive to microwave pulses than were units with higher characteristic frequencies, a point to be addressed later.

Fig. 6.18 Auditory (eighth cranial) nerve responses of cat irradiated with acoustic click and microwave pulses

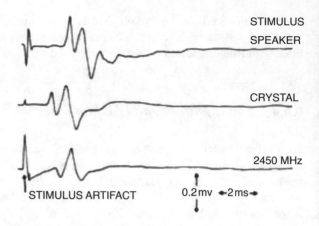

6.6.5   Cochlear Round Window

In a series of experiments, a high resistance carbon electrode like that employed in the cortical recordings was applied to the round window of the cochlea to record activity evoked by acoustic clicks and microwave pulses [Guy et al., 1972, 1975a, b]. In this experiment, before the cats were placed in the stereotaxic instrument, the lateral and ventral surface of the auditory bulla was revealed by reflection and removal of the overlying soft tissue. The lateral wall of the bulla was perforated with a drill and was enlarged with a small rongeur until the round window of the cochlea could be clearly visualized. A carbon electrode was cemented to the round window and connected to a low-pass microwave filter for further signal processing (Fig. 6.10). The remaining surgical procedure was like that performed for the medial geniculate experiments, including the attachment of the piezoelectric transducer.

It can be seen from Fig. 6.19 that both acoustic clicks and microwave pulses elicited activity at the round window. The first trace shows the composite cochlear microphonic and N_1 and N_2 auditory nerve responses from an animal elicited by a loudspeaker pulse. The cochlear microphonic was quite strong in amplitude. When the auditory system of the same animal was stimulated by microwave pulses, a microwave artifact pulse and clear N_1 and N_2 auditory nerve responses were elicited, but there was no evidence of a cochlear microphonic as seen from the second trace in Fig. 6.19. The cochlear microphonic in this case is either extremely brief and lost in the microwave artifact, greatly attenuated, or absent completely.

The role of the cochlea in microwave-induced auditory phenomena has been discounted in earlier reports, partly based on not observing a microphonic in either cats or guinea pigs [Frey, 1967, 1971] (see Sect. 6.6.3). However, the later studies have found in some animals the cochlear microphonic is considerably reduced (see third trace in Fig. 6.19) or not present at all (see fourth trace in Fig. 6.19) when the auditory system of the animal is stimulated by an acoustic click [Guy et al., 1972,

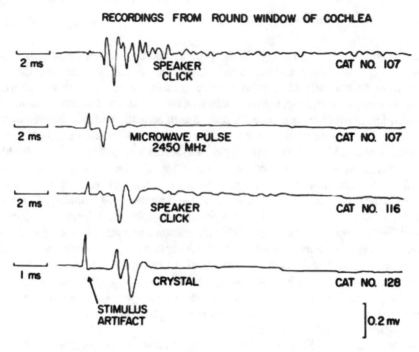

Fig. 6.19 Carbon electrode recordings from the round window of the cat cochlea elicited by acoustic click and microwave pulse stimuli

1975a,b]. Several factors could prevent the observance of a cochlear microphonic, especially at low stimulus intensity [Wever, 1966]. Some reports referred to studies in which the auditory thresholds of cats, as determined by behavioral tasks, were established as being 40 dB below the first stimulus level effective in eliciting cochlear microphonics of enough amplitude to be observed with the conventional oscilloscopes. Thus, considering the fact that the microwave pulse generator used was capable of only providing 10–17 dB gain in peak power over that corresponding to the threshold of evoked auditory responses, the absence of a microwave-evoked cochlear microphonic does not necessarily rule out the hypothesis that microwave-induced auditory sensation is mediated at the periphery as are conventional acoustic stimuli.

A peripheral site of interaction should involve displacement of the tissues in the head with resultant dynamic consequence in the cochlear fluids and nervous system correlates that have been well described for the acoustic case. However, cochlear microphonics, the signature of mechanical modification of cochlear hair cells, eluded researchers for some time. This had led to the persistent speculation that microwave pulses, in contrast to conventional acoustic pulses, might not act on any tissue structure before acting directly on the inner ear sensory apparatus or the more centrally located auditory nervous nuclei.

6.6.6   Cochlear Microphonics

The electrophysiological findings in the primary auditory cortex, brainstem, auditory eighth cranial nerve, and the round window described above indicate that the
microwave auditory effect is exerted on the animal in a manner like that of conventional acoustic stimuli. Also, the elimination of the first stage of sound transduction
affected the central nervous system's response to acoustic and microwaves in the
same way, i.e., the evoked electrical activities of all three sites (eighth cranial nerve,
the brain stem, and the primary auditory cortex) were abolished by cochlear disablement, suggesting that the locus of initial interaction of pulse-modulated microwave
energy with the auditory system resides at the cochlea or peripherally with respect
to the cochlea. This interpretation was augmented by the observations made in systematic studies of loci involved, through production of lesions in ipsilateral auditory
nuclei and bilateral ablation of the cochlea, the known first stage of transduction for
acoustic energy into nerve impulses. On the other hand, cochlear microphonic, the
signature of mechanical disturbance of cochlear hair-cell, needs to be observed
under experimental situations involving microwave pulses. The absence of any
recording of cochlear microphonics could lead to a proposition that pulsed microwaves, in contrast to conventional acoustic stimuli, might not act on any sensor prior
to acting directly on the inner ear apparatus.

As mentioned in the previous section, failure to observe any microwave-induced
cochlear microphonic in experimental animals may have been due to limitations of
the output of the microwave pulse generator or a large microwave pulse artifact
which concealed the cochlear microphonic. In fact, subsequent experiments [Chou
et al., 1975, 1976, 1977, 1982] have successfully demonstrated the existence of
microwave-induced cochlear microphonics in laboratory animals with clearly visible acoustically evoked cochlear microphonics by alleviating the difficulties just
mentioned.

In these experiments, guinea pigs (400–600 g) were anesthetized with sodium
pentobarbital (40 mg/kg) and allowed to breathe normally through a tracheal can-
nula. After clearing either the left or right auditory bulla, a fine carbon electrode was
inserted against the round window and cemented onto the bulla. The animals were
then screened based on whether the amplitude of the cochlear microphonic evoked
by an acoustic click exceeded 0.5 mV. If the answer was positive, the guinea pig was
chosen for the experiment. In this case, its head was then placed in the cylindrical
cavity through an opening on the side of the waveguide (Fig. 6.20). The head was
supported by a microwave-transparent polystyrene foam block inside the cavity.
With the animal's head inside, the cavity was tuned for maximum power to the head
by adjusting the position of a sliding short located on one end and the depth of penetration of the animal's head. Since only 0.1% of the input power was detected to be
leaking around the neck of the guinea pig, the available power was assumed to be
completely absorbed by the subject. The microwave pulse artifacts were greatly
reduced by locating the microwave source (AML model PH 40 K), the cavity, and
the animal in a shielded room and recording the cochlear potentials via coaxial
cables connected to differential amplifiers outside the shielded room.

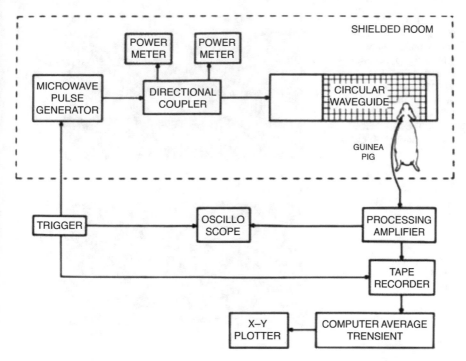

Fig. 6.20 Schematic of experimental apparatus for recording microwaveinduced cochlear micro-phonics in guinea pigs

The animals were intermittently exposed to 918 MHz microwave pulses, 1–10µs in duration, for 90 s intervals at a pulse repetition frequency of 100 Hz and at peak powers up to 10 kW. The evoked electrical activities were stored on a magnetic tape system having a frequency response to 80 kHz. The responses were then processed either online or offline using a signal averaging computer. Figure 6.21 illustrates the evoked potentials recorded from the round window of a guinea pig. The responses due to single acoustic clicks derived from a loudspeaker driven at 10 kHz consisted of a cochlear microphonic which preceded the N_1 and N_2 auditory nerve responses. The polarity of the cochlear microphonic changed with a change in the polarity of the electrical energy driving the loudspeaker, confirming the authenticity of the cochlear microphonics observed. When the same guinea pig was exposed to pulsed microwave, in addition to the well-defined N_1 and N_2 nerve responses, a high-frequency (50 kHz) oscillation was seen preceding and immediately following the microwave stimulus artifact. Clearly, cochlear microphonic responses like that evoked by conventional acoustic stimuli can be induced by pulse-modulated microwave energy.

Figure 6.22 compares the cochlear microphonic induced by microwave pulses of 1µs, 5µs, and 10µs at the same peak power (10 kW). Each trace is the average of 400 responses played back from tape. While the frequency of the cochlear microphonic remained constant, its amplitude increased as pulse width increased, and the energy

Fig. 6.21 Cochlear microphonics recorded from round window of the guinea pig. (**a**) Acoustic click stimulus. (**b**) Single 918 MHz microwave pulse 10μs wide including in (**b**) time expansion of initial 200μs. The absorbed energy density is 1.33 J/ kg

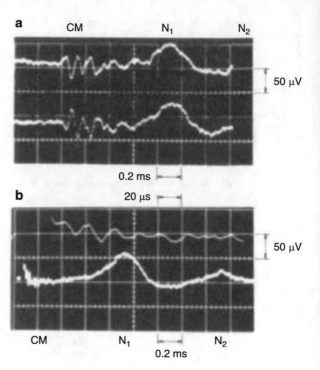

Fig. 6.22 Cochlear microphonics evoked by 918 MHz microwave pulses at a peak power of 10 kW, but variable pulse width

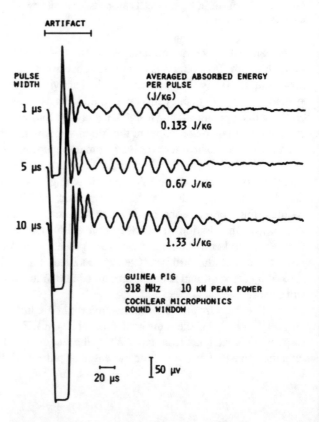

absorption correspondingly increased. Furthermore, latency of cochlear microphonic occurrence was nearly the same for all three cases. Following the death of the animal, whether by anoxia or by drug overdose, microwave-evoked nerve responses disappeared before the cochlea microphonic. Similar disappearances occurred during acoustic stimulation of the dead animal. After many minutes, the cochlear microphonics also disappeared, but the artifact persisted, indicating that the 50 kHz oscillatory signal is a genuine physiological response. In some experiments, 38 kHz cochlear microphonics were recorded from the round window of cats.

6.6.7   Brainstem Nuclei Ablations on Auditory Evoked Responses

The electrophysiological evidence indicates that an auditory sensation can be induced in laboratory animals by pulse-modulated microwave energy. Those studies suggest that microwave-induced auditory sensation is detected by the animal in a manner very similar to conventional sound detection and that the site of conversion from microwave to sonic energy resides somewhere peripheral to the cochlea. This interpretation finds support in systematic studies of responses from brainstem nuclei after successive production of coagulative ablations in the central auditory loci [Lin et al., 1978, 1979]. Also, as already mentioned in Sect. 6.6.3, disablement of the cochlea led to total loss of the medial geniculate body's brainstem-evoked responses to both acoustic and microwave stimuli (Fig. 6.14).

The effects of brainstem lesions on microwave pulse-evoked responses at fixed peak power and pulse-repetition frequency are shown in Figs. 6.23, 6.24, and 6.25. Figure 6.23 shows decreased amplitude of all responses recorded from each of the electrode sites with successive lesion production in the inferior colliculus, lateral lemniscus, and superior olive nuclei is readily apparent. The reduction in amplitude is most pronounced for the inferior colliculus following lesion production in

Fig. 6.23 Effect of brainstem ablative lesions on electrical potentials from the inferior colliculus nucleus (control) evoked by 2450 MHz microwave pulses in the cat. 25μs pulse; 100–200 averages. Inferior colliculus (IC), lateral lemniscus (LL), superior olive (SO) nuclei

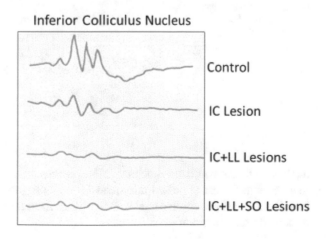

Inferior Colliculus Nucleus

Control

IC Lesion

IC+LL Lesions

IC+LL+SO Lesions

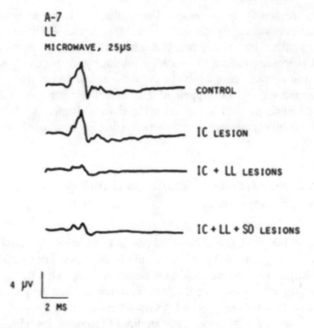

Fig. 6.24 Responses in brainstem potentials recorded in lateral lemniscus nucleus (control) following ablative lesion production. 2450 MHz microwave pulses in the cat. 25μs pulse; 100–200 averages. Inferior colliculus (IC), lateral lemniscus (LL), superior olive (SO) nuclei

Fig. 6.25 Effect of brainstem ablative lesions on electrical potentials from the superior olive nucleus (control) evoked by 2450 MHz microwave pulses in the cat. 25μs pulse; 100–200 averages. Inferior colliculus (IC), lateral lemniscus (LL), superior olive (SO) nuclei

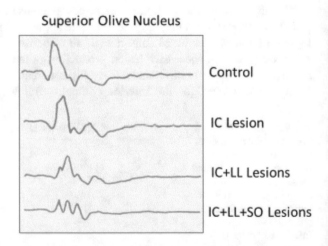

itself – inferior colliculus nucleus. Note severe loss of the inferior colliculus responses in the cat to pulsed microwave stimulation following ablation in lateral lemniscus and superior olive nuclei. Recordings from the vertex and all other nuclei remained practically unaltered.

The effect of brainstem ablative lesions on electrical potentials recorded from the lateral lemniscus (LL) nucleus in response to microwave pulse stimulation is shown in Fig. 6.24. Lesions in proximal nucleus (inferior colliculus – IC) had negligible influence on the response recorded from lateral lemniscus nucleus. However, the response disappeared after ablative lesion was made in the distal nucleus (superior olivary – SO) thus confirming the peripheral nature of the primary site of transduction.

A further example is given in Fig. 6.25, where the effect of brainstem ablative lesions on electrophysiological signals recorded from the more distal SO nucleus in response to microwave pulse stimulation is shown. Lesions in the proximal nuclei, IC and LL, had insignificant impact on the response recorded from SO nucleus. However, the response disappeared after a coagulative lesion of the SO nucleus, thus confirming again the peripheral nature of the primary site of transduction.

In these studies, cats (2.0–3.9 kg) were anesthetized intraperitoneally by sodium pentobarbital (30 mg/kg) and kept under a surgical level of anesthesia by intravenous supplementation. The animals were placed in a stereotaxic apparatus with hollow ear bars. The animal's body temperature was monitored rectally and was maintained between 36 and 39 °C by a heating pad.

The dorsal aspect of the skull was revealed through a skin incision and by partially separating the nuchal muscle. Insulated stainless-steel electrodes (100µm in diameter) were inserted and advanced stereotaxically to various brainstem nuclei. When the optimal electrode position was reached (as determined by advancing the electrode until some component of the evoked potential was maximal), the electrode was fixed to the skull with dental acrylic. When the desired number of electrodes has been fixed to the skull, the entire occipital area was reinforced with dental acrylic. An indentation was then made at the vertex, and a stainless-steel screw electrode was fastened to the skull. A reference gold pin electrode was always located near the lowest part of the pinna.

Acoustic pulses were generated by feeding 0.1 ms of electric current from a Grass model S-4 stimulator into a pair of commercial earphones. The resulting pulses were presented binaurally at the rate of 10–100 Hz and were approximately 70 dB above threshold sound level.

Short rectangular microwave pulses were produced by an Applied Microwave Laboratory pulsed signal source at 2450 MHz and were applied to the dorsal or frontal surface of the head via a small (15 mm dia) direct-contact applicator, which restricted microwave exposure to a small near-zone region. The pulses were 0.5 to 25 µs wide and were delivered at the rate of 10–100 Hz. The peak power of the incident microwave was 10 kW. Incident and reflected power were monitored through a bidirectional coupler by a power meter.

Bioelectric activity recorded at the vertex and from each depth electrode was amplified with a bandpass filter of 80 Hz to 10 kHz and summed online by a signal averaging computer. The first 10 ms of the averaged response was displayed on a video display and was then photographed.

Brainstem lesions were produced at the tips of the depth electrodes by a Grass LM-3 RF lesion maker; lesions were successively made in the IC nucleus, LL nucleus, and SO nucleus. After each lesion, a new series of averaged evoked responses was recorded from the vertex and from each of the depth electrodes; these responses were compared to those made prior to production of lesions. As mentioned earlier, decreased amplitude of all responses recorded from each of the electrode sites with successive lesion production in the inferior colliculus, lateral lemniscus, and superior olive nuclei was readily observed. Moreover, Fig. 6.8 shows that the amplitude of responses recorded at the vertex is clearly a function of the integrity of brainstem nuclei along the central auditory pathway.

Note that at the end of each of the experiments, the animals were euthanized by an overdose of sodium pentobarbital. The brain was fixed by intracarotid injection of formalin. The fixed brainstem was removed, imbedded in paraffin, and the serial sections were made to ascertain the exact locations of lesion and electrode track.

6.6.8 Manipulation of Middle and Outer Ears on Brainstem Evoked Responses

The surgical maneuver described in Sects. 6.6.1 and 6.6.7 offered a useful way to explore the role of the cochlea in the auditory response to pulsed microwaves. Ablation of central auditory nuclei and damage to the cochlea in the cat resulted in total loss of the response signal. Those results show that the same pathway through the central nervous system is activated by both microwave and acoustic pulses. Furthermore, the cochlea has an essential role in the auditory perception of pulsed microwaves.

As shown in Fig. 6.26, when the external auditory meatus of guinea pigs was blocked by cotton balls socked in mineral oil, the amplitude of microwave-evoked

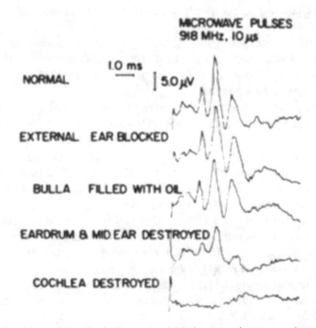

Fig. 6.26 Effect of external and middle ear manipulations on microwave pulse-evoked auditory brainstem potential.

auditory brainstem potentials, recorded from an electrode placed on the scalp at the vertex of the head, remained unaltered. Similarly, filling the middle ear cavity with mineral oil, which impeded ossicular movement, had no effect on microwave-evoked response. In addition, disablement of both tympanic membrane and middle ear ossicles led only to a reduced brainstem potential [Taylor and Ashleman, 1974, Chou and Galambos, 1979].

These findings indicate that the external and middle ears are not the route used by microwave-induced sound and that the decrease in brainstem potential probably arose from a small relative motion between the stapes and the cochlear round window. Furthermore, they suggest that the interaction of pulse-modulated microwave energy with the auditory system is not mediated by any airborne sound reception mechanism. Furthermore, the site of initial sound detection is inside the cochlea.

6.6.9   Brain Tissue as Site of Interaction

Briefly, these are the succinct facts. The microwave pulse-induced auditory nerve responses disappeared after ablative lesions made in the central auditory nuclei, pointing to the peripheral nature of the primary site of transduction. Cochlear microphonics, the signature of mechanical disturbance inside the cochlea of guinea pigs, have been observed in response to microwave pulse stimulation. Destruction of the cochlea by perforating the round window leads to complete abolishment of any microwave-induced brainstem evoked response. These results clearly implicate the mode of interaction for pulse-induced microwave auditory effect as mechanical in nature, and it does not involve the middle ear apparatus. Thus, the site of initial interaction is peripheral to the cochlea and in brain matters or soft tissues of the head.

The involvement of brain matters has been demonstrated in laboratory animal studies through visualization of functional activity in the brain and its associated glucose utilization [Wilson et al., 1980]. An autoradiographic technique in which (^{14}C)2-deoxy-D glucose is used to map in vivo metabolic auditory activity in the brain of rats exposed to acoustic clicks and microwave pulses. Prior to exposure, one middle ear was destroyed to block transmission of sound in one side of the head. As expected, asymmetry of radioactive tracer uptake was observed in the central auditory nuclei of rats exposed to acoustic clicks. In contrast, a symmetrical uptake of tracer was found in the auditory nuclei of animal brains exposed to microwave pulses. Aside from confirming that the microwave auditory effect does not require an intact middle ear, the autoradiographic results authenticate the involvement of brain tissues in the microwave pulse-induced auditory response.

6.7   Animal Thresholds, Noise, and BER Potentials

6.7.1   Electrophysiologic Threshold Determination in Animals

In Sect. 6.3.2, an account of psychophysical efforts to establish the threshold of microwave-induced auditory sensation in humans was given. Several investigators attempted to ascertain the minimally effective magnitudes of pulsed microwave energy for evoking auditory system responses in laboratory animals. These "threshold" determinations, however, must be considered incomplete because measurements were usually attempted with too few subjects and at only a single frequency in most cases.

Potentials recorded from the medial geniculate body of the cat were used to estimate the threshold of pulse-modulate, microwave-evoked auditory response in animals [Guy et al., 1973, 1975a, b]. The experimental protocols were analogous to those described in Sect. (6.6.3). Tables 6.6 and 6.7 present the threshold of 918 MHz and 2450 MHz microwave pulse-evoked thalamic responses. The peak absorbed

Table 6.6 Threshold of evoked auditory responses in cat exposed to 918 MHz microwave pulses at 1 Hz. Background noise 64 dB

Pulse width	Peak power density	Average power density	Peak energy per pulse	Peak SAR
μs	W/m²	W/m²	J/m²	W/kg
3	5.80	17.4	17.4	4.1
5	3.88	19.4	19.4	2.76
10	2.26	22.6	22.6	1.6
15	1.37	20.6	20.6	0.97
20	1.17	20.6	20.6	0.83
25	0.97	24.3	24.3	0.69
32	0.80	28.3	28.3	0.63

Table 6.7 Threshold evoked auditory responses in cat exposed to 2450 MHz microwave pulses at 1 Hz. Background noise 64 dB

Pulse width	Peak power density	Average power density	Peak energy per pulse	Peak SAR
μs	W/m²	W/m²	J/m²	W/kg
0.5	35.6	17.8	17.8	20.2
1	7.8	1.8	17.8	10.1
2	10.0	20.3	20.3	5.3
4	55.0	20.3	20.3	2.4
5	4.0	20.3	20.3	2.32
10	2.2	21.6	21.6	1.23
15	1.9	28.0	28.0	1.06
20	1.7	33.0	33.0	0.94
25	0.6	15.2	15.2	0.35
32	1.5	47.0	47.0	0.83

energy density per pulse in these tables was measured with a thermographic method [Guy, 1971] and the results compared favorably with that calculated using a spherical model of the head [Johnson and Guy, 1972].

The data in these tables show that as the pulse width was increased, the peak incident power density required to elicit an auditory response in the cat decreased proportionately, except at the longer pulse width of 32μs for the 2450 MHz case. Although the incident energy density per pulse also increased with pulse width, the increases were not as clear cut. This observation has led to the conclusion that the threshold for the pulsed microwave-evoked auditory response was related to the incident energy density per pulse, at least for pulse duration shorter than 10μs [Guy et al., 1973, 1975a, b]. The incident energy density per pulse appeared to be at a level about one-half of that which produced audible sensations in humans (Sect. 6.3.2). On the other hand, one cannot easily rule out the possible linking between the pulsed microwave-evoked auditory responses and the peak incident power density and the pulse width of the incident microwave pulses. Note that threshold of sensation is inversely proportional to perceived loudness; the lower the threshold, the higher the perceived loudness. In fact, the data listed in Tables 6.6 and 6.7 are consistent with the thermoelastic expansion theory, as shown later in this book.

Exposure of guinea pigs to 2450 MHz microwave pulses in a circular waveguide cavity showed that the threshold peak absorbed power density for producing an identifiable cochlear microphonic response was nearly 2000 W/kg for a 10μs wide square pulse [Chou et al., 1975]. The peak absorbed power density was determined by measuring the induced temperature in the guinea pig's head using a thermographic procedure. One would expect the threshold value to be higher than those determined using evoked responses from the thalamus. It is known, at least in cats, that the auditory threshold determined by behavioral tasks is 40 dB below the sound levels first effective in producing cochlear microphonic potentials of sufficient amplitude to be identified with conventional oscilloscopes [Wever, 1966].

Microwave-induced auditory thresholds in several different laboratory animals, specifically cat, chinchilla, and dog [Rissmann and Cain, 1975; Cain and Rissmann, 1978]. The experimental protocol was similar to that employed by others [Guy et al., 1973, 1975a, b], with the exception that the recording electrode implanted in the inferior colliculus and with scalp electrodes on the and side of the head of the animals. The threshold peak incident power densities were determined as a function of the pulse width of the impinging 3000 MHz microwaves for two cats, two chinchillas and one dog. The results of this study are presented in Table 6.8. The peak incident power density required to elicit an auditory response decreased as the pulse width increased in all cases, although not proportionately. The threshold energy density per pulse seemed to stay relatively constant for cats and chinchillas in agreement with the results reported by others [Guy et al., 1973, 1975a, b] for the pulse widths used. There was, however, no apparent relationship between audible threshold and energy density per pulse for the dog, although in this case an increase in pulse width was accompanied by a decrease in peak power density required.

The threshold parameters required to elicit a response from the medial geniculate body of the cat were assessed in two animals for X-band, 32μs pulses at frequencies

Table 6.8 Threshold of 3000 MHz pulsed microwave-evoked auditory responses in cat, chinchilla, and dog

	Cat		Chinchilla		Dog	
Pulse width	Peak power density	Energy per pulse	Peak power density	Energy per pulse	Peak power density	Energy per pulse
μs	kW/m^2	mJ/m^2	kW/m^2	mJ/m^2	kW/m^2	mJ/m^2
5	25	125	25	125	18	90
10	13	130	15	150	3	33
15	5.8	87	5.4	81	2	30

between 8670 and 9160 MHz [Guy et al., 1973, 1975a, b]. The results showed that the incident power density and energy density per pulse required were approximately 148–388 kW/m^2 and 4720–12,400 mJ/m^2, respectively. These values are much higher than those required for the lower frequencies tested to date.

The results from the above studies suggest a threshold in microwave-induced auditory sensation. Note the nonlinear nature of threshold as a function of pulse width, which implies the existence of a minimum or optimum pulse width for conversion of microwave to acoustic energy in the mammalian cranial structure. This observation is consistent with the thermoelastic expansion theory, as shown later in this book.

6.7.2   Effect of Ambient Noise Level

The effect of ambient noise level or sound masking on the threshold of the microwave-induced auditory effect in animals was quantitatively evaluated using evoked response from the medial geniculate body [Guy, et al., 1975a, b]. Adult cats (2.2–3.3 kg) were prepared under alpha chloralose anesthesia for recording electrical responses from the medial geniculate body evoked by pulsed microwaves and conventional acoustic inputs. The electrode used was glass pipette filled with Ringer's solution with a tip diameter of 80–100μm. The experimental details for procedures were like those described earlier in Sect. 6.6.3. A loudspeaker located 17 cm to the right of the cat's head centerline was used to deliver the air-conducted acoustic clicks. Each cat was fitted with a piezoelectric transducer for providing sound stimuli via the bone conduction route (see Fig. 6.10). Microwave pulses were provided by a 2450 MHz horn antenna placed 8 cm away from the occipital pole of the cat and driven by a pulse power generator (AML Model PH40K). The average incident power density was measured in the eat's absence using a power meter, and the peak power density was calculated from the known duty factor. A noise generator was used in combination with an audio amplifier and a loudspeaker to provide 50 Hz to 15 kHz artificial ambient noise levels up to 90 dB, as measured with a sound level meter.

The averaged thresholds of evoked responses due to the three stimulating modalities, microwave pulse, air-borne sound, and acoustic bone conduction, are presented in Fig. 6.27 as a function of ambient noise level. Each point represents the threshold averaged over 3–5 cats. There was no noticeable increase in the threshold

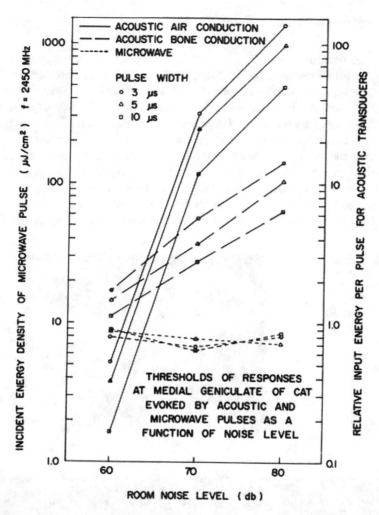

Fig. 6.27 Microwave pulse-evoked medial geniculate response thresholds as a function of background noise in young adult cats

for the microwave stimuli as the ambient noise level was increased. (This observation reinforces the notion that the conversion of microwave energy to acoustic energy takes place in brain tissue.) A moderate rise, however, was seen in the threshold for the piezoelectric transducer attached to the skull. There was also a large increase in the threshold response evoked by the loudspeaker. These results suggest that the acoustic energy produced by pulse-modulated microwaves probably lies predominately in the frequency range above 15 kHz in cats, since the cat's "threshold of audibility" for microwave pulses was not raised by the presence of masking background noise (50 Hz to 15 kHz). This estimate is consistent with the observation that 38 kHz cochlear microphonic oscillations were induced in cats by pulse modulated microwaves. It is also in agreement with the theory of microwave pulse-induced thermoelastic pressure waves discussed later in this book.

6.7.3   Microwave BER Recordings

Section 6.6.1 referred to a series of experiments involving recording microwave-evoked brainstem response (BER) potentials from laboratory cats that employed, for the first time, a small direct-contact microwave antenna applicator (15 mm in diameter, Elmed, Addison, IL) for selective microwave irradiation of an animal's head. At the time, the use of a small direct-contact applicator or antenna was innovative (Fig. 6.28). It has several advantages over conventional microwave exposure techniques, such as open-ended waveguides or horns in an anechoic chamber, in many laboratory situations. It did not require any specialized components or equipment such as carbon-loaded, high-resistance electrodes and leads or microwave filters designed specifically for microwave exposure studies [Guy, et al., 1972; Lin, 1978]. Figure 6.29 shows that the interference due to direct pickup of microwave

Fig. 6.28 Selective microwave irradiation of animal brain using a small contact antenna applicator. The illustration shows the relative size of applicator antenna to a rat immobilized in a stereotaxic device

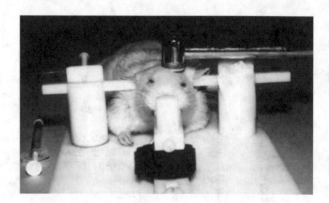

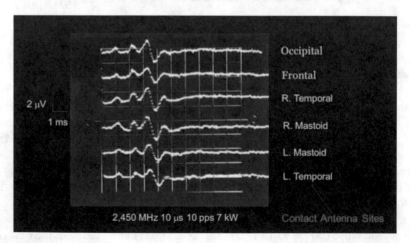

Fig. 6.29 Microwave pulse-evoked brainstem potentials via a small contact antenna sited around a cat's head

artifacts by the recording electrode was substantially reduced (or practically nonexistent) for these experimental protocols (see Sect. 6.6.7). It can be seen that as the small narrow beam, direct-contact microwave antenna was moved around the head of a cat from the occipital, frontal, mastoid, and temporal regions, the recordings from vertex indicated similar evoked response characteristics for each antenna location. The brainstem-evoked responses show remarkable preservation of amplitude and temporal relations from antenna location to location. Furthermore, these results clearly demonstrate that microwave energy does not need to be deposited in the cochlea to initiate the microwave auditory effect.

References

Airborne Instruments Laboratory (1956) An observation on the detection by the ear of microwave signals. Proc IRE 44:10 (Oct), p 2A

Cain CA, Rissmann WJ (1978) Mammalian auditory response to 3.0 GHz microwave pulses. IEEE Trans Biomed Eng 25:288–293

Chou CK, Galambos R (1979) Middle ear structure contribute little to auditory perception of microwaves. J Microwave Power 14:321–326

Chou CK, Guy AW (1979) Microwave-induced auditory responses in guinea pigs: relationship of threshold and microwave-pulse duration. Radio Sci 14:193–197

Chou CK, Guy AW, Galambos R, Lovely R (1975) Cochlear microphonics generated by microwave pulses. J Microwave Power 10:361–367

Chou CK, Guy AW, Galambos R (1976) Microwave induced cochlear microphonics in cats. J Microwave Power 11:171–173

Chou CK, Guy AW, Galambos R (1977) Characteristics of microwave-induced cochlear microphonics. Radio Sci 12:221–227

Chou CK, Guy AW, Galambos R (1982) Auditory perception of radiofrequency electromagnetic fields. J Acoust Soc Am 71:1321–1334

Chou CK, Yee KC, Guy AW (1985) Auditory response in rats exposed to 2,450 MHz electromagnetic fields in a circularly polarized waveguide. Bioelectromagnetics 6:323–326

Constant PC (1967) Hearing EM waves. In: Digest of the 7th international conference on medical and biological engineering, 7th international conference on medical and biological engineering, pp 24–27

Frey AH (1961) Auditory system response to radio frequency energy. Aerospace Med 32:1140–1142

Frey AH (1962) Human auditory system response to modulated electromagnetic energy. J Appl Physiol 17:689–692

Frey AH (1963) Some effects on human subjects of ultra-high-frequency radiation. Am J Med Electron 2:28–31

Frey AH (1965) Behavioral biophysics. Psychol Bull 63:322–337

Frey AH (1966) A restraint device for cats in a UHF electromagnetic energy field. Psychophysiology 2:381–383

Frey AH (1967) Brain stem evoked responses associated with low intensity pulse UHF energy. J Appl Physiol 23:984–988

Frey AH (1971) Biological function as influenced by low-power modulated RF energy. IEEE Trans Microwave Theory Tech 19:153–164

Frey AH, Eichert ES (1985) Psychophysical analysis of microwave sound perception. J Bioelectricity 4:1–14

Frey AH, Feld S (1972) Perception and avoidance of illumination with low-power pulsed UHF electromagnetic energy. In: Proceeding of the International Microwave Power Symposium, Ottawa, May 1972, pp 130–138

Frey AH, Feld SR (1975) Avoidance by rats of illumination with low power nonionizing electromagnetic energy. J Comp Physiol Psychol 89:183–188

Frey AH, Messenger R Jr (1973) Human perception of illumination with pulsed ultra-high-frequency, electromagnetic energy. Science 181:356358

Frey AH, Fraser A, Siefert E, Brish T (1968) A coaxial pathway for recording from the cat brain stem during illumination with UHF energy. Physiol Behav 3:363–365

Frey AH, Feld SR, Frey B (1975) Neural function and behavior: defining the relationship. Ann NY Acad Sci 247:433–439

Guy AW (1971) Analysis of electromagnetic fields induced in biological tissues by thermographic studies on equivalent phantom models. IEEE Trans Microwave Theory Tech, Special issue on biological effects of microwaves 19:205–214

Guy AW, Lin JC, Harris FA (1972) The effect of microwave radiation on evoked tactile and auditory CNS responses in cats. In: Proceeding of the international microwave power symposium, Canada, International Microwave Power Institute, pp 120–129

Guy AW, Taylor EM, Ashleman B, Lin JC (1973, June) Microwave interaction with the auditory systems of humans and cats. In: Proceeding of the IEEE International Microwave Symposium, Boulder, pp 321–323

Guy AW, Lin JC, Chou CK (1975a) Electrophysiological effects of electromagnetic fields on animals. In: Fundamentals and applied aspects of nonionizing radiation. Plenum Press, New York, pp 167–211

Guy AW, Chou CK, Lin JC, Christensen D (1975b) Microwave induced acoustic effects in mammalian auditory systems and physical materials. Ann NY Acad Sci 247:194–218

Ingalls CE (1967) Sensation of hearing in electromagnetic fields. NY State J Med 67:2992–2997

Jewett DL (1970) Volume-conducted potentials in response to auditory stimuli as detected by averaging in the cat. Electroencephalogr Clin Neurophysiol 28(6):609–618

Jewett DL, Williston JS (1971) Auditory-evoked far fields averaged from the scalp of humans. Brain 94(4):681–696

Johnson CC, Guy AW (1972) Nonionizing electromagnetic wave effects in biological materials and systems. Proc IEEE 60:692–718

Johnson RB, Myers D, Guy AW, Lovely RH, Galambos R (1976) Discriminative control of appetitive behavior by pulsed microwave in rats. In: Johnson CC, Shore ML (eds) Biological effects of electromagnetic waves, HEW publication (FDA) 77-8010, pp 238–247

Lebovitz RM, Seaman RL (1977) Single auditory unit responsesto weak, pulsed microwave radiation. Brain Res 126(2):370–375

Lin JC (1977a) On microwave-induced hearing sensation. IEEE Trans Microwave Theory Tech 25(7):605–613

Lin JC (1977b) Further studies on the microwave auditory effect. IEEE Trans Microwave Theory Tech 25(11):938–943

Lin JC (1978) Microwave auditory effects and applications. Springfield, Charles C Thomas

Lin JC (1980) The microwave auditory phenomenon. Proc IEEE 68:67–73

Lin JC (1981) The microwave hearing effect. In: Illinger KH (ed) Biological effects of nonionizing radiation, American Chemical Society, pp 317–330

Lin JC (1990) Chapter 12: Auditory perception of pulsed microwave radiation. In: Gandhi OP (ed) Biological effects and medical applications of electromagnetic fields. Prentice-Hall, New York, pp 277–318

Lin JC (2004) Studies on microwaves in medicine and biology: from snails to humans. Bioelectromagnetics 25:146–159

Lin JC, Wang ZW (2010) Acoustic pressure waves induced in human heads by RF pulses from high-field MRI scanners. Health Phys 98(4):603–613

Lin JC, Wu C (1976) Scattering of microwaves by dielectric materials used in laboratory animal restrainers. IEEE Trans Microwave Theory Tech 24(4):219–223

Lin JC, Guy AW, Caldwell LR (1974) Behavioral changes of rats exposed to microwave radiation. Paper presented at the IEEE international microwave symposium, Atlanta, Georgia, June 1974

Lin JC, Guy AW, Caldwell LR (1977a) Thermographic and behavioral studies of rats in the near field of 918-MHz radiations. IEEE Trans Microwave Theory Tech 25:833–836

Lin JC, Bassen HI, Wu C-L (1977b) Perturbation effect of animal restraining materials on microwave exposure. IEEE Trans Biomedical Eng BME-24(1):80–83

Lin JC, Meltzer RJ, Redding FK (1978) Microwave-evoked brainstem auditory responses. Proc Diego Biomed Symp 17:451–466

Lin JC, Meltzer RJ, Redding FK (1979) Microwave-evoked brainstem potential in cats. J Microwave Power 14:291–296

Lin JC, Meltzer RJ, Redding FK (1982) Comparison of measured and predicted characteristics of microwave-induced sound. Radio Sci 17:159S–163S

Picton TW, Hillyard SA, Krausz HI, Galambos R (1974) Human auditory evoked potentials. I. Evaluation of components. Electroencephalogr Clin Neurophysiol 36(2):179–190

Rissmann WJ, Cain CA (1975) Microwave hearing in mammals. Proc Natl Elec Cong 30:239–244

Roschmann P (1991) Human auditory system response to pulsed radiofrequency energy in RF coils for magnetic resonance at 2.4 to 170 MHz. Magn Reson Med 21:197–215

Seaman RL, Lebovitz RM (1989) Thresholds of cat cochlear nucleus neurons to microwave pulses. Bioelectromagnetics 10(2):147–160

Sheridan CL (1971) Fundamentals of experimental psychology. HR&W, New York

Snider RS, Niemer WI (1961) A stereotaxic atlas of the cat brain. University of Chicago Press, Chicago

Stevens SS (1961) The psychophysics of sensory function. In: Rosenblith WA (ed) Sensory communication. MIT Press, Cambridge

Taylor EM, Ashleman BT (1974) Analysis of central nervous involvement in the microwave auditory effect. Brain Res 74:201–208

Tyazhelov VV, Tigranian RE, Khizhnian EP, Akoev IG (1979) Some pecularities of auditory sensations evoked by pulsed microwave fields. Radio Sci 14(supp 6):259–263

Wever EG (1966) Electrical potentials of the cochlea. Physiol Rev 46:102–127

Wilson BS, Zook JM, Joines WT, Casseday JH (1980) Alterations in activity at auditory nuclei of the rat induced by exposure to microwave radiation: autoradiographic evidence using [14C]2-deoxy-D-glucose. Brain Res 187(1980):291–306

Zwislocki J (1957) In search of the bone-conduction threshold in a free sound field. J Acoust Soc Am 29:795–804

Chapter 7
Mechanisms for Microwave to Acoustic Energy Conversion

The microwave auditory effect has been generally accepted as one of the most interesting and significant biological effects from microwave exposure. Research conducted to date has shown that a cascade of events takes place when a beam of microwave pulses is aimed at a human or animal subject's head. Absorption of pulsed microwave energy creates a rapid expansion of brain matter and launches an elastic wave of pressure that travels inside the head to the inner ear. There, it activates the nerve cells in the cochlea, and the neural signals are then relayed through the central auditory system to the cerebral cortex for sound perception. This is the microwave thermoelastic theory which is widely recognized as the mechanism of interaction for the microwave auditory effect. This chapter will begin with a discussion of some of the mechanisms that have been suggested whereby auditory responses might be induced by pulse-modulated microwave radiation. Details of the thermoelastic theory will then follow.

7.1   Site of Microwave to Acoustic Energy Transduction

Psychophysical studies in humans and laboratory animals have confirmed that auditory perception of pulse-modulated microwaves transmitted through air to the head is a genuine phenomenon. The vast amount of electrophysiological evidence accumulated to date demonstrates scientifically that auditory responses to microwave pulses are like those evoked by conventional acoustic pulses. The findings of brainstem ablative lesions on electrical potentials recorded from central auditory nuclei confirm the peripheral nature of the primary site of microwave to acoustic transduction. Manipulations on middle and outer ears of animals showed that the site of initial sensation of pulse-modulated microwave energy by the auditory system is inside the cochlea. These results strongly indicate that microwave auditory effect is mediated by a microwave-to-mechanical transduction occurring in the brain tissue.

© Springer Nature Switzerland AG 2021
J. C. Lin, *Auditory Effects of Microwave Radiation*,
https://doi.org/10.1007/978-3-030-64544-1_7

A peripheral or distal interaction should involve displacements in tissue and initiation of pressure waves in the head that propagate into the inner ear, where pressure waves are converted into electrical nerve potentials and then relayed into the central nervous system, ultimately to the auditory cortex for recognition. The question is then what anatomical structure in the head would be the site for primary transduction from microwave energy to acoustic energy? A hint comes in the need to enlarge the hole in the skull and lay bare the soft brain tissues for an auditory response to pulsed microwaves to be observed in the brainstem central auditory nuclei study in animals (Sect. 6.6.3). As mentioned in the previous chapter, a direct evidence for it is provided by visualization of glucose utilization through autoradiography (Sect. 6.6.9), which clearly indicates anatomically the brain tissues as the site of primary initial interaction in the microwave auditory effect. So, without any question, the site of interaction and transduction is the soft brain tissue in the head.

7.2   Possible Mechanisms of Interaction

Many investigators have attempted to account for the mechanism of microwave auditory effect from physical and physiological considerations [Lin, 1978, 1981]. The suggestions include microwave forces acting on cochlea proper [Frey and Coren, 1979], the tympanic membrane, oval or round window in the cochlea, basilar membrane, organ of Corti [Joines and Wilson, 1981], caloric vestibulocochlear stimulation, or an alleged sensitive area (Fig. 7.1) for detecting microwave pulse-induced sound over the temporal lobe of the brain [Frey, 1962, 1963]. All were readily ruled out as they should. While these anatomical structures are found on or linked to both sides of the head, their position on the head is fixed in space with respect to the incident microwave. The sensitivity to microwave auditory effect

Fig. 7.1 An alleged sensitive area for detecting microwave pulse-induced sound over the temporal lobe of the brain

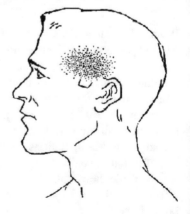

would have to be orientation dependent. However, experimental evidence indicates that the perceived loudness remains the same regardless of subject's position or orientation in the microwave field. Moreover, as the data from a small (1.5 cm) direct-contact microwave antenna moving around, the head of a cat showed, the amplitude and temporal relations of brainstem evoked responses (BER) were remarkably preserved (Fig. 6.28). Microwave energy does not need to be deposited in the cochlea, or the inner, middle, or outer ears to initiate the microwave auditory effect. Of course, results from the cat experiment also demonstrated the auditory response's independence of the microwave source location or orientation.

Several other peripheral transduction mechanisms for a pulsed microwave-induced auditory effect were also offered. Most of these hypotheses were qualitative and lacked specific details; consequently, they remain as highly speculative conjectures in want of experimental verification and theoretical substantiation. One example is the well-known piezoelectric effect in bones, which possibly could result from small bone deformation caused by microwave pulses as a candidate for electrically mediated response [Sharp et al., 1974]. This is contrary to the experiments on cochlear microphonics and the observations showing a need to remove some of the skull bone to allow the applied microwave direct access to the soft brain tissue to detect an evoked auditory response at higher microwave frequencies.

Another hypothesis regarding possible mechanisms is waveguide tuning of the external auditory meatus [Lebovitz, 1975]. Aside from the requirement of subject orientation for optimal detection sensitivity, the waveguide tuning hypothesis neglected the physical fact of cut-off frequency. For a mean external auditory meatus diameter of 7.5 mm, assuming the skin and musculature are ideal electrical conductors at microwave frequencies, the waveguide theory predicts a lowest cut-off frequency for an air-filled waveguide of about 23.45 GHz [Ramo et al., 1965]. That is, microwaves with a frequency below 23.45 GHz would not be able to propagate within the auditory meatus. Conversely, for the waveguide hypothesis to hold, the impinging microwave energy must be above 23.45 GHz, which is in direct contradiction to available experimental data.

7.2.1   Radiation Pressure

Radiation pressure or Maxwell stress exerted by the pulse-modulated microwaves on the surface of the head was assessed for induced vibrations in tissue, which may then launch an acoustic signal of sufficient amplitude to be detected by the inner ear through bone conduction [Sommer and von Gierke, 1964]. Although the radiation pressure computed at the level of microwave perception threshold was found to be slightly above the free-field air conduction threshold, it was almost two orders of magnitude below the sound pressure ($20\mu Pa$) required for threshold bone conduction hearing (see Fig. 7.2). Nevertheless, considering the observations point toward bone

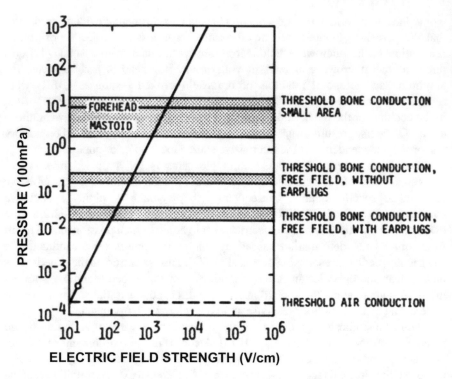

Fig. 7.2 Radiation pressure on the head in an electric field. Approximate pressure ranges for the human auditory thresholds for air and bone conduction at 1000 Hz are indicated. The straight line indicates the radiation pressure on a conducting sphere in a plane wave with a wave length that is small compared to the diameter of the sphere

conduction mediation, the paper concluded that there was no evidence of any direct stimulation which cannot be explained on the basis of microwave-induced vibrations in tissue and normal bone conduction reception in the cochlea.

In attempting to elucidate the mechanism responsible for microwave-induced auditory sensation, a microwave exposure system was set up in 1973 at the Walter Reed Army Institute of Research in Maryland. Investigators found that carbon-impregnated polyurethane microwave absorber (Eccosorb WG4, Emerson and Cuming) acted as a transducer from microwave energy to acoustic energy [Sharp et al., 1974]. If the microwave absorber was placed between the human subject and the pulsed microwave source, the apparent locus of the audible click moved from the observer's head to the absorber. Using a microphone and sound level meter, sounds produced by pulsed microwaves in absorbers of different sizes and shapes were detectable for absorber sizes as small as 4 mm square by 2 mm thick. Several other types of microwave absorbers also produced audible sound. However, aluminum foil had to be crumbled before audible sound was detected from it. Thus, the observed sonic phenomenon implicated a connection of microwave absorption to pulse-induced auditory sensation in humans, but also provided a counterexample against radiation pressure as an operating interaction mechanism.

Physical requirements and simple calculations indicated that the radiation pressure exerted on highly reflective smooth surfaces such as a sheet of aluminum foil would be greater than that exerted on the surfaces of microwave absorbers or tissue materials, in opposition to the observed results. Since the microwave penetration or skin depth is very small, on the order of one micrometer for aluminum, compared with microwave absorbers or tissue materials, one would therefore expect the acoustic energy generated in aluminum foil to be much smaller than that generated in the other materials used. On the other hand, crumpling the aluminum foil presumably increases the "effective penetration depth" and amount of microwave energy absorption to allow production of an audible sound when it is exposed to pulse-modulated microwaves. The critical features were microwave absorbers and crumbled aluminum foils both offered sufficiently high microwave absorption cross sections. Thus, microwave absorbers became a substitute for tissues in the human head for sound wave production. In the case of crumbling aluminum foil, crumbling acted to increase microwave absorption to enable audible sound production and detection.

7.2.2   Electrostrictive Force

A microwave-induced electrostrictive force was also introduced [Guy et al., 1973, 1975]. It was based on a mathematical theory on the elastic deformation of a dielectric body in response to an applied electrostatic field [Stratton, 1941]. At the frequencies where the auditory effect can be easily detected, microwaves can penetrate deeply and are absorbed in tissues of the head. The absorbed energy may produce a volumetric strictive force (elastic deformation) in biological materials, which sets up a pressure wave that travels in the cranial tissue structure and initiates movement of the cochlear partition. It was assumed that, by extension, the equations derived from electrostatics were also valid for microwaves and calculated the pressure within the brain that would be expected. The acoustic pressure calculated in this way was of the same order of magnitude as the computed internal threshold pressure or a little below the threshold of hearing [Guy et al., 1973, 1975a, b].

7.2.3   Thermoelastic Pressure Wave Generation

The conversion of electromagnetic to elastic pressure waves by the transient surface heating of liquid materials had been reported [White, 1963; Gournay, 1966]. In this process, a portion of the incident pulse radiation is converted into heat which generates a temperature gradient normal to the surface. The temperature gradient produces strains in the liquid dielectric material, which results in rapid thermal expansion occurring within a few µs and the generation of pressure waves which propagate away from the surface.

The observation that microwave absorbers acted as a transducer from microwave energy to audible acoustic energy appeared to suggest a similar conversion mechanism – production of elastic pressure wave via absorption of pulsed microwave energy in a brief duration over a meaningful distance inside the surface of an exposed object [Sharp et al., 1974].

While the above publications were indicative of a pulsed microwave-induced thermoelastic pressure wave generation mechanism for the microwave auditory effect, there was no direct physiological evidence confirming the existence of pulsed microwave to sound conversion in the human or animal head, at that time. Taking a cue from the above findings, a bench-top experiment was initiated and demonstrated that microwave pulses impinging on water produced acoustic pressure transients with peak amplitude, which were within the human auditory frequency range of 200 Hz to 20 kHz [Foster and Finch, 1974].

A large polystyrene container is filled with 0.15 N KCI solution at 25 °C and exposed to pulse-modulated 2450 MHz microwaves at a constant energy density per pulse with pulse width from 2 to 25µs in a microwave anechoic chamber. Microwave energy was derived from an Applied Microwave Laboratory pulse source, which fed into a standard gain horn antenna. The average incident power density was measured with an isotropic radiation monitor. The peak sound pressure generated in the solution was measured using a sensitive, electrically shielded hydrophone. Acoustic transients were recorded in water, physiological saline, blood, muscle, and brain samples irradiated with pulsed 2450 MHz microwave energy. Also, using distilled water, the experiment showed that between 0 and 4 °C the recorded pressure wave was inverted from that at higher temperatures and at 4 °C the pressure wave signal disappeared completely. This agrees with the known behavior of water as a function of temperature. At 4 °C, the coefficient of thermal expansion of water is zero. This observation argues strongly for a microwave-induced thermoelastic mechanism of sound wave generation in water. Since similar pressure signals were observed in biological tissues exposed to pulse-modulated microwaves with pressure amplitudes approximately 90 dB, which is above the estimated threshold for sound perception by bone conduction, it is reasonable to suppose that a similar mechanism may be at work when humans and animals sense pulsed microwaves impinging on their heads.

7.3   Analysis of Possible Transduction Mechanisms

The foregoing section discussed many potential transduction mechanisms suggested by various investigators. First-order calculations are given to compare the three possible physical mechanisms which are the most likely to be involved in the interaction of microwave pulses with the auditory systems of animals and humans.

Several investigators [Guy et aI., 1975; Lin, 1976a, b, 1978; Borth and Cain, 1977] have reported comparative data on the amplitude of the acoustic energy generated through radiation pressure, electrostrictive force, and thermoelastic stress. The results indicated that thermoelastic waves greatly exceed radiation pressure.

While the strictive forces are high compared to radiation pressure, they are much smaller than those generated by rapid thermoelastic expansion, based on a planar model of biological materials. Moreover, the amplitude of the induced thermoelastic stress wave is clearly above the established threshold of hearing in humans via bone conduction. Thus, while all three mechanisms may be operating in an exposure situation, the large pressure due to thermoelastic stress may completely mask the impact of others.

Consider a simple one-dimensional model in which a plane wave impinges normally on the boundary of a semi-infinite region of homogeneous tissue material shown in Fig. 7.3. For microwave absorptions at and near the surface of the dielectric tissue medium and a power density at the surface I_0, the power density at a distance z from the surface is given by

$$
\begin{aligned}
I &= I_0 e^{-2\alpha z}, \quad 0 < t < t_0 \\
I &= 0, \qquad\qquad \text{otherwise}
\end{aligned}
\tag{7.1}
$$

where α is the attenuation coefficient which describes the absorbing characteristics of the material medium. This microwave energy corresponds to a rectangular pulse with pulse width t_0. It may exert a radiation pressure on the surface of the absorbing medium and launch a pressure wave, or it may generate body forces via dielectric volume change, or it may be absorbed by the lossy dielectric tissue and converted to a pressure wave as a result of rapid thermoelastic expansion. It is assumed that the tissue dielectric medium possesses linear, isotropic, and elastic properties characterized by Lame's constant λ and μ and a volume mass density ρ. Allowing particle displacement u only along the z direction in the medium, the equation of motion of the particles responding to an applied force or pressure obtained from Newton's second law of motion is [Love, 1927; Sokolnikoff, 1956],

$$
\frac{\partial^2}{\partial t^2} u(z,t) - c^2 \frac{\partial^2}{\partial z^2} u(z,t) = G(z,t)
\tag{7.2}
$$

where $c^2 = [(\lambda + 2\mu)/\rho]$ is the square of bulk velocity of pressure wave propagation in the medium and $G(z,t)$ is the generating function proportional to force per unit

Fig. 7.3 A plane wave impinging normally on a semi-infinite tissue medium

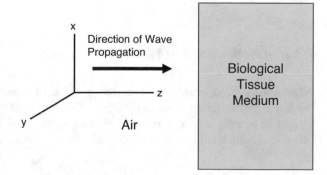

mass in newtons per kilogram (Nt/kg). For the development presented here, the temperature variations of ρ and c are neglected. Although a sound pressure wave is, in general, attenuated as it propagates through the medium, attenuation will be neglected in formulating the mathematical description of the response. That is, the fraction of sound energy dissipated in the medium is assumed to be relatively small. However, in analyzing the data, attenuation will be taken into consideration.

In what follows, the D'Alembert's method of solution [Tychonov and Samarski, 1964] of the governing differential Eq. (7.2) is used to obtain displacements and pressures for the one-dimensional response of a homogeneous isotropic half-space due to power deposition. The method is very useful for obtaining results for many types of volume-force excitation. Much of the developments follow those given in Lin [1978].

Only the case of a rigidly constrained surface is considered here because the resulting pressure for a constrained surface is greater than that given by a stress-free surface. So, if Eq. (7.2) is solved by assuming that the surface is rigidly constrained and is initially at rest, that is, at $z = 0$,

$$u(0,t) = 0 \tag{7.3}$$

and

$$u(z,0) = \frac{\partial}{\partial t} u(z,0) = 0 \tag{7.4}$$

then the displacement as a function of z and t for a generating function $G(z,t)$ is given by

$$u(z,t) = \frac{1}{2c} \int_0^t dt' \int_{z-ct+ct'}^{z+ct-ct'} G(x,t')dx, \qquad t < z/c$$

$$= \frac{1}{2c} \int_0^t dt' \int_{z-ct+ct'}^{z+ct-ct'} G(x,t')dx, \qquad t < z/c \tag{7.5}$$

It is readily verified by substitution that Eq. (7.5) formally satisfies the equation of motion and the auxiliary conditions.

7.3.1   Computation of Radiation Pressure

When a plane wave impinges on an infinite plane surface, a pressure is exerted by the impinging microwave on the medium [Stratton, 1941; Smythe, 1968]. If the surface is entirely within a medium that supports essentially no shearing stress ($\mu/\lambda << 1$), which is the case in soft tissues, the total pressure P is given by

$$P(z,t) = \sqrt{\mu_0 \varepsilon_0 \varepsilon_r}\, I(z,t) \tag{7.6}$$

where $I(z,t)$ is that expressed by Eq. (7.1), ε_r is the relative dielectric constant of the medium, and ε_0 and μ_0 are the vacuum permittivity and permeability, respectively. The net force (pressure) acting on a differential tissue volume is $\{A \, \Delta z \, P(z,t)\}$ for area A and depth or thickness, Δz. Therefore, the total force per unit mass is

$$G(z,t) = -\frac{1}{\rho}\frac{\partial}{\partial z} P(z,t) \tag{7.7}$$

or equivalently,

$$G(z,t) = 2\alpha I_0 \frac{1}{\rho}\sqrt{\mu_0\varepsilon_0\varepsilon_r}\,e^{-2\alpha z}, \quad 0 < t < t_0$$
$$G(z,t) = 0, \qquad\qquad\qquad\qquad \text{otherwise} \tag{7.8}$$

where ρ is the density of the medium. The displacement obtained by evaluating the integrals in Eq. (7.5) with the generating function given by Eq. (7.8) for radiation pressure is

$$u(z,t) = \frac{I_0\sqrt{\mu_0\varepsilon_0\varepsilon_r}}{\alpha\rho c^2} F e^{-2\alpha z}, \tag{7.9}$$

where $F = F1, F2, F3, F4,$ or $F5$, and are given as

$$F1 = \sinh^2 \alpha ct, \qquad t < t_0; \, t < z/c \tag{7.10}$$

$$F2 = \sinh \alpha ct_0 \sinh \alpha c(2t - t_0), \qquad t > t_0; \, t < z/c \tag{7.11}$$

$$F3 = \frac{1}{2}\left[1 - e^{-2\alpha z} - \sinh 2\alpha z e^{-2\alpha ct}\right], \qquad t < t_0 \,; t > z/c \tag{7.12}$$

$$F4 = \frac{1}{2}\left[1 - \sinh 2\alpha z e^{-2\alpha ct} - \cosh 2\alpha c(t - t_0)e^{-2\alpha z}\right], \, t_0 < t < t_0 + z/c; \, t > z/c \tag{7.13}$$

$$F5 = \sinh 2\alpha z \sinh \alpha ct_0 \, e^{-\alpha c(2t - to)}, \qquad t > t_0 + z/c; \, t > z/c \tag{7.14}$$

The local pressure distribution, $p(z,t)$ is given by

$$p(z,t) = (\lambda + 2\mu)\frac{\partial}{\partial z} u(z,t) \tag{7.15}$$

The complete expressions of local pressure distribution following indicated differentiation process, in terms of the functionals, $F6$, $F7$, and $F8$, along with the F functionals $F1$ and $F2$, take the form

$$p(z,t) = \frac{2(\lambda + 2\mu)I_0\sqrt{\mu_0\varepsilon_0\varepsilon_r}}{\rho c^2} F e^{-2\alpha z}, \tag{7.16}$$

where $F = -F1$ or $-F2$, as specified in Eqs. (7.10) and (7.11), and the functionals $F6$, $F7$, or $F8$ given by

$$F1 = \sinh^2 \alpha ct, \qquad t < t_0; \; t > z/c \qquad (7.10)$$

$$F2 = \sinh \alpha ct_0 \sinh \alpha c(2t - t_0), \qquad t > t_0; \; t < z/c \qquad (7.11)$$

$$F6 = \frac{1}{2}\left[1 - \cosh 2\alpha z e^{-2\alpha ct + 2\alpha z}\right], \qquad t < t_0; \; t > z/c \qquad (7.17)$$

$$F7 = \frac{1}{2}\left[\cosh 2\alpha c(t - t_0) - \cosh 2\alpha z e^{-2\alpha ct + 2\alpha z}\right], \qquad t_0 < t < t_0 + z/c; \; t > z/c \quad (7.18)$$

$$F8 = \cosh 2\alpha z \sinh \alpha ct_o \; e^{-\alpha c(2t - to) + 2\alpha z}, \qquad t > t_0 + z/c; \; t > z/c \qquad (7.19)$$

7.3.2 Electrostrictive Force Calculation

A dielectric body exhibits tendencies to contract or expand in an applied electromagnetic field. The force associated with the elastic deformation is called strictive force. Although a complete derivation of the strictive force in a microwave field is difficult to obtain, an approximate expression may be obtained by considering the pressure increase in a fluid (which is an approximation of most soft tissue materials) exposed to a microwave field. The pressure increase at any interior point due to microwave exposure, according to Stratton [1941] and Smythe [1968], is given by

$$p(z,t) = \frac{1}{3}(\varepsilon_r - 1)(\varepsilon_r + 2)I_0 \sqrt{\frac{\mu_0 \varepsilon_0}{\varepsilon_r}} e^{-2\alpha z}, \qquad (7.20)$$

Following the descriptions above, the total strictive force per unit mass inside the dielectric fluid is

$$G(z,t) = \frac{2\alpha}{3\rho}(\varepsilon_r - 1)(\varepsilon_r + 2)I_0 \sqrt{\frac{\mu_0 \varepsilon_0}{\varepsilon_r}} e^{-2\alpha z}, \qquad (7.21)$$

The generating function corresponding to an incident rectangular microwave pulse with pulse width t_0 is therefore given by

$$G(z,t) = \frac{2\alpha}{3\rho}(\varepsilon_r - 1)(\varepsilon_r + 2)I_0 \sqrt{\frac{\mu_0 \varepsilon_0}{\varepsilon_r}} e^{-2\alpha z}, \qquad 0 < t < t_0 \qquad (7.22)$$

$$G(z,t) = 0, \qquad \qquad \text{otherwise}$$

The displacement due to strictive force, after substituting Eq. (7.22) into the integral solution of Eq. (7.5), is

$$u(z,t) = \frac{1}{\alpha \rho c^2}(\varepsilon_r - 1)(\varepsilon_r + 2)I_0\sqrt{\frac{\mu_0 \varepsilon_0}{\varepsilon_r}} Fe^{-2\alpha z}, \qquad (7.23)$$

where $F = F1, F2, F3, F4,$ or $F5$ is given in Eqs. (7.10) to (7.14). The pressure due to forces of electrostriction using Eqs. (7.15) and (7.23) becomes

$$p(z,t) = \frac{2}{3\rho c^2}(\lambda + 2\mu)(\varepsilon_r - 1)(\varepsilon_r + 2)I_0\sqrt{\frac{\mu_0 \varepsilon_0}{\varepsilon_r}} Fe^{-2\alpha z}, \qquad (7.24)$$

where $F = -F1, -F2, F6, F7,$ or $F8$ are specified by Eqs. (7.10) or (7.11) and (7.17) to (7.19), respectively.

7.3.3   Thermoelastic Pressure Generation

In the process of microwave energy absorption, a portion of the incident radiation is converted into heat which generates a temperature gradient normal to the surface. As a result of thermoelastic expansion occurring within a few μs, this temperature gradient produces strains in the dielectric material and leads to the generation of stress waves which propagate away from the surface.

For the power distribution described by Eq. (7.1), the energy absorption occurs only during the short microwave pulse application between $t = 0$ and $t = t_0$. Neglecting any heat loss due to conduction and radiation, a solution of the equation of heat conduction [Carslow and Jaeger, 1959; Gournay, 1966] gives a simple approximate temperature distribution $v(z,t)$ inside the medium as

$$v(z,t) = 2\alpha t I_0 \frac{1}{\rho c'} e^{-2\alpha z}, \qquad 0 < t < t_0 \qquad (7.25)$$

where c' is the specific heat of the medium. It is interesting to note that Eq. (7.25) predicts potentially rapid temperature rises and steep temperature gradients when the peak input power is close to the magnitude observed for microwave-induced auditory sensation. Using the property values listed in Tables 7.1 and 7.2, for an incident power density of 10 kW/m² and a 10μs microwave pulse, Eq. 7.25 yields a calculated temperature rise of less than 10^{-6} °C at the surface of brain equivalent tissues.

In biological materials, as in many nonmetallic and dielectric media, the cooling curve for $t > t_0$ is a slowly varying function of time and becomes appreciable only for times greater than ms. Moreover, the times for production and propagation of stress waves are short compared with temperature equilibration. Thus, for $t > t_0$

$$v(z,t) = 2\alpha t_0 I_0 \frac{1}{\rho c'} e^{-2\alpha z}, \qquad (7.26)$$

Table 7.1 Acoustic or elastic properties of biological materials

Biological material	Temperature °C	Mass density Kg/m³	Bulk velocity m/s	Shear velocity m/s	Lame's constant, λ [b]Pa (Nt/m²)	Lame' constant, μ [b]Pa (Nt/m²)	Absorption coefficient 1/m	Frequency
Fat	37	970	1440	1.78			5–13	1 MHz
Bone	37	1700	3360	1576	2.25×10^{11}	5.5×10^{10}	90	800 kHz
Teeth	37	1700	3380	1576	2.25×10^{11}	5.5×10^{10}	40	600 kHz
Muscle	37	1070	1580	1.78			13–25	1 MHz
Brain	37	1040	1500				11	1 MHz
Brain	37		1523					25 kHz
Brain	37			0.10	2.24×10^{9}	1.052×10^{3}		10–100 Hz
[a]Liver	25	1070		9–100			$(2–30) \times 10^{5}$	2–14 MHz
Saline	40	998.4	1539					1 MHz
Saline	20	1005	1493					1 MHz
Water	40	992.2	1529				2.5×10^{-2}	1 MHz
Water	20	998.2	1483				2.5×10^{-2}	1 MHz

[a]Average of liver, kidney, muscle, and red blood cells at 2.0, 6.5, and 14 MHz
[b]1 Pascal = 1 Nt/m²

Table 7.2 Thermophysical properties of biological materials

Biological material	Thermal conductivity W/(m °C)	Specific heat capacity J/(kg °C)	Thermal diffusivity 10^{-7} m²/s	Thermal expansion 10^{-5}(°C)$^{-1}$
Fat	0.18	2595	0.873	2.76
Bone	0.35	2051	4.20	2.76
Muscle	0.85	3139	1.52	4.14
Brain	0.55	3684	1.38	4.14
Liver	0.75	3192	1.95	4.14
Water	0.63	4177	1.50	6.9

In the brain medium, the small but rapid temperature rise produces a strain given by

$$\varepsilon_z(z,t) = \frac{\partial}{\partial z} u(z,t) = \beta v(z,t), \tag{7.27}$$

where $u(z,t)$ is the particle displacement and β is the coefficient of linear thermal expansion. Negligible strains were assumed along the x and y directions. The strain of Eq. (7.27) could also be produced, in the absence of any temperature elevation, by a mechanical stress of

$$P_z(z,t) = (3\lambda + 2\mu)\beta v(z,t), \tag{7.28}$$

In the presence of both temperature increase and stress, the stress-strain relationship [Love, 1927; Sokolnikoff, 1956] requires

$$P(z,t) = (\lambda + 2\mu)\frac{\partial}{\partial z} u(z,t) - (3\lambda + 2\mu)\beta v(z,t), \tag{7.29}$$

where $P(z,t)$ is pressure or stress.

The net force due to rapid temperature rise acting on the differential volume is $\{A\Delta z P(z,t)\}$. Thus, the total force per unit mass as a result of rapid temperature rise in the elastic dielectric medium is

$$G(z,t) = -\frac{1}{\rho}\frac{\partial}{\partial z} P_z(z,t), \tag{7.30}$$

Substituting Eq. (7.28) in Eq. (7.30) produces

$$G(z,t) = -(3\lambda + 2\mu)\frac{\beta}{\rho}\frac{\partial}{\partial z} v(z,t) \tag{7.31}$$

By combining Eqs. (7.25), (7.26), and (7.31), the generating function due to rapid temperature increase by a short microwave pulse with pulse width t_0 under the approximation of negligible heat transfer becomes

$$G(z,t) = (3\lambda + 2\mu)\beta(2\alpha)^2 tI_0 \frac{1}{\rho^2 c}, e^{-2\alpha z}, \quad t < t_0$$

$$= (3\lambda + 2\mu)\beta(2\alpha)^2 t_0 I_0 \frac{1}{\rho^2 c}, e^{-2\alpha z}, \quad t > t_0 \qquad (7.32)$$

Now substituting Eq. (7.32) into Eq. (7.5) and performing the simple integrations to obtain for the displacement,

$$u(z,t) = (3\lambda + 2\mu)\beta I_0 \frac{1}{2\alpha c' \rho^2 c^3} M e^{-2\alpha z}, \qquad (7.33)$$

where $M = M1, M2, M3, M4,$ or $M5$ is given by

$$M1 = \sinh 2\alpha ct - 2\alpha ct, \qquad\qquad t < t_0; \quad t < z/c \qquad (7.34)$$

$$M2 = \sinh 2\alpha ct - \sinh 2\alpha c(t - t_0) - 2\alpha ct_0 \cosh 2\alpha c(t - t_0), \quad t > t_0; t < z/c \qquad (7.35)$$

$$M3 = \sinh 2\alpha z e^{-2\alpha ct + 2\alpha z} + 2\alpha c\left[(t - z/c)e^{-2\alpha z} - t\right], \qquad t < t_0; t > z/c \qquad (7.36)$$

$$M4 = \sinh 2\alpha z e^{-2\alpha ct + 2\alpha z} + 2\alpha c(t - z/c)e^{2\alpha z} - \sinh 2\alpha c(t - t_0) - 2\alpha ct_0 \cosh \\ 2\alpha c(t - t_0), \; t_0 < t < t_0 + z/c; \; t > z/c \qquad (7.37)$$

$$M5 = 2\sinh 2\alpha z e^{-2\alpha c(t - z/c) + \alpha ct_0}\left(\alpha ct_0 e^{\alpha cto} - \sinh \alpha ct_0\right), \quad t > t_0 + z/c; \; t > z/c \qquad (7.38)$$

From Eq. (7.29), using the results of Eq. (7.33), the pressure or stress is found to be

$$p(z,t) = (3\lambda + 2\mu)\beta I_0 \frac{1}{\rho cc'} Q e^{-2\alpha z}, \qquad (7.39)$$

where $Q = Q1, Q2, Q3, Q4,$ or $Q5,$ and they are defined as

$$Q1 = -\sinh 2\alpha ct - 2\alpha ct, \qquad t < t_0; \; t < z/c \qquad (7.40)$$

$$Q2 = \sinh 2\alpha c(t - t_0) - \sinh 2\alpha ct - \sinh 2\alpha c(t - t_0) - 2\alpha ct_0\left[1 - \cosh 2\alpha c(t - t_0)\right], \\ t > t_0; \; t < z/c \qquad (7.41)$$

$$Q3 = \left(\cosh 2\alpha z e^{-2\alpha ct} - 1\right)e^{2\alpha z}, \quad t < t_0; t > z/c \qquad (7.42)$$

$$Q4 = 2\alpha ct_0 \left[\cosh 2\alpha c(t-t_0)-1\right] + \sinh 2\alpha c(t-t_0) + \left(\cosh 2\alpha z e^{-2\alpha ct} -1\right)e^{2\alpha z},\ (7.43)$$
$$t_0 < t < t_0 + z/c; t > z/c$$

$$Q5 = 2\cosh 2\alpha z \left(\alpha ct_0\, e^{\alpha ct_0} - \sinh \alpha ct_0 \right) e^{-2\alpha c(t-z/c)+\alpha ct_0} - 2\alpha ct_0,\ \ t > t_0 + z/c; t > z/c$$

$$(7.44)$$

The above expressions represent a complete analysis of the displacement and pressure generated by microwave pulse-induced thermoelastic expansion in a semi-infinite dielectric medium. A few special cases have previously been obtained using the usual transform technique in solving the equation of motion [White, 1963; Gournay, 1966]. As mentioned earlier, it is considerably simpler to solve the problem using the integral solution of Eq. (7.5) obtained through D'Alembert's method.

7.4   Biophysical Properties of Biological Materials

Several proposed mechanisms for microwave to sonic energy transduction have been presented in the preceding section. To place the results in proper perspective, the derivations should be quantitatively examined with appropriate biophysical properties of biological materials. The microwave properties of tissue materials, dielectric permittivity and conductivity, including those for brain matters, are found in Chap. 4 under the title "Microwave Properties of Biological Materials."

The relevant mechanical properties of biological materials are sonic, elastic, and thermal properties and mass density of materials. The elastic properties may be also characterized in terms of Lame's constants, absorption, and velocity of sound wave propagation in tissue. In general, the velocity of sound wave propagation in biological materials is frequency independent, while the absorption coefficient varies strongly as a function of acoustic or sonic frequency [Schwan, 1965].

The values of acoustic or elastic properties for various tissues have been reported by several investigators [Goldman and Hueter, 1956; Dunn et al., 1969; Fallenstein et al., 1969; Lang, 1970, Frizzell et al., 1977; Lin et al., 1988; Madsen et al., 1983]. A summary is given in Table 7.1. Hard tissues such as bone and teeth absorb 5–10 times more acoustic energy than soft tissues like brain and muscle. The velocity of propagation in soft tissues is nearly that of physiological saline and varies only slightly among different tissues, except for fat tissue which has a velocity about 5% less than the soft tissue average of 1520 m/s. In contrast, the velocity in hard tissue is about 2.2 times that of the average for soft tissues. Note that aside from tissue geometry and interfaces, the propagation of sonic energy in tissue medium is determined primarily by the acoustic absorption coefficient and sound velocity.

The thermophysical properties of biological materials, namely, specific heat capacity and thermal conductivity, are required to predict the transient and steady-state temperature elevation and distribution due to microwave energy absorption. The thermal properties for various tissues have been summarized in detail by Chato [1966, 1969, 1984], and other data were presented by Lehmann [1965] and Cooper and Trezek [1971, 1972]. Table 7.2 provides a short collection of thermophysical properties of biological materials. Since water is a major component of biological materials and it has the highest thermal conductivity, 0.63 (W/m °C), its value may be considered as an upper limit for tissue thermal conductivity [Chato, 1984]. The thermophysical properties of human and animal tissues depend on detailed composition and structure, which are known to vary considerably among tissue types and preparations; the data presented should be considered as good approximations for the properties of similarly structured tissues with similar composition.

The coefficients of thermal expansion for biological materials are also included in the table. Because values for biological materials and structures do not seem to have been measured, the values for tissues with high water content, i.e., brain and muscle, were assumed to be 60% of the corresponding value for water [Weast, 1974], whereas bone and fat were assumed to have a value approximately 40% of that for water, reflecting their lower water content.

7.5   Comparison of Possible Transduction Mechanisms

It is instructive to examine quantitatively the explicit mathematical expressions describing the induced acoustic pressure waves in a semi-infinite biological medium. In what follows, the mathematical derivations for three of the most likely microwave-to-sound converting schemes will be applied to calculate and compare with known thresholds for sound perception in humans. In this case, the analytical expressions presented in Sect. (7.3) serve as a useful platform for computation and comparison, especially since they have been developed based on a common foundation and used shared mathematical formulation.

For radiation pressure, the explicit expressions of displacement (Eq. 7.9) and induced pressure (Eq. 7.16) along with the biophysical parameters given above in the Sect. (7.4) are used for a 2450 MHz microwave pulse ($t_0 = 10\mu s$ and $I_0 = 10$ kW/ m^2) impinging normally on the surface of a half-space layer of brain material. The calculated results are shown in Figs. 7.4, 7.5, 7.6 and 7.7, as a function of time, t and depth, z. Note that the displacement is zero, while the pressure is the highest at $z = 0$ (constrained surface). The displacement close to the surface is characterized by a rapid rise time and a slightly slower fall time. The temporal behavior becomes increasingly symmetric as the distance into the brain medium increases.

The pressure wave is initially monophasic and becomes diphasic with increasing depth into the brain medium. Both displacement and pressure attain the asymptotic traveling waveform after passing out of the region of most effective power deposition (Eq. 2.53, depth of penetration δ, or $z = 1/\alpha$). This is shown in Figs. 7.6 and 7.7 where the maximum displacement and pressure are plotted as a function of distance.

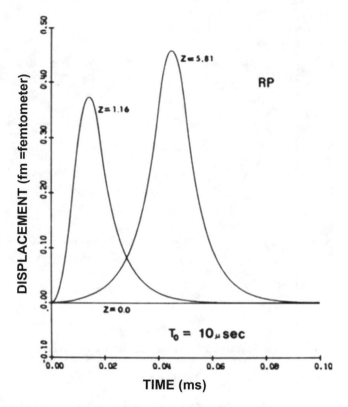

Fig. 7.4 Radiation pressure induced displacement at z = 0, 1.16, and 5.81 cm in a planar brain model exposed to pulsed 2450 MHz microwave radiation: $t_0 = 10\mu s$ and $I_0 = 10$ kW/m²

It is seen that the maximum displacement increases to a limiting value (0.45×10^{-15}m or 0.45 fm) and the maximum pressure decreases from 0.14 mPa. to a minimum value of about 0.065 mPa, after z = 2.32 cm.

The results of calculations made for forces of electrostriction using Eqs. (7.23 and 7.24) are given in Figs. 7.8 to 7.11. As expected, the waveforms as functions of time and the dependence of maximum displacement and maximum pressure on distance have similar characteristics as for radiation pressure, except that the magnitudes are greater for electrostriction by approximately a factor of ten. In the case of electrostriction, the maximum displacement and pressure are 4.9 fm and 15 mPa, respectively.

For thermoelastic stress, calculated results using the mathematical Eqs. (7.33) and (7.39) for displacement and pressure, respectively, are presented in Figs. (7.12, 7.13, 7.14 and 7.15). These figures show that the displacement elicited in the planar brain model by microwave-induced thermoelastic stress has qualitative features like those obtained from radiation pressure and electrostriction, except that the peak displacement is greater by a factor of one thousand. Figure (7.13) depicts typical pressures developed because of thermoelastic stress. In this case, only the traveling

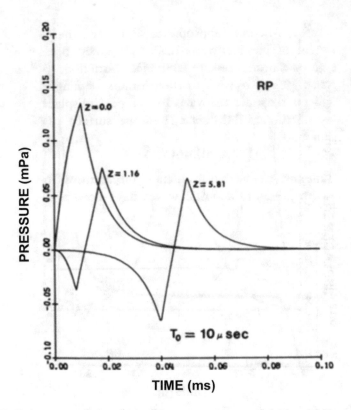

Fig. 7.5 Radiation pressure induced acoustic pressure wave at z = 0, 1.16, and 5.81 cm in a planar brain model exposed to pulsed 2450 MHz microwave radiation: $t_0 = 10\mu s$ and $I_0 = 10$ kW/m²

component of the pressure wave is plotted by removing from Eq. (7.39) terms proportional to $2act$ or $2act_0$ wherever appropriate. It is seen that the traveling component of the thermoelastic stress-generated pressure waves have characteristics like radiation pressure and electrostriction, but with a peak pressure value of 65 mPa which is greater by two to three orders of magnitude. Figures (7.14 and 7.15) illustrate the variation of peak displacement and pressure as functions of distance from the surface of the half-space brain model. The associated maximum displacement is 0.19 pm (pico meter).

It can be seen from the above results that the peak pressure, compressive, or tensile stress always occurs at the surface and decreases with depth into the tissue, regardless of which mechanism is involved in its development. A direct comparison of the maximum pressures shown in Figs. (7.7, 7.9, and 7.13) indicates that microwave-induced thermoelastic stress exceeds radiation pressure by almost three orders of magnitude; pressure generated by electrostriction is about ten times as high as that produced by the radiation pressure mechanism. Thus, while the electrostriction mechanism may be effective in comparison with radiation pressure, it is much less effective than thermoelastic stress in developing pressure waves in brain matters.

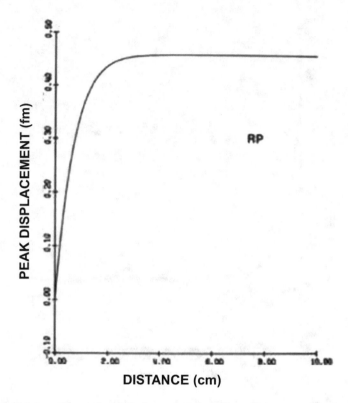

Fig. 7.6 Spatial dependence of peak displacement induced by radiation pressure in a planar brain model exposed to pulsed 2450 MHz microwave radiation: $t_0 = 10\mu s$ and $I_0 = 10$ kW/m^2

In fact, simple relationships among the pressure magnitudes predicted by these three mechanisms for producing pressure waves may be obtained from a consideration of the equations of pressure development. Comparing Eqs. (7.24) and (7.16) yields

$$\frac{\text{Pressure Amplitude due to Electrostriction}}{\text{Pressure Amplitude due to Radiation Pressure}} = \frac{\varepsilon_r}{3}, \qquad (7.42)$$

where dielectric permittivity, $\varepsilon_r >> 1$ is assumed. Also, a comparison of Eqs. (7.39) and (7.16) gives

$$\frac{\text{Pressure Amplitude due to Thermoelastic Stress}}{\text{Pressure Amplitude due to Radiation Pressure}} = \frac{3}{2}\frac{c\beta}{c'\sqrt{\mu_0\varepsilon_0\varepsilon_r}}, \qquad (7.43)$$

where the Lame's constants are such that $\mu/\lambda << 1$. The ratios of relative pressure amplitude calculated for impinging 2450 MHz microwave pulses via these three mechanisms are tabulated in Table 7.3 for brain, muscle, and water. It is readily

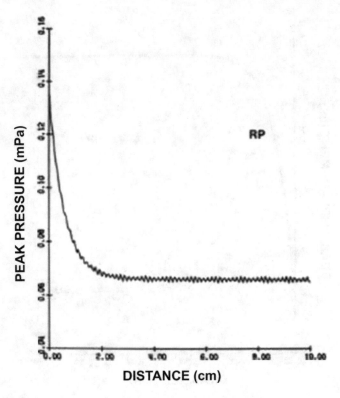

Fig. 7.7 Spatial dependence of peak pressure induced by radiation pressure wave in a planar brain model exposed to pulsed 2450 MHz microwave radiation: $t_0 = 10\mu s$ and $I_0 = 10$ kW/m^2

observed that, in all cases, thermoelastic expansion-generated stress exceeds the others by a large margin. A closer examination of the pressures generated by electrostriction force, radiation pressure, and thermoelastic stress in brain tissue shows that the thermoelastic pressures are two to three orders of magnitude greater than the other mechanisms [Lin, 1976a, b, 1978]. Thus, while all three mechanisms may be operating at given incident power density, the large values due to thermoelastic expansion may completely mask the effect of the others.

Note that a much later publication [Bennett, 1998] claimed that "microwave hearing" was primarily due to a radiation pressure mechanism. The claim was based on a misreading or misinterpretation of the comparative data given in Table 7.3, as previously published [Lin, 1976a, b, 1978]. The conclusion is in significant disagreement with the present understanding of the way in which sensation of sound induced by pulsed microwave in humans and animals.

With a 60% coupling coefficient and a peak incident power density of 40 kW/m^2 (which was found to be the minimally effective value for 2450 MHz microwave radiation to produce audible signals in an adult with normal hearing acuity), the calculated maximum thermoelastically generated pressure is approximately 15 mPa

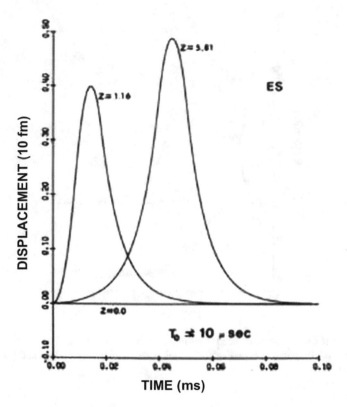

Fig. 7.8 Electrostriction force induced displacement at z = 0, 1.16, and 5.81 cm in a planar brain model exposed to pulsed 2450 MHz microwave radiation: $t_0 = 10\mu s$ and $I_0 = 10$ kW/m^2

for the brain, which is clearly above the established threshold of perception by bone conduction (see Fig. 7.1). Thus, if the acoustic pressure wave is produced in a human head, it would reach the inner ear via bone conduction causing a distinct auditory sensation.

It should be mentioned that the human head is not a planar half-space medium of brain material, but rather a heterogeneous spheroidal body with a semirigid surface. The calculations made in this chapter are therefore not expected to be accurate predictions of the precise magnitude of microwave-induced acoustic pressure waves in the human head, except where exposure geometry and scenario are suited for a planar model, for example, the side of a human dead in the field of a microwave horn antenna. The results do, however, show the importance and applicability of various transduction mechanisms. In particular, the analysis indicates that the amplitude of a thermoelastic expansion-generated pressure signal is of such magnitude that it is the most attractive physical mechanism to explain the microwave-induced auditory effect in humans. In fact, the thermoelastic theory has been accepted scientifically as the mechanism of interaction for the microwave auditory effect.

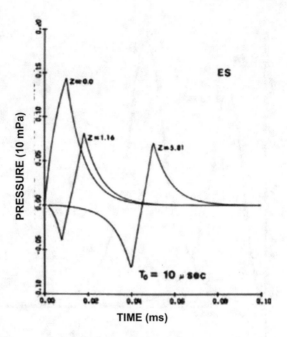

Fig. 7.9 Electrostriction force induced acoustic pressure wave at z = 0, 1.16, and 5.81 cm in a planar brain model exposed to pulsed 2450 MHz microwave radiation: $t_0 = 10\,\mu s$ and $I_0 = 10\,kW/m^2$

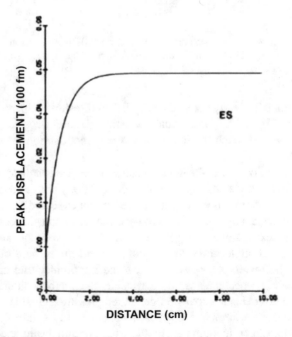

Fig. 7.10 Spatial dependence of peak displacement induced by electrostriction force in a planar brain model exposed to pulsed 2450 MHz microwave radiation: $t_0 = 10\mu s$ and $I_0 = 10\,kW/m^2$

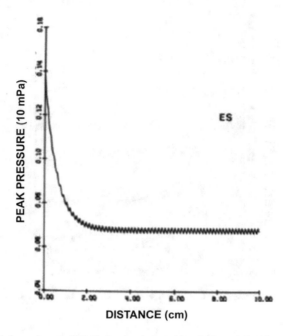

Fig. 7.11 Spatial dependence of peak pressure induced by electrostriction force wave at in a planar brain model exposed to pulsed 2450 MHz microwave radiation: $t_0 = 10\mu s$ and $I_0 = 10$ kW/m^2

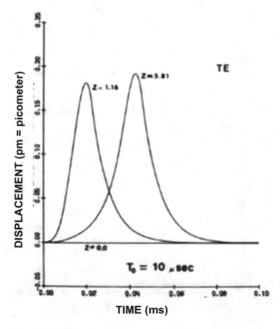

Fig. 7.12 Thermoelastic stress induced displacement at z = 0, 1.16, and 5.81 cm in a planar brain model exposed to pulsed 2450 MHz microwave radiation: $t_0 = 10\mu s$ and $I_0 = 10$ kW/m^2

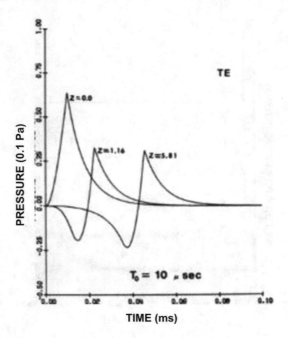

Fig. 7.13 Thermoelastic stress induced acoustic pressure wave at $z = 0$, 1.16, and 5.81 cm in a planar brain model exposed to pulsed 2450 MHz microwave radiation: $t_0 = 10\mu s$ and $I_0 = 10$ kW/m^2

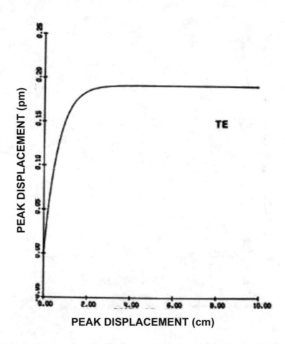

Fig. 7.14 Spatial dependence of peak displacement induced by thermoelastic stress in a planar brain model exposed to pulsed 2450 MHz microwave radiation: $t_0 = 10\mu s$ and $I_0 = 10$ kW/m^2

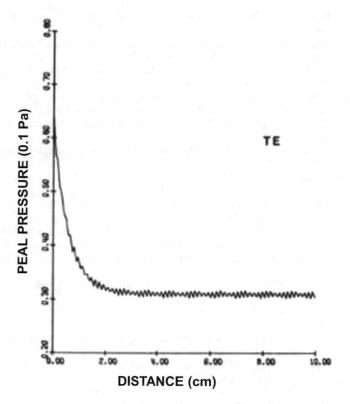

Fig. 7.15 Spatial dependence of peak pressure induced by electrostriction force wave in a planar brain model exposed to pulsed 2450 MHz microwave radiation: $t_0 = 10 \, \mu s$ and $I_0 = 10 \, kW/m^2$

Table 7.3 A comparison of three physical mechanisms of acoustic pressure produced in planar models of biological materials exposed to short 2450 MHz microwave pulses

Material	Relative amplitude		
	Electrostriction/radiation pressure	Thermoelastic stress/ electrostriction	Thermoelastic stress/ radiation pressure
Brain	10.67	122	1301
Muscle	15.67	82	1290
Water	26.0	47	1225

References

Bennett WR (1998) Radio frequency hearing: electrostrictive detection and bone conduction. J Acoust Soc Am 103:2111–2116

Borth DE, Cain CA (1977) Theoretical analysis of acoustic signal generation in materials irradiated with microwave energy. IEEE Trans Microwave Theory Tech 25:944–954

Carslow HS, Jaeger JC (1959) Conduction of heat in solids, 2nd edn. Oxford University Press, London

Chato JC (1966) A survey of thermal conductivity and diffusivity data on biological materials. ASME winter annual meeting paper No. 66-WA/HT 37

Chato JC (1969) Heat transfer in bioengineering. In: Chao BJ (ed) Advanced heat transfer. University Illinois Press, Urbana, pp 395–414

Chato JC (1984) Selected thermophysical properties of biological materials. In: Shitzer A, Eberhart RC (eds) Heat transfer in medicine and biology. Plenum, New York, pp 413–418

Cooper TE, Trezek GJ (1971) Correlation of thermal properties of some human tissues with water content. Aerosp Med. 42:24–27

Cooper TE, Trezek GJ (1972) A probe technique for the determination of thermal conductivity of tissue. J Heat Transfer 94:133–138

Dunn F, Edmonds PD, Fry WJ (1969) Absorption and dispersion of ultrasound in biological media. In: Schwan HP (ed) Biological engineering. McGraw, New York, pp 205–332

Fallenstein GT, Hulce VD, Melvin JW (1969) Dynamic mechanical properties of human brain tissue. J Biomechanics 2:217–226

Foster KR, Finch EE (1974) Microwave hearing: Evidence for thermoacoustical auditory stimulation by pulsed microwaves. Science 185:256–257

Frey AH (1962) Human auditory system response to modulated electromagnetic energy. J Appl Physiol 17:689–692

Frey AH (1963) Some effects on human subjects of ultra-high-frequency radiation. Am J Med Electron 2:28–31

Frey AH (1971) Biological function as influenced by low-power modulated RF energy. IEEE Trans Microwave Theory Tech 19:153–164

Frey AH, Coren E (1979) Holographic assessment of hypothesized microwave hearing mechanism. Science 206:232–234

Frizzell LA, Carstensen EL, Dyro JF (1977) Shear properties of mammalian tissues at low megahertz frequencies. J Acoust Soc Am 60(6):1409–1411

Goldman DE, Hueter TF (1956) Tabular data of the velocity and absorption of high-frequency sound in mammalian tissues. J Acoust Soc Am 28:35–37

Gournay LS (1966) Conversion of electromagnetic to acoustic energy by surface heating. J Acoust Soc Am 40:1322–1330

Guy AW, Taylor EM, Ashleman B, Lin JC (1973) Microwave interaction with the auditory systems of humans and cats. In: Proceeding of the international microwave symposium, Boulder, June 1973, pp 321–323

Guy AW, Chou CK, Lin JC, Christensen D (1975) Microwave induced acoustic effects in mammalian auditory systems and physical materials. Ann NY Acad Sci 247:194–218

Guy AW, Lin JC, Chou CK (1975a) Electrophysiological effects of electromagnetic fields on animals. In: Michaelson S et al (eds) Fundamentals and applied aspects of nonionizing radiation. Plenum Press, New York, pp 167–211

Joines WT, Wilson BS (1981) Field-induced forces at dielectric interfaces as a possible mechanism of RF hearing effects. Bull Math Biol 43(4):401–413

Lang SB (1970) Ultrasonic method for measuring elastic coefficients of bone and results on fresh and dried bovine bones. IEEE Trans Biomed Eng 17:101–105

Lebovitz RM (1975) Detection of weak electromagnetic radiation by the mammalian vestibulocochlear apparatus. Ann NY Acad Sci 247:182–193

Lehmann JF (1965) Ultrasound therapy. In: Licht S (ed) Therapeutic heat and cold, New Haven, pp 321–386

Lin JC (1976a) Microwave auditory effect – a comparison of some possible transduction mechanisms. J Microwave Power 11:77–81

Lin JC (1976b) Microwave induced hearing sensation: some preliminary theoretical observations. J Microwave Power 11:295–298

Lin JC (1978) Microwave auditory effects and applications. CC Thomas, Springfield

Lin JC (1981) The microwave hearing effect. In: Illinger KH (ed) Biological effects of nonionizing radiation. American Chemical Society, pp 317–330

Lin JC, Su J-L, Wang Y (1988) Microwave-induced thermoelastic pressure wave propagation in the cat brain. Bioelectromagnetics 9(2):141–147

Love AEH (1927) The mathematical theory of elasticity. Cambridge University Press, New York

Madsen EL, Sathoff HJ, Zagzebski JA (1983) Ultrasonic shear wave properties of soft tissues and tissue like materials. J Acoust Soc Am 74(5):1346–1355

Ramo S, Whinnery JR, Van Duzer T (1965) Fields and waves in communication electronics. Wiley, New York

Schwan HP (1965) Biophysics of diathermy. In: Licht S (ed) Therapeutic heat and cold, New Haven

Sharp JC, Grove HM, Gandhi OP (1974) Generation of acoustic signals by pulsed microwave energy. IEEE Trans Microwave Theory Tech 22:583–584

Smythe WR (1968) Static and dynamic electricity. McGraw, New York

Sokolnikoff IS (1956) Mathematical theory of elasticity. McGraw, New York

Sommer HC, Von Gierke HE (1964) Hearing sensations in electric fields. Aerosp Med 35:834–839

Stratton JA (1941) Electromagnetic theory. McGraw, New York

Tychonov AN, Samarski AS (1964) Partial differential equations of mathematical physics. Holden-Day, San Francisco

Weast RC (ed) (1974) Handbook of chemistry and physics. CRC Press, Cleveland

White RM (1963) Generation of elastic waves by transient surface heating. J Appl Physics 34:3559–3567

Chapter 8
Thermoelastic Pressure Waves in Canonical Head Models

The preceding chapters have discussed and documented that an audible sound occurs when human subjects are exposed to pulse-modulated microwave radiation which originates from within the head. Also, auditory detection of pulsed microwaves occurring in laboratory animals has been confirmed both in behavioral and electrophysiological investigations. The site of microwave to sound energy conversion is in the brain tissues. The primary mechanism of interaction is pulsed microwave-induced thermoelastic expansion of brain matter, which is the most effective mechanism, since pressures generated by thermoelastic stress are two to three orders of magnitude greater than by any other possible mechanisms [Lin, 1980, 1990; Lin and Wang, 2007].

Thus, the microwave auditory effect arises from the miniscule (on the order of 10^{-6} °C for a 10μs pulse), but rapid elevation of temperature in the brain due to absorption of pulsed microwave radiation. The sudden rise of theoretically tiny temperature and practically unmeasurable currently available instruments occurring in a very short time creates a thermoelastic expansion of the brain matter, which then launches a stress or pressure wave that travels through the structures in the head, that is detected by the hair cells in the cochlea. It is then relayed to the central auditory system for perception and recognition at the cerebral auditory cortex. (See Fig. 1.2 for an illustration of the cascade of events.)

The human head is a heterogeneous spheroidal body with a semirigid surface not a planar half-space medium of only brain material. The calculations mentioned in earlier chapters are therefore not expected to accurately predict the precise magnitude of microwave-induced acoustic pressure waves in the human head (except at higher microwave frequencies where absorption is concentrated near the surface). The results do, however, indicate the magnitude of a thermoelastic expansion-generated pressure signal is such that it is the dominant physical mechanism to explain the microwave auditory effect in humans and animals.

This chapter presents rigorous multidisciplinary, mathematical analyses of the thermoelastic pressure wave generated in canonical or spherical human and animal heads exposed to pulsed microwave radiation. The results include a variation of

© Springer Nature Switzerland AG 2021
J. C. Lin, *Auditory Effects of Microwave Radiation*,
https://doi.org/10.1007/978-3-030-64544-1_8

induced sound pressure frequency and strength on microwave pulse characteristics. They also provide information on the attributes of sound generated in the head as perceived by humans. More precise computer simulations of the properties of microwave-pulse-induced sound pressure wave using realistic anatomic head structures are presented in the chapter that follows.

The first mathematical models for analyzing the sound pressure waves inside the head due to a microwave pulse using a homogeneous spherical model of the head were developed by Lin [1976a, b; 1977a, b, c], which had assumed a spherically symmetric SAR pattern that peaked at the sphere center. The same analytic method was applied by Shibata et al. [1986], but with a slightly modified SAR pattern which included a uniform SAR offset. A generalization of the methodology, again using a homogeneous sphere head model, but with an arbitrary SAR pattern of spherical symmetry, was presented by Yitzhak et al. [2009].

8.1 Analytic Formulation of Microwave-Induced Thermoelastic Pressure

The mathematical analysis considers a spherical head model consisting of homogeneous brain materials [Lin, 1976a, b; 1978]. Using a multidisciplinary approach, the source of the impinging radiation is assumed to be a plane wave of pulsed microwave radiation. The resulting microwave energy absorption (or SAR) inside the head can be obtained from Maxwell's equations (Chap. 2). The ensuing small but finite temperature elevation is then derived, and lastly, the thermoelastic equation of motion is solved for the stress pressure waves generated inside the spherical head. The analytical solution of the thermoelastic equation of motion was presented for two types of boundary conditions: (1) stress-free boundary, where the normal stress vanishes at the surface of the spherical head [Lin, 1977a] and (2) rigidly constrained boundary, where the displacement vanishes at the sphere's surface [Lin, 1977b].

8.1.1 Microwave Absorption

Examples of microwave absorption or SAR distribution patterns calculated from analytical formulations for the homogeneous spherical model of the human child (5-cm radius) or adult (9-cm radius) head exposed to 400–915 MHz plane waves are given in Figs. 5.7, 5.8 and 5.9. It is seen for these calculations, both human child- and adult-sized heads show peak absorption occurring near the center and inside of both brain spheres. SAR distributions along the three coordinate axes exhibit standing-wave-like oscillations along the outer portion of the spherical head and reach a maximum near the center. For mathematical simplicity, instead of invoking rigorous calculations from Maxwell's equations, which have been done elsewhere [Johnson and Guy, 1972; Kritikos and Schwan, 1975; Lin, et al., 1973; Lin, 1976a, b; Shapiro et al., 1971], a spherically symmetric absorption pattern is adopted to approximate the SAR distribution $I(r)$, inside the head using the spherically symmetric function,

$$I(r) = I_0 \left(\frac{1}{\rho}\right) \frac{\sin\left(\dfrac{N\pi r}{a}\right)}{\left(\dfrac{N\pi r}{a}\right)}, \tag{8.1a}$$

where peak rate of energy absorption or peak absorbed power density is I_0, r is the radial variable, and a is the radius of the spherical head. The parameter N specifies the number of oscillations in the approximated spatial dependence of absorbed energy. For some frequencies and sphere size combinations, the integer N may be adjusted to account for the difference in absorption patterns. For instance, $N = 6$ may be chosen to approximate the absorption pattern inside a 3- or 7- cm radius spherical head of a cat or human adult exposed to 2450 or 915 MHz microwave radiations. Figure 8.1a gives the approximated energy absorption pattern for $N = 6$, showing a specific spherically symmetric microwave energy absorption pattern which peaks at the sphere center.

It was mentioned that the basic results would be the same [Lin, 1978, p.142] if the pattern of absorbed energy distribution takes the form of

$$I(r) = I_0 \left(\frac{1}{\rho}\right) \frac{\sin\left(\dfrac{N\pi r}{a}\right)}{\left(\dfrac{N\pi r}{a}\right)} + I_c, \tag{8.1b}$$

where I_c is a constant offset included to account for the uniform component of the pattern of absorbed energy distribution, so long as I_c is small compared with I_0. Indeed, this was done by Shibata et al. [1986] for a modified specific absorption pattern with $N = 4$ and a 60 and 40% split between I_0 and I_c (Fig. 8.1b), but still spherically symmetric and centrally peaked, under a stress-free boundary condition.

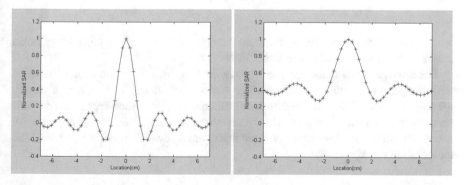

Fig. 8.1 Spherically symmetric microwave energy absorption or SAR distribution according to: Left (**a**) for Eq. (8.1a) and Right (**b**) for Eq. (8.1b)

8.1.2   Temperature Elevation

The temperature elevation $T(r,t)$ resulting from microwave absorption is specified by the heat conduction equation [Carslaw and Jaeger, 1959]. Namely,

$$\nabla^2 T(r,t) - \frac{1}{k}\frac{\partial}{\partial t}T(r,t) = -\frac{1}{K}I(r,t), \tag{8.2}$$

where $T(r,t)$ is temperature elevation; k and K are the thermal diffusivity and conductivity, respectively; and $I(r,t)$ is the heat production rate, which is the same as the absorbed microwave energy and is assumed to be rectangular in shape or constant during the short incident microwave pulse. Given the spherical symmetry of absorbed energy distribution, the spatial dependence of heat conduction equation may be expressed as a function of r alone. That is,

$$\frac{1}{r^2}\frac{\partial}{\partial r}r^2\frac{\partial}{\partial r}T(r,t) - \frac{1}{k}\frac{\partial}{\partial t}T(r,t) = -\frac{1}{K}I(r,t). \tag{8.3}$$

Since the incident microwave pulse duration is much shorter than the time constant of heat conduction in biological tissues, the temperature elevation due to microwave absorption would take place under adiabatic conditions. There will not be any temperature decay and the spatial derivatives in Eq. (8.3) may be neglected such that

$$\frac{1}{k}\frac{\partial}{\partial t}T(r,t) = \frac{1}{K}I(r,t), \tag{8.4}$$

By setting the initial temperature equal to zero, Eq. (8.4) may be integrated to give the temperature variation as

$$T(r,t) = I_0\left(\frac{1}{\rho c'}\right)\frac{\sin\left(\dfrac{N\pi r}{a}\right)}{\left(\dfrac{N\pi r}{a}\right)}t, \tag{8.5}$$

where ρ and c' are respectively, the mass density and specific heat capacity of brain matter or biological materials in general. Note that $\rho c' = K/k$.

For biological materials, the stress-wave development times are short compared with temperature equilibrium times. The temperature decay is therefore a slowly varying function of time and becomes significant only for times greater than milliseconds (ms). We may thus assume for a rectangular pulse of microwave energy with t_0 = pulse width that, immediately after microwave power is removed at $t = t_0$, the temperature remains as

$$T(r,t) = I_0 \left(\frac{1}{\rho c'}\right) \frac{\sin\left(\dfrac{N\pi r}{a}\right)}{\left(\dfrac{N\pi r}{a}\right)} t_0, \tag{8.6}$$

8.1.3 Thermoelastic Equation of Motion

Considering the spherical head with homogeneous brain matter as an isotropic, linear, elastic medium without viscous damping, i.e., lossless, and taking advantage of the spherical symmetry, the thermoelastic equation of motion may be expressed in spherical coordinates when $\lambda >> \mu$, as follows [Love, 1927; Sokolnikoff, 1956]:

$$\frac{\partial^2 u(r,t)}{\partial r^2} + \frac{2}{r}\frac{\partial u(r,t)}{\partial r} - \frac{2}{r^2}u(r,t) - \frac{1}{c^2}\frac{\partial^2 u(r,t)}{\partial t^2} = \frac{3\lambda + 2\mu}{\lambda + 2\mu}\beta \frac{\partial T(r,t)}{\partial r}, \tag{8.7}$$

where $u(r,t)$ is the radial displacement of brain matter, β is the coefficient of linear thermal expansion, λ and μ are Lame's elasticity constants, and $c^2 = [(\lambda + 2\mu)/\rho]$ is the square of bulk velocity c of pressure wave propagation in the brain medium. Note that the mathematical curl of u equals zero since u is in the radial direction only. The right-hand side of Eq. (8.7) is the sudden change in temperature which gives rise to the displacement. It takes on different values according to Eqs. (8.5) and (8.6), so that

$$\frac{3\lambda + 2\mu}{\lambda + 2\mu}\beta \frac{\partial T(r,t)}{\partial r} = I_0 \left(\frac{1}{\rho c'}\right)\frac{3\lambda + 2\mu}{\lambda + 2\mu}\beta \frac{d}{dr}\left[\frac{\sin\left(\dfrac{N\pi r}{a}\right)}{\left(\dfrac{N\pi r}{a}\right)}\right] F_t(t), \tag{8.8}$$

where the time function,

$$F_t(t) = t \quad \text{for} \quad 0 < t < t_0,$$

or

$$F_t(t) = t_0, \quad \text{for} \quad t > t_0. \tag{8.9}$$

The initial conditions are

$$u(r,0) = \frac{\partial}{\partial t}u(r,0) = 0, \tag{8.10}$$

For a stress-free spherical surface, the boundary condition at $r = a$ is given by,

$$(\lambda + 2\mu)\frac{\partial u(a,t)}{\partial r} + 2(\lambda)\frac{u(a,t)}{r} - (3\lambda + 2\mu)\beta T(a,t) = 0. \qquad (8.11)$$

If the surface of the sphere is rigidly constrained, the boundary condition at the surface requires the displacement,

$$u(a,t) = 0. \qquad (8.12)$$

The above derivations can be applied to obtain solutions for the time function, $F_t(t)$.

8.2   Pressure Wave in Stress-Free Brain Spheres

In the following sections, the derivations above will be applied to obtain a solution for the case of $F_t(t) = 1$, corresponding to a constant input, and extending the solution to a rectangular pulse using Duhamel's principle [Lin, 1976a, b; 1977a, b; 1978].

8.2.1   Thermoelastic Pressure for $F_t(t) = 1$

Given the linearity of the thermoelastic equation of motion of Eq. (8.7), the displacement $u(r,t)$ may be written in the form of a linear combination [Lin, 1977a],

$$u(r,t) = u_s(r,t) + u_t(r,t). \qquad (8.13)$$

Substituting Eq. (8.13) into Eq. (8.7) and for $F_t(t) = 1$, the equation of motion becomes two differential equations: a stationary one and a time-varying one. They are

$$\frac{d^2 u_s(r)}{dr^2} + \frac{2}{r}\frac{du_s(r)}{dr} - \frac{2}{r^2}u_s(r) = I_0\left(\frac{1}{\rho c'}\right)\frac{3\lambda + 2\mu}{\lambda + 2\mu}\beta\frac{d}{dr}\left[\frac{\sin\left(\frac{N\pi r}{a}\right)}{\left(\frac{N\pi r}{a}\right)}\right], \qquad (8.14)$$

and

$$\frac{\partial^2 u_t(r,t)}{\partial r^2} + \frac{2}{r}\frac{\partial u_t(r,t)}{\partial r} - \frac{2}{r^2}u_t(r,t) = \frac{1}{c^2}\frac{\partial^2 u_t(r,t)}{\partial t^2}, \qquad (8.15)$$

The corresponding boundary conditions at $r = a$ for a stress-free surface are

$$(\lambda + 2\mu)\frac{du_s}{dr} + 2\lambda\frac{u_s}{r} = 0, \tag{8.16}$$

and

$$(\lambda + 2\mu)\frac{\partial u_t}{\partial r} + 2\lambda\frac{u_t}{r} = 0, \tag{8.17}$$

A solution to the second-order ordinary differential equation (8.14) may be obtained by writing

$$u_s(r) = u_p(r) + Dr. \tag{8.18}$$

where u_p is a particular solution of (8.14) and is obtained by integrating (8.14) from 0 to r. Thus,

$$u_p(r) = u_0\left(\frac{a}{N\pi}\right)j_1\left(\frac{N\pi r}{a}\right), \tag{8.19a}$$

where the function $j_1\left(\dfrac{N\pi r}{a}\right)$ is the first-order spherical Bessel function of the first kind. The coefficient D of (8.18) is evaluated by applying the boundary condition given in (8.16) and it is

$$D = \pm u_0\left(\frac{1}{N\pi}\right)^2\frac{4\mu}{3\lambda + 2\mu}, \quad \text{for} \quad N = \begin{cases} 1,3,5,\dots \\ 2,4,6,\dots, \end{cases} \tag{8.19b}$$

The solution to (8.14) is therefore given by

$$u_s(r) = u_0\left(\frac{1}{N\pi}\right)\left[aj_1\left(\frac{N\pi r}{a}\right) \pm \left(\frac{r}{N\pi}\right)\frac{4\mu}{3\lambda + 2\mu}\right] \quad \text{for} \quad N = \begin{cases} 1,3,5,\dots \\ 2,4,6,\dots, \end{cases} \tag{8.20}$$

Now let

$$u_t(r,t) = R(r)T(t) \tag{8.21}$$

and use the method of separation of variables to solve Eq. (8.15) for the time-varying component. Inserting Eq. (8.21) into Eq. (8.17) yields

$$\frac{d^2R(r)}{dr^2} + \frac{2}{r}\frac{dR(r)}{dr} + \left[k^2 - \frac{2}{r^2}\right]R(r) = 0 \tag{8.22}$$

and

$$\frac{d^2T(t)}{dt^2} + k^2c^2T(t) = 0 \tag{8.23}$$

where k is the yet undetermined constant of separation, Eq. (8.22) is the Bessel's equation [Stratton, 1941] and its solution is given by

$$R(r) = B_1 j_1(kr) + B_2 y_1(kr) \tag{8.24}$$

where $j_1(kr)$ and $y_1(kr)$ are the first-order spherical Bessel functions of the first and second kind, respectively. Since $R(r)$ is finite at $r = 0$, B_2 must be zero. Combining Eq. (8.24) and the boundary condition of Eq. (8.17), we obtain a transcendental equation for k, the constant of separation,

$$\tan(ka) = \frac{ka}{\left[1 - \dfrac{(\lambda + 2\mu)(ka)^2}{4\mu} \right]} \tag{8.25}$$

The solution of this equation is an infinite sequence of eigenvalues, k_m; each corresponds to a characteristic mode of vibration inside the spherical head. Using the values for brain matters, it can be shown that $k_m a = m\pi$, $m = 1,2,3 \ldots$ to within an accuracy of 10^{-7}. Moreover, since Eq. (8.23) is harmonic in time, a general solution for $u_t(r,t)$ takes the form

$$u_t(r,t) = \sum_{m=1}^{\infty} A_m j_1(k_m r)\cos\omega_m t \tag{8.26}$$

where

$$\omega_m = k_m c = \frac{m\pi c}{a} \qquad \text{or} \qquad f_m = \frac{mc}{2a} \tag{8.27}$$

and $\omega_m = 2\pi f_m$ is the angular or radian frequency of vibration (or acoustic pressure wave) inside the head sphere for $m = 1, 2, 3 \ldots$. Note that the frequency of vibration $f_m = \omega_m/(2\pi) = mc/(2a)$ is independent of the patterns of absorbed microwave energy distribution. It is only a function of the head size and speed of pressure wave propagation, which depends on the elastic properties of the brain matter with negligible shear stress. This result indicates that the frequency of sound perceived by a subject irradiated by rectangular pulses of microwave energy is the same regardless of the frequency of the impinging radiation. As an example, a graph of calculated fundamental frequency of acoustic pressure wave for $m = 1$ as a function of head radius is given in Fig. 8.2. The frequency exceeds 80 kHz for a radius of 1 cm and decreases rapidly to a value of 25 kHz at $a = 3$ cm [Lin, 1977a, c]. For larger head sizes, between 7 and 10 cm, it gradually decreases to about 7.3–10.4 kHz.

The constants A_m are determined by using the initial condition given in Eq. (8.10). Thus,

$$A_m = -u_0 \frac{\left\{ \left(\dfrac{a}{N\pi}\right) \int_0^a r^2 j_1\left(\dfrac{N\pi r}{a}\right) j_1(k_m r)\,dr \pm \left(\dfrac{1}{N\pi}\right)^2 \dfrac{4\mu}{3\lambda + 2\mu} \int_0^a r^3 j_1(k_m r)\,dr \right\}}{\int_0^a r^2 \left[j_1(k_m r) \right]^2 dr},$$

$$\text{for} \quad N = \begin{cases} 1,3,5,\ldots \\ 2,4,6,\ldots, \end{cases} \tag{8.28}$$

Fig. 8.2 Calculated
fundamental frequency of
acoustic pressure wave in a
brain equivalent brain
sphere with stress-free and
constrained surfaces as a
function of head radius

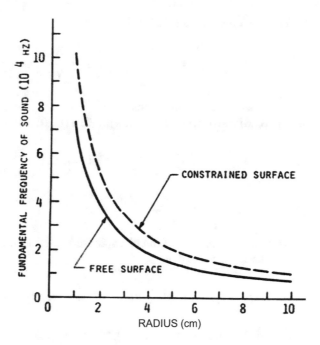

The integrals in the above equation may be evaluated with the help of Jahnke and
Emde [1945] to give

$$\int_0^a r^2 j_1\left(\frac{N\pi r}{a}\right) j_1(k_m r)\, dr = \pm\left(\frac{ka}{N\pi}\right) j_0(k_m a)\frac{a^3}{(N\pi)^2 - (k_m a)^2},$$

$$\text{for}\quad N = \begin{cases} 1,3,5,\dots \\ 2,4,6,\dots, \end{cases} \tag{8.29}$$

$$\int_0^a r^3 j_1(k_m r)\, dr = \frac{a^3}{k_m} j_2(k_m a) \tag{8.30}$$

$$\int_0^a r^2 \left[j_1(k_m r)\right]^2 dr = \frac{a^3}{2}\left\{ \left[j_1(k_m a)\right]^2 - j_0(k_m a) j_2(k_m a)\right\} \tag{8.31}$$

where $j_0(k_m a)$ and $j_2(k_m a)$ are the zeroth- and second-order spherical Bessel function
of the first kind. Using these relations, Eq. (8.28) becomes

$$A_m = \pm u_0 a \left(\frac{1}{N\pi}\right)^2 \left\{\frac{2}{j_0(k_m a) j_2(k_m a) - \left[j_1(k_m a)\right]^2}\right\}$$

$$\left\{\frac{4\mu}{3\lambda + 2\mu}\frac{1}{k_m a} j_2(k_m a) + k_m a\, j_0(k_m a)\frac{1}{(N\pi)^2 - (k_m a)^2}\right\}$$

$$\text{for}\quad N = \begin{cases} 1,3,5,\dots \\ 2,4,6,\dots \end{cases} \tag{8.32}$$

For $k_m a = m\pi = N$, Eq. (8.32) simplifies to

$$A_m = -u_0 a \left(\frac{1}{N\pi} \right) \left[1 + \left(\frac{1}{N\pi} \right)^2 \frac{24\mu}{3\lambda + 2\mu} \right] \qquad (8.33)$$

The displacement in response to $F_t(t) = 1$ is now obtained by introducing Eq. (8.32) in Eq. (8.26) and then combining it with Eqs. (8.20) and (8.13), and can be written as

$$u(r,t) = u_0 \left[\left(\frac{a}{N\pi} \right) j_1 \left(\frac{N\pi r}{a} \right) \pm \left(\frac{1}{N\pi} \right)^2 \left(\frac{4\mu}{3\lambda + 2\mu} \right) r \right]$$

$$+ \sum_{m=1}^{\infty} A_m j_1 (k_m r) \cos \omega_m t, \quad \text{for} \quad N = \begin{cases} 1,3,5,\dots \\ 2,4,6,\dots \end{cases} \qquad (8.34)$$

The radial stress (pressure) can be expressed in terms of displacement [Love, 1927; Sokolnikoff, 1956], so that

$$p_r(r,t) = (\lambda + 2\mu) \frac{\partial}{\partial r} u(r,t) + 2\lambda \left(\frac{1}{r} \right) u(r,t) - (3\lambda + 2\mu) \beta T(r,t) \qquad (8.35)$$

Therefore, by substituting Eqs. (8.5), (8.6), and (8.34) into the above expression, the thermoelastic pressure for $F_t(t) = 1$ is found as

$$p_r(r,t) = (4\mu) u_0 \left[\pm \left(\frac{1}{N\pi} \right)^2 - \frac{j_1 \left(\frac{N\pi r}{a} \right)}{\frac{N\pi r}{a}} \right]$$

$$+ \sum_{m=1}^{\infty} A_m k_m [(\lambda + 2\mu) j_0 (k_m r) - 2\mu \frac{j_1 \left(\frac{N\pi r}{a} \right)}{\frac{N\pi r}{a}}] \cos \omega_m t,$$

$$\text{for} \quad N = \begin{cases} 1,3,5,\dots \\ 2,4,6,\dots \end{cases} \qquad (8.36)$$

Equations (8.34) and (8.36) can be used to obtain solutions of displacement and pressure in response to any time-varying input. The process is illustrated in the following section for rectangular pulses.

8.2.2 Rectangular Pulse-Induced Pressure in Stress-Free Sphere

The displacement and pressure in stress-free spherical head models from exposure to a rectangular pulse of microwave radiation can be obtained by applying Duhamel's principle [Churchill, 1958), that is

$$u(r,t) = \frac{\partial}{\partial t} \int_0^t u'(r,t') F_t(t-t') dt',$$ (8.37)

where the displacement $u'(r,t)$ is the solution given by Eq. (8.34) for the case of $F_t(t) = 1$. An equivalent expression can be written for the pressure using Eq. (8.36). Therefore, by substituting Eqs. (8.9) and (8.34) into Eq. (8.37), the displacement for $0 < t < t_0$ is given by

$$u(r,t) = u_0 \left[\left(\frac{a}{N\pi} \right) j_1 \left(\frac{N\pi r}{a} \right) \pm \left(\frac{1}{N\pi} \right)^2 \left(\frac{4\mu}{3\lambda + 2\mu} \right) r \right] t$$

$$+ \sum_{m=1}^{\infty} A_m j_1(k_m r) \frac{\sin \omega_m t}{\omega_m},$$

$$\text{for} \quad N = \begin{cases} 1,3,5,\dots \\ 2,4,6,\dots \end{cases}$$ (8.38)

and for $t > t_0$,

$$u(r,t) = u_0 \left[\left(\frac{a}{N\pi} \right) j_1 \left(\frac{N\pi r}{a} \right) \pm \left(\frac{1}{N\pi} \right)^2 \left(\frac{4\mu}{3\lambda + 2\mu} \right) r \right] t_0$$

$$+ \sum_{m=1}^{\infty} A_m j_1(k_m r) \left[\frac{\sin \omega_m t}{\omega_m} - \frac{\sin \omega_m (t-t_0)}{\omega_m} \right],$$

$$\text{for} \quad N = \begin{cases} 1,3,5,\dots \\ 2,4,6,\dots \end{cases}$$ (8.39)

Similarly, the pressure for $0 < t < t_0$,

$$p_r(r,t) = (4\mu) u_0 \left[\pm \left(\frac{1}{N\pi} \right)^2 - \frac{j_1 \left(\frac{N\pi r}{a} \right)}{\frac{N\pi r}{a}} \right] t$$

$$+ \sum_{m=1}^{\infty} A_m k_m [(\lambda + 2\mu) j_0(k_m r)$$

$$- 2\mu \frac{j_1 \left(\frac{N\pi r}{a} \right)}{\frac{N\pi r}{a}}] \frac{\sin \omega_m t}{\omega_m},$$

$$\text{for} \quad N = \begin{cases} 1,3,5,\dots \\ 2,4,6,\dots \end{cases}$$ (8.40)

and for $t > t_0$,

$$p_r(r,t)=(4\mu)u_0\left[\pm\left(\frac{1}{N\pi}\right)^2-\frac{j_1\left(\dfrac{N\pi r}{a}\right)}{\dfrac{N\pi r}{a}}\right]t_0$$

$$+\sum_{m=1}^{\infty}A_mk_m[(\lambda+2\mu)j_0{}^{(k_mr)}$$

$$-2\mu\frac{j_1\left(\dfrac{N\pi r}{a}\right)}{\dfrac{N\pi r}{a}}]\left[\frac{\sin\omega_mt}{\omega_m}-\frac{\sin\omega_m(t-t_0)}{\omega_m}\right],$$

$$\text{for}\quad N=\begin{cases}1,3,5,\ldots\\2,4,6,\ldots\end{cases}$$

(8.41)

where k_m and A_m are as given in Eqs. (8.27) and (8.32), respectively. Equations (8.38) to (8.41) represent the general solutions for the displacement and pressure in a spherical head exposed to rectangular-pulse-modulated microwave radiation as functions of the microwave, thermoelastic, and geometric parameters of the model in the absence of shear and surface stresses.

Since u_0 and A_m are directly proportional to I_0, both the displacement and the pressure are proportional to the peak absorption. It is easily seen that the displacement and radial pressure also depend linearly on the peak incident power density. At the center of the sphere, $r = 0$, both Eqs. (8.38) and (8.39) reduce to zero and there is no displacement at the center of the model. On the other hand, at the surface of the brain sphere ($r = a$), the first terms of Eqs. (8.36, 8.40, and 8.41) all reduce to zero. The radial pressure is given by the summation of the harmonic time functions alone, indicating an oscillatory pressure response or reverberation phenomenon inside the brain sphere. Examples of numerically calculated displacements and pressures are provided in a later section.

8.3   Pressure in Brain Spheres with Constrained Surface

If the surface of the head sphere is rigidly constrained, the boundary condition at the surface requires that the displacement,

$$u(a,t)=0.$$

(8.42)

The corresponding solutions for pressure and displacement are outlined below.

8.3.1 Induced Thermoelastic Pressure for $F_t(t) = 1$

Following the procedures set forth in the previous section, a solution for $F_I(t) = 1$ is first found by expressing the displacement $u(r, t)$ as a summation of two terms, such that from Eq. (8.13),

$$u(r,t) = u_s(r,t) + u_t(r,t)$$ (8.43)

where $u_s(r)$ is found by using the boundary condition $u_s(a) = 0$ to be

$$u_s(r) = u_0 \left(\frac{1}{N\pi} \right) \left[aj_1 \left(\frac{N\pi r}{a} \right) \pm \left(-\frac{r}{N\pi} \right) \right], \quad \text{for} \quad N = \begin{cases} 1,3,5,\ldots \\ 2,4,6,\ldots, \end{cases}$$ (8.44)

and the time-varying component is given by

$$u_t(r,t) = \sum_{m=1}^{\infty} A_m j_1 (k_m r) \cos \omega_m t$$ (8.45)

where $k_m a$ is the positive zeroes of the first-order spherical Bessel function of the first kind,

$$j_1(ka) = 0$$ (8.46)

and

$$\omega_m = k_m c$$ (8.47)

where c is the bulk speed of pressure wave propagation in the brain medium.

The solution of Eq. (8.46) is an infinite sequence of eigenvalues, k_m. Since $\omega_m = 2\pi f_m$, thus $f_m = \frac{\omega_m}{2\pi}$, where f_m represents the frequency of acoustic vibration inside the spherical head model. Therefore, there are an infinite number of modes of acoustic vibration of the spherical head exposed to pulse-modulated microwave radiation. The values of k_m a for $m = 1\text{–}10$ are listed in Table 8.1. The frequency of sound generated inside a spherical head with constrained boundary without shear stress is then given by

$$f_m = \frac{\omega_m}{2\pi} = \frac{(k_m a)c}{2\pi a}$$ (8.48)

The acoustic vibration frequency under constrained surface condition is significantly higher than that given in the stress-free case. As before, the frequency of the acoustic pressure wave generated is a function of the elastic property (speed of propagation) and size of the sphere. An example of calculated fundamental frequency of acoustic pressure wave in a constrained spherical head model for $m = 1$ is

Table 8.1 Zeroes of the first-order spherical Bessel function of the first kind, $j_1(ka)$

m	$k_m a$
1	4.493411
2	7.725252
3	10.904122
4	14.066194
5	17.220755
6	20.371303
7	23.519453
8	26.666054
9	29.811598
10	32.956389

plotted in Fig. 8.2, as a function of head radius where the frequencies predicted by the stress-free formulation are shown for comparison. The constrained head model predicts a fundamental frequency of about 50% higher than those calculated based on stress-free boundary conditions.

The coefficient A_m in Eq. (8.45) is determined by inserting Eqs. (8.54) and (8.55) into Eq. (8.43) and making use of the initial condition of Eq. (8.10). Because the function $j_1(k_m r)$ is orthogonal, after integration over r from 0 to a,

$$A_m = \pm u_0 a \left(\frac{1}{N\pi}\right)^2 \left\{ \frac{2}{\left[j_1(k_m a)\right]^2 - j_0(k_m a) j_2(k_m a)} \right\}$$

$$\left\{ \frac{1}{k_m a} j_2(k_m a) \pm k_m a \, j_0(k_m a) \frac{1}{(k_m a)^2 - (N\pi)^2} \right\},$$

$$\text{for} \quad N = \begin{cases} 1,3,5,\ldots \\ 2,4,6,\ldots \end{cases} \tag{8.49}$$

The displacement inside the spherical model is therefore given by

$$u(r,t) = u_0 \left(\frac{1}{N\pi}\right)\left[aj_1\left(\frac{N\pi r}{a}\right) \mp \left(\frac{r}{N\pi}\right)\right]$$

$$+ \sum_{m=1}^{\infty} A_m j_1(k_m r)\cos \omega_m t,$$

$$\text{for} \quad N = \begin{cases} 1,3,5,\ldots \\ 2,4,6,\ldots \end{cases} \tag{8.50}$$

The pressure p_r (r, t) for $F_t(t) = 1$ is obtained as

$$p_r(r,t) = u_0 \left[\left[(-4\mu) \frac{j_1\left(\frac{N\pi r}{a}\right)}{\frac{N\pi r}{a}} \mp (3\lambda + 2\mu) \left(\frac{1}{N\pi}\right)^2 \right] \right.$$

$$\left. + \sum_{m=1}^{\infty} A_m k_m [(\lambda + 2\mu) j_0(k_m r) - \frac{4\mu}{k_m r} j_1(k_m r)] \cos \omega_m t, \right.$$

$$\text{for} \quad N = \begin{cases} 1,3,5,\dots \\ 2,4,6,\dots \end{cases}$$

$$(8.51)$$

8.3.2   Rectangular Pulse-Induced Pressure for Constrain Surfaces

The solutions expressed by Eqs. (8.50) and (8.51) are applied to the case of a rectangular pulse using Duhamel's principle (Eq. 8.37). Therefore, by letting $u'(r,t)$ be equal to Eq. (8.50), the displacement in the spherical head with a constrained surface in response to a rectangular pulse of microwave energy with pulse width t_0, in the absence of shear stress for $0 < t < t_0$, is given by

$$u(r,t) = u_0 \left(\frac{1}{N\pi}\right) \left[aj_1\left(\frac{N\pi r}{a}\right) \mp \left(\frac{r}{N\pi}\right) \right] ts$$

$$+ \sum_{m=1}^{\infty} A_m j_1(k_m r) \frac{\sin \omega_m t}{\omega_m},$$

$$\text{for} \quad N = \begin{cases} 1,3,5,\dots \\ 2,4,6,\dots \end{cases}$$

$$(8.52)$$

and for $t > t_0$

$$u(r,t) = u_0 \left(\frac{1}{N\pi}\right) \left[aj_1\left(\frac{N\pi r}{a}\right) \mp \left(\frac{r}{N\pi}\right) \right] t_0$$

$$+ \sum_{m=1}^{\infty} A_m j_1(k_m r) \left[\frac{\sin \omega_m t}{\omega_m} - \frac{\sin \omega_m (t - t_0)}{\omega_m} \right],$$

$$\text{for} \quad N = \begin{cases} 1,3,5,\dots \\ 2,4,6,\dots \end{cases}$$

$$(8.53)$$

Similarly, the thermoelastic pressure for $0 < t < t_0$

$$p_r(r,t) = u_0 \left[\left[(-4\mu) \frac{j_1\left(\frac{N\pi r}{a}\right)}{\frac{N\pi r}{a}} \right] \mp (3\lambda + 2\mu)\left(\frac{1}{N\pi}\right)^2 \right] t$$

$$+ \sum_{m=1}^{\infty} A_m k_m [(\lambda + 2\mu) j_0(k_m r)$$

$$- 2\mu \frac{j_1\left(\frac{N\pi r}{a}\right)}{\frac{N\pi r}{a}}] \frac{\sin \omega_m t}{\omega_m},$$

$$\text{for} \quad N = \begin{cases} 1,3,5,\dots \\ 2,4,6,\dots \end{cases} \tag{8.54}$$

and for $t > t_0$

$$p_r(r,t) = u_0 \left[\left[(-4\mu) \frac{j_1\left(\frac{N\pi r}{a}\right)}{\frac{N\pi r}{a}} \right] \mp (3\lambda + 2\mu)\left(\frac{1}{N\pi}\right)^2 \right] t_0$$

$$+ \sum_{m=1}^{\infty} A_m k_m [(\lambda + 2\mu) j_0(k_m r)$$

$$- 2\mu \frac{j_1\left(\frac{N\pi r}{a}\right)}{\frac{N\pi r}{a}}] \left[\frac{\sin \omega_m t}{\omega_m} - \frac{\sin \omega_m (t - t_0)}{\omega_m} \right],$$

$$\text{for} \quad N = \begin{cases} 1,3,5,\dots \\ 2,4,6,\dots \end{cases} \tag{8.55}$$

It is seen that the displacement becomes zero both at the center and at the surface of the spherical head. Because u_0 and A_m are directly related to I_0, both the displacement and pressure are proportional to peak absorption, as in the stress-free boundary case. The dependence of sound pressure on pulse width is more complex in character. However, the temporal variation of induced pressure is governed in general by the summation of harmonic time functions, suggesting an oscillatory pressure response or reverberation phenomenon inside the head like that for brain spheres with stress-free surface boundaries.

8.4 Other Absorption Patterns and Pulse Waveforms

Following the analytical approach and mathematical technique introduced by Lin [1976a, b; 1977a, b; 1978], extensions and generalizations have been published by others. A special case of the problem, as suggested, was discussed in detail by Shibata et al. [1986]. In this case, the spherically symmetric microwave absorption (SAR) pattern which peaked at the sphere center like that given by Eq. (8.1b) was employed, in which $N = 4$ and a modified specific absorption pattern with $N = 4$ and a 60 and 40% split between I_0 and I_c. Note that depending on the I_0 and I_c split, the introduction of a uniform offset term may affect the specific details of the waveform of the pressure waves.

Furthermore, the acoustic pressure wave generation inside a lossy dielectric sphere from an incident microwave pulse was analyzed using a Green's function theoretic approach [Uzunoglu and Polychronopoulos, 1988]. The SAR distribution was computed by applying the exact Mie theory solution for the dielectric sphere [Stratton, 1941]. In this case, the result also showed a transient burst pressure wave, plus an infinite set of damped oscillations related to the normal acoustic modes of the spherical resonator.

The analytical approach and mathematical technique introduced earlier have been generalized and extended to an arbitrary spherically symmetric and centrally peaked microwave absorption pattern using a Taylor series expansion [Yitzhak, et al., 2009]. For $f(t)$ representing the specific pulse shape, the microwave absorption or SAR pattern can be written as,

$$I(r,t) = f(t) \sum_{n=0}^{\infty} I_n \left(\frac{r}{a}\right)^n \tag{8.56}$$

where $I(r,t)$ is the SAR at radial distance r, from the sphere center at time t, and I_n is the nth Taylor series expansion coefficient. In addition to the rectangular pulse, the cases considered include sawtooth pulse defined by $f(t) = \beta t$ for $0 < t < t_0$, and zero otherwise, and half-sine pulse defined by $f(t) = \sin\left(\frac{\pi}{t_0} t\right)$, for $0 < t < t_0$, and zero otherwise. The dependence of induced acoustic pressures on the shape of the microwave pulse was explored in this paper for a rectangular pulse, half-sine, and sawtooth waveforms. The results showed that the generalized solution reduces to the published results [Lin, 1976a, b; 1977a, b; 1978] for the centrally peaked absorption patterns associated with rectangular pulses.

Also, the generalized model was applied to the case in which the microwave absorption or SAR pattern is concentrated near the sphere surface, in this case, it is approximated by a single-term spherically symmetric function,

$$I(r,t) = f(t) I_6 \left(\frac{r}{a}\right)^6 \tag{8.57}$$

Note that in this case, SAR is maximum at the sphere's surface but zero at the center. It was found that, for equal average SAR or equal whole-body SAR, the acoustic or sound pressure amplitude generated by an SAR pattern that concentrates absorption more toward the sphere's surface is comparable to that generated by SAR patterns that are peaked at the center [Yitzhak, et al., 2009]. The generalized mathematical extensions have also been applied to examine the effects of repeated pulses and varying pulse repetition rate.

8.5   Calculated and Measured Frequencies for Spherical Head Models

Some detailed results calculated from these mathematical developments including those of the generalized solutions are presented in this section. The results are compared with available experimental data where appropriate. Since the frequencies of sound pressure wave generated in a spherical head with stress-free and constrained boundary are given by two simple Eqs. (8.27 and 8.48), the discussion will begin with calculated frequencies for pressure waves inside the head model.

8.5.1   Theoretically Predicted Frequency of Acoustic Pressure Waves

The thermoelastic theory predicted that the frequencies of microwave pulse-induced sound pressure waves are only functions of the size and elastic properties of tissue for both the constrained and stress-free spherical head models. As noted above the frequency is also independent of microwave absorption patterns. This observation reveals an important finding that the frequency of sound perceived by a subject whose head is exposed to pulses of microwave energy will be the same regardless of the frequency of the impinging radiation.

The predicted fundamental sound frequencies in the mammalian heads are plotted as a function of equivalent spherical head size in Fig. 8.2. The specific values are calculated using Eqs. (8.27 and 8.48) and properties of the brain materials previously listed (e.g., $v = 1600$ m/s). It is seen that the sound frequency in various subjects differs according to their equivalent spherical head sizes, i.e., the smaller the head size, the higher the frequency. For example, the average head radius for guinea pigs is about 1.5–2.5 cm. Figure 8.2 yields a range of 40–70 kHz for the corresponding fundamental sound frequency. The average head radius for cats is approximately 2.5–3.5 cm; the corresponding fundamental sound frequency is between 30 and 40 kHz. Note that the frequencies predicted by the constrained-surface formulation are higher than those calculated based on stress-free boundary conditions. Since most mammalian heads are neither entirely stress-free nor rigidly constrained, thus the actual fundamental sound frequency may fall somewhere between those predicted under these two conditions.

The predicted fundamental and second harmonic frequencies are shown as functions of spherical head size in Fig. 8.3. As expected, the frequencies of acoustic vibration under a constrained surface condition are significantly higher than that of the stress-free surface case. Observe that the frequency of the higher order harmonics in different subjects differs according to their equivalent spherical radius of the head. Specifically, the frequency of the second harmonic ($m = 2$) is nearly twice that of the fundamental mode ($m = 1$). While the microwave auditory effect arises predominantly from the excitation of the fundamental mode, in cases like the human subjects with a fundamental frequency of 8 kHz, a second harmonic frequency near 16 kHz would be present and in principle audible although at a much lower

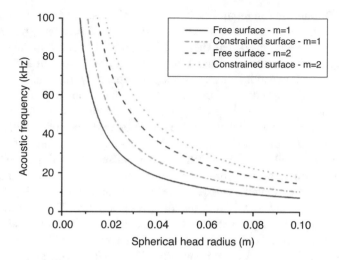

Fig. 8.3 Fundamental and second harmonic frequencies as functions of spherical head size

amplitude. The other higher order modes are typically above the high-frequency auditory limits for animals and humans with much lower amplitudes.

8.5.2   Measured Frequency in Spherical Head Models

Experimentally, a small hydrophone was used to measure microwave pulse-induced pressure waves in spherical head models filled with brain-tissue equivalent dielectric materials [Olsen and Lin, 1981]. The phantom head models were composed of hemispherical cavities machined in 20.3 × 20.3 × 7.6 cm blocks of foamed polystyrene plastics. The foamed polystyrene material provided a stress-free boundary for the brain model. The dielectric phantom material has electromagnetic, mechanical, and thermal properties akin to those of brain tissues. It consisted of finely granulated polystyrene powder, sodium chloride, water, and a gelling agent. It has an acoustic propagation speed of 1600 m/s at room temperature.

A spherical, barium-titanate piezoelectric hydrophone, 1 cm in diameter, was placed in the center of the model. Its output signal was displayed on an oscilloscope and photographed on film. The barium-titanate piezoelectric element (Edo Western, Model 6600) had a response of 50.1 pA/mV for the range of frequencies (5–40 kHz) encountered in this study. The model was exposed to microwave pulses at 1.1 GHz with 4 kW peak power obtained from a microwave generator (Epsco PG5KB) delivered through an open-ended waveguide (WR 650), placed in contact with the foam surface, which has the same dielectric property as air.

A series of three types of experiments were performed which began with the application of a single microwave pulse to elicit acoustic responses. The ringing in the response after microwave application corresponds to the fundamental mode frequency. During the single-pulse test, an optimal pulse width that produced the highest amplitude response was also determined. Brief pulse trains or pulses consisting of three microwave pulses were then applied to the spherical phantom model. The applied pulse repetition frequency equals the frequency of the fundamental mode. For each pulse train combination, the maximum post-artifact hydrophone output voltage was recorded and graphed as a function of pulse repetition frequency. Figure 8.4 shows the output voltage of the hydrophone as a function of pulse repetition frequency for a 6-cm diameter model. A pulse repetition frequency of 25.5 kHz gave the highest acoustic pressure amplitude, indicating a resonance frequency of 25.5 kHz.

In the 10-cm diameter sphere, a single 14-μs pulse produced a 16 kHz ringing signal with maximum acoustic pressure, which indicates a resonance frequency of 16 kHz. This is also revealed in the response "tuning curve" shown in Fig. 8.5, which peaked at 16 kHz. In the 14-cm diameter model, a single 35-μs pulse exposure yielded a ringing frequency slightly above 10 kHz with maximum acoustic pressure. Figure 8.6 gives the results of exposing 14-cm diameter model with various combinations of three-pulse bursts. A pulse frequency of 11.5 kHz provided the highest value in pressure amplitude, indicating a resonance pressure wave frequency of 11.5 kHz.

These results demonstrate that microwave pulses can indeed generate measurable acoustic pressures in spherical models of animal- and human-size heads. Furthermore, they indicate that appropriately selected pulse repetition frequencies can promote resonance that may elevate the microwave-induced acoustic pressure by several folds. For example, the acoustic pressure signal produced by a three-pulse burst was increased by threefold over the response to a single pulse. In general, the hydrophone response gradually increases from a low value to a peak amplitude at the resonance frequency and then falls off rapidly as pulse repetition frequency increases further.

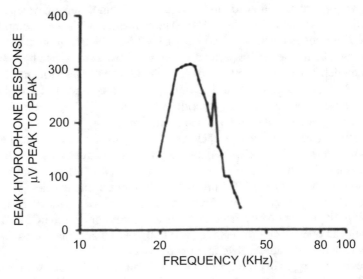

Fig. 8.4 Amplitude of hydrophone response in a 6-cm diameter, spherical brain model exposed to 10μs wide, 1100 MHz microwave pulses at different repetition frequencies

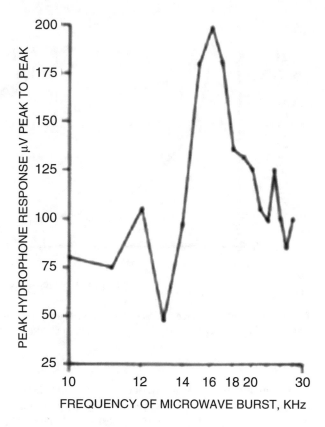

Fig. 8.5 Hydrophone response amplitude in a 10 cm diameter brain sphere exposed to 14μs wide, 1100 MHz microwave pulses at different tuning frequencies

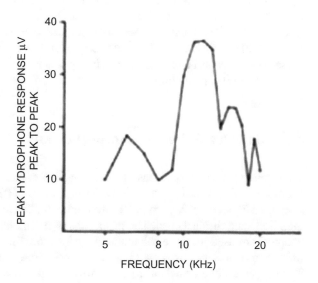

Fig. 8.6 Amplitude of hydrophone response in a 14 cm diameter, spherical head model exposed to 10μs wide, 1100 MHz microwave pulses at different repetition frequencies

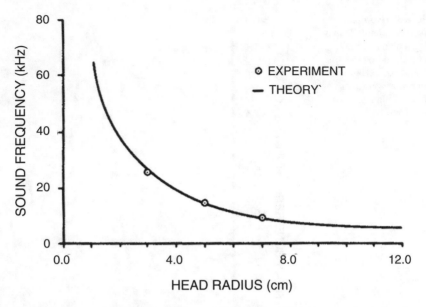

Fig. 8.7 Comparison of predicted and measured frequencies of microwave pulse-induced pressure waves in brain equivalent head models with stress-free boundary

The measured resonance frequencies of pressure waves in the spherical models compare favorably with those predicted by the thermoelastic theory based on homogeneous brain-tissue equivalent spheres with stress-free boundaries (Fig. 8.7). Specifically, measured resonance frequency for 6-, 10-, and 14-cm diameter brain spheres were 25.5, 16, and 11.5 kHz, respectively. The corresponding fundamental frequencies of sound pressure predicted from the thermoelastic theory are 26.6, 16, and 11.4 kHz, respectively. The measured results clearly indicate the microwave pulse-induced acoustic pressure waves in the spherical brain model, thus confirming earlier theoretical predictions.

8.6 Calculated Pressure Amplitude and Displacement

The explicit analytical solutions for the acoustic pressure waves generated in spherical models of the head exposed to rectangular pulses of RF and microwave radiation presented in Sects. (8.2.2 and 8.3.2) are applied to calculate induced pressure wave amplitude and displacement. The properties of brain material provided in previous chapters are used for these calculations. Aside from the calculated pressure amplitudes and displacements shown below for both stress-free and constrained surface brain spheres, additional data are available elsewhere [Lin, 1977a, b, c; 1978; 1990].

8.6.1   Stress-Free Boundary

Figures 8.8 and 8.9 show pressure waves induced in a 6-cm diameter brain sphere (cat size head) and a 14-cm human-sized spherical brain exposed to pulsed 2450 MHz and 915 MHz microwave radiation, respectively. The pulse width is 10μs and the peak SAR is set at 1 kW/kg. The calculations obtained at radial distances, $r = 0, 1.5,$ and 3.0 cm for the 6-cm diameter stress-free sphere (Fig. 8.8) show the pressures are the highest in the center of the spherical brain and zero at the surface, but initially goes negative as a function of time. After a transient buildup, the pressure oscillates at a constant level in the absence of elastic loss assumed for brain tissue. Similar results are shown in Fig. 8.9 for a 14-cm brain sphere under 915 (918) MHz plane wave exposure, except for slower temporal variations, associated with afore-mentioned inverse dependence of pressure frequency on size of brain sphere.

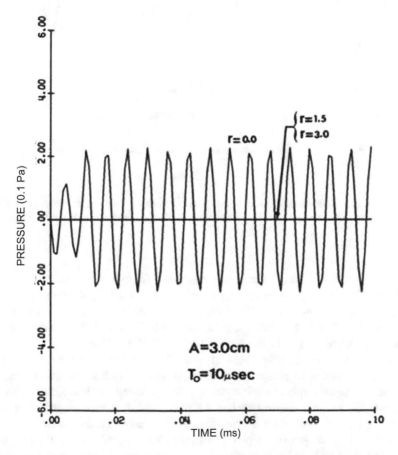

Fig. 8.8 Thermoelastic pressure (radial stress) generated in a 6-cm diameter spherical brain with stress-free boundary exposed to 2450 MHz plane wave. The peak absorption is 1 kW/kg in 10μs

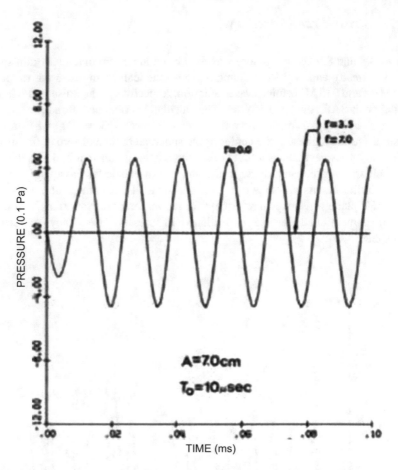

Fig. 8.9 Thermoelastic pressure (radial stress) generated in a 14-cm diameter spherical brain with stress-free boundary exposed to 915 MHz plane wave. The peak absorption is 1 W/kg in 10μs

   Table 8.2 provides the peak pressures calculated for guinea pig- and human-sized heads for 10μs rectangular pulses. For a peak SAR of 1 kW/kg the pressures generated at the center of the sphere are 70–90 dB above 0.02 mPa. At this SAR, the rate of temperature rise at the center of both spheres is 0.26 °C/s. The calculated temperature elevation in 10μs is about 2.6×10^{-6} °C from Eq. (8.6).

   The corresponding displacements are illustrated in Figs. 8.10 and 8.11. At the center of the sphere, displacement is zero. At more peripheral locations, the displacement increases, either positively or negatively, but almost linearly as a function of time until the time elapse matches the pulse width; it then begins to oscillate around a constant level. For the cases shown, the maximum displacements are on the order of 10^{-13} meters or 0.1 picometer (pm). The displacements stay sinusoidal after a transient buildup because of the lossless assumption for the elastic media.

Table 8.2 Calculations of peak pressure and displacement in stress-free spherical models at SAR of 1 kW/kg for 6-cm and 14-cm diameter brains

Pulse width	Incident power density	SAR	Pressure	dB
μs	kW/m²	kW/kg	mPa	Re: 0.02 mPa
6-cm diameter brain at 2450 MHz				
0.1	5.89	1	12	55.6
0.5	5.89	1	59	69.4
1.0	5.89	1	115	75.2
5.0	5.89	1	140	76.9
10.0	5.89	1	230	81.2
20.0	5.89	1	135	76.6
30.0	5.89	1	150	72.5
40.0	5.89	1	220	80.8
50.0	5.89	1	120	75.6
14-cm diameter brain at 915 (918) MHz				
0.1	21.83	1	12	55.5
0.5	21.83	1	60	69.5
1.0	21.83	1	119	75.5
5.0	21.83	1	490	87.8
10.0	21.83	1	470	87.4
20.0	21.83	1	510	88.1
30.0	21.83	1	280	82.9
40.0	21.83	1	410	86.2
50.0	21.83	1	540	88.6

8.6.2   Constrained Boundary

The calculated pressure in a 6-cm diameter spherical brain exposed to 2450 MHz radiation as a function of time for a 10μs pulse is shown in Fig. 8.12. The values are evaluated at $r = 0$, 1.5, and 3.0 cm. Like that for the stress-free boundaries, the pressure is the highest in the center of the spherical brain but can begin either with positive or negative values. After a transient buildup, which lasts for the duration of the pulse, the pressure wave oscillates at varying levels as a function of time, which is different from the behavior under stress-free conditions. It is also important to note that the high-frequency oscillation is modulated by a low-frequency envelope whose frequency is the same as the fundamental sound frequency of a 6-cm (dia.) spherical brain. The pressure generated at the center of the spherical head is 369 mPa for a peak SAR of 1 kW/kg. The pressure is considerably higher than that obtained from the surface stress-free model in comparison (230 mPa). The fundamental frequency of pressure variation agrees with the prediction in Fig. 8.2. The displacement in the cat-sized head is about 0.151 pm (Fig. 8.13). Similar results for pressure and displacement in a 14-cm brain sphere under 915 (918) MHz exposure are shown in Figs. 8.14 and 8.15.

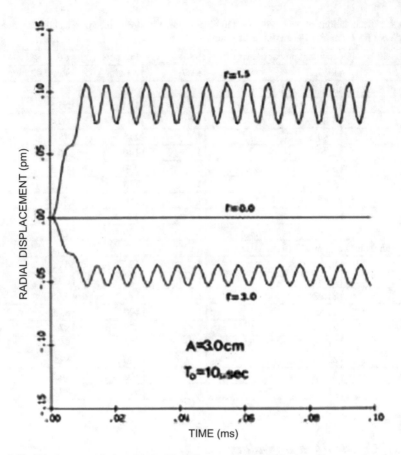

Fig. 8.10 Displacement induced in a 6-cm diameter spherical brain with stress-free boundary exposed to 2450 MHz plane wave. The peak absorption is 1 kW/kg in 10μs

Table 8.3 gives the calculated peak pressure and displacement in these spherical mammalian brain models exposed to 10-μs pulses at the same SAR level. Although SAR and pulse width are the same, the peak pressure and tissue displacement differ in this case. Note the microwave frequency and incident power density differ according to the species or size of spherical brain model involved as well.

8.6.3 Pressure Wave Dependence on Pulse Width and Pulse Shape

Results of pressure calculation for spherical brain models with stress-free boundary exposed to rectangular pulses from 0.1 to 100μs are illustrated in Figs. 8.16 and 8.17 for a 6-cm diameter sphere exposed to 2450 MHz and a 14-cm diameter sphere

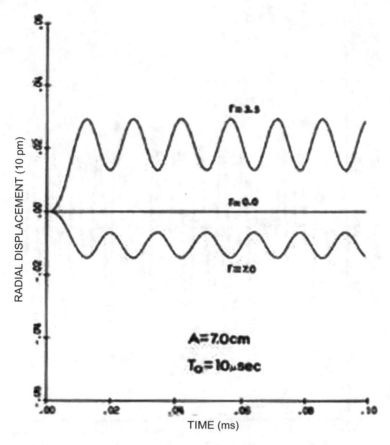

Fig. 8.11 Displacement induced in a 14-cm diameter spherical brain with stress-free boundary exposed to 915 MHz plane wave. The peak absorption is 1 kW/kg in 10μs

exposed to 918 MHz radiation, respectively. The peak SAR in the stress-free sphere is 1.0 kW/kg. In general, sound pressure first rises rapidly to a maximum and then oscillates around a constant value. The oscillatory behavior was a surprise when first predicted [Lin 1977a, b, c] and has since been verified in laboratory studies with animal and human subjects. Microwave-induced sound pressure amplitudes clearly depend on the pulse width of the impinging microwave radiation. Moreover, there is apparently a pulse width for more efficient sound pressure generation which varies according to the head size and frequency of the impinging microwave radiation, among other factors. For the cases displayed in Figs. 8.16 and 8.17, the induced pressure peaks around 3 and 6μs, respectively.

The calculated pressures for a 6-cm diameter (3-cm radius) constrained brain sphere exposed to 2450 MHz radiation and for a 14-cm diameter (7-cm radius) brain sphere exposed to 918 MHz radiation are shown in Figs. 8.18 and 8.19, respectively, as a function of pulse width [Lin, 1977b, c; 1978]. The curves were evaluated at an absorbed microwave energy rate of 1.0 kW/kg and showed a clear pulse width effect on the pressure wave. These figures are similar in quality and character to

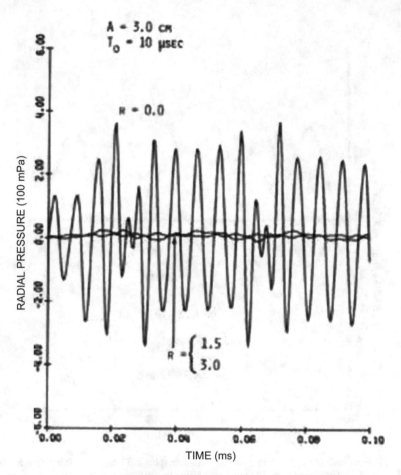

Fig. 8.12 Thermoelastic pressure (radial stress) generated in a 6-cm diameter spherical brain with constrained surface exposed to 2450 MHz plane wave. The peak absorption is 1 kW/kg in 10μs

results from the stress-free calculations, although the detailed dependence on pulse width differs somewhat. The microwave-induced pressure peaks around 2μs for the 6-cm case and around 5μs for the 14-cm case. Thus, these values may be considered as the rectangular pulse widths to provide for more efficient conversion in spherical head models with SAR patterns characterized by Eq. (8.1a), in which the absorption of microwave radiation peaks at the center of the spherical brain.

Note that the prediction of a longer pulse width could lead to an increase in thermoelastic pressure response was confirmed independently in a recent experiment [Nan and Arbabian, 2017]. The relationship between the magnitude of the pressure measured with 0.5- and 1-MHz transducers, whose bandwidths are between 50% and 60% in response to short microwave pulses with a carrier frequency of 2.1 GHz

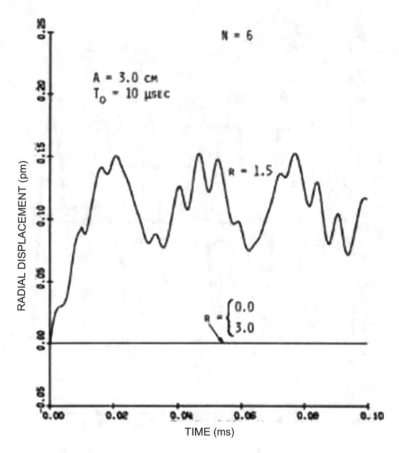

Fig. 8.13 Displacement induced in a 6-cm diameter spherical brain with constrained surface exposed to 2450 MHz plane wave. The peak absorption is 1 kW/kg in 10μs

under constant peak power, is shown in Fig. 8.20. The measured pressure first increases with the applied pulse width and then appears to reach a plateau. The bending points in Fig. 8.20a, b are approximately 3.5 and 2.5μs, respectively, which are also functions of the transducer's bandpass filters.

As mentioned, a generalized solution was applied to obtain acoustic pressure generated by a surface concentrated absorption pattern of Eq. (8.57), in which microwave absorption occurs mainly near the surface of the sphere. The same was done for a spherically symmetric but centrally concentrated absorption pattern of Eq. (8.1b). An average SAR of 0.4 W/kg was chosen for both cases to normalize calculated results. A comparison was performed between the results obtained with Eq. (8.1b) for a sphere of 14-cm diameter at 918 MHz and those obtained with Eq. (8.57) for a sphere of 20-cm diameter at 2450 MHz [Yitzhak, et al., 2009]. The

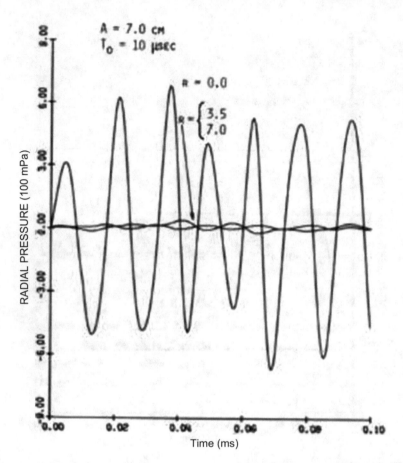

Fig. 8.14 Thermoelastic pressure (radial stress) generated in a 14-cm diameter spherical brain with constrained surface exposed to 915 MHz plane wave. The peak absorption is 1 kW/kg in 10μs

dependence of the acoustic pressure amplitude of the fundamental mode of sonic vibration at the center of the sphere on pulse width is given in Fig. 8.21. The results show that for equal average SAR, the pressure generated by a surface localized SAR pattern (2450 MHz and Eq. 8.57) is comparable to that generated by an SAR pattern that is peaked at the sphere center (918 MHz and Eq. 8.1b). The behavior of the responses as a function of pulse width resembles those of Figs. 8.16, 8.17, 8.18 and 8.19, calculated with the SAR model of Eq. (8.1a). In each of these cases, the sound pressure first rises to a maximum and then oscillates to some constant value, except for the offset in average amplitude. Note that according to the SAR model of Eq. (8.1a), the first pressure peak occurs at a pulse width of about 6μs. However, the results in Fig. 8.21 suggest a pulse width that maximizes the pressure varying

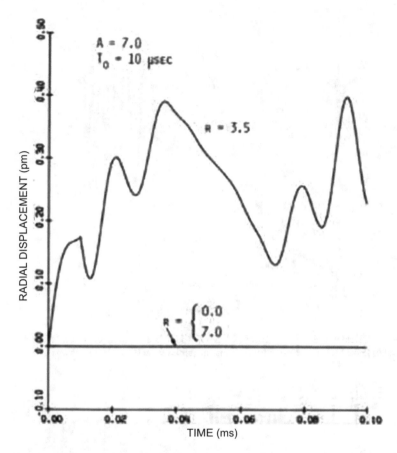

Fig. 8.15 Displacement induced in a 14-cm diameter spherical brain with constrained surface exposed to 915 MHz plane wave. The peak absorption is 1 kW/kg in 10μs

Table 8.3 Calculated pressure amplitude and displacement in constrained spherical brain models exposed at SAR of 1 kW/kg for 10μs rectangular pulses

Brain model	Sphere diameter	Microwave frequency	Incident power density	SAR	Pressure	Displacement (picometer)
	cm	MHz	kW/m²	kW/ kg	mPa	pm
Guinea pig	4	2450	4.45	1	408	0.216
Cat	6	2450	5.89	1	369	0.151
Human child	10	915 (918)	12.82	1	961	0.934
Human adult	14	915 (918)	21.83	1	682	0.397

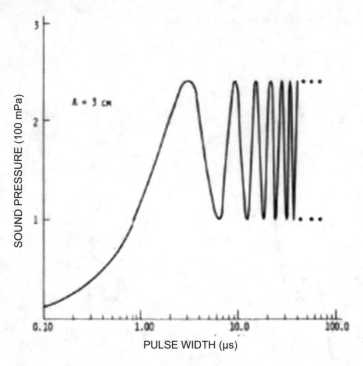

Fig. 8.16 Calculated acoustic pressure amplitude in a 6-cm diameter stress-free brain sphere exposed to 2450 MHz plane wave for rectangular pulses from 0.1 to 100μs. Peak SAR is 1 kW/kg

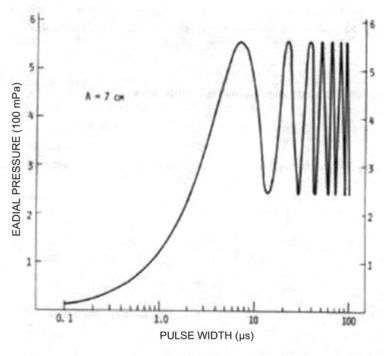

Fig. 8.17 Calculated acoustic pressure amplitude in a 14-cm diameter stress-free brain sphere exposed to 915 MHz plane wave for rectangular pulses from 0.1 to 100μs. Peak SAR is 1 kW/kg

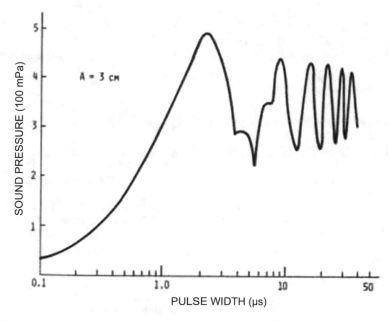

Fig. 8.18 Calculated acoustic pressure amplitude in a 6-cm diameter constrained brain sphere exposed to 2450 MHz plane wave for rectangular pulses from 0.1 to 100μs. Peak SAR is 1 kW/kg

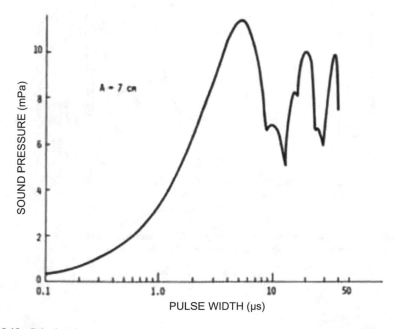

Fig. 8.19 Calculated acoustic pressure amplitude in a 14-cm diameter constrained brain sphere exposed to 915 MHz plane wave for rectangular pulses from 0.1 to 100μs. Peak SAR is 1 kW/kg

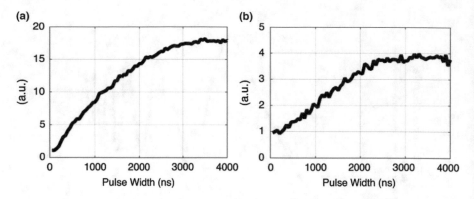

Fig. 8.20 Relationship between measured pressure amplitude (in arbitrary unit) and pulse width under constant peak-power constraint: (**a**) 0.5- and (**b**) 1-MHz transducer bandwidth

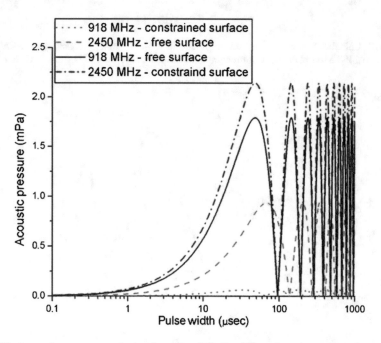

Fig. 8.21 Acoustic pressure, at the head center, of the fundamental mode as a function of pulse width for different models having the same average SAR of 0.4 W/kg

between 50 and 70μs for surface localized and centrally peaked SAR patterns with offset. In all the rectangular pulses shown in Fig. 8.21 [Yitzhak, et al., 2009], the amplitude of the acoustic pressure oscillates between zero and maximum for longer pulse widths.

In addition to rectangular pulses, the effect of other pulse shapes such as saw-tooth and half-sine waves have been examined by using the generalized solution [Yitzhak, et al., 2009]. A comparison of rectangular, sawtooth, and half-sine pulse shapes is presented in Fig. 8.22. The acoustic pressure at the center of a 20-cm diameter brain sphere, with a constrained surface, was calculated for an incident power density of 10 W/m^2 (~0.19 W/kg SAR) at 2450 MHz. As seen in the other pulse scenarios, in each case, the sound pressure first rises to a maximum and then oscillates for longer pulse widths. However, there are some differences in detailed behaviors. The rectangular pulse provided the highest maximum pressure, while the sawtooth pulse gave the lowest pressure. The pulse width with the maximum pres-sure appears to depend on pulse shape, specifically, 48μs for the rectangular pulse, 62μs for the sawtooth pulse, and 65μs for the half-sine pulse. Unlike the rectangular pulses, for the sawtooth pulse, the amplitude of the acoustic pressure does not reduce to zero at longer pulse widths. For the half-sine pulse, the amplitude of the acoustic pressure oscillates to an intermediate lower value, but not zero, but then reduces to zero for wider pulses. Interestingly, the pressure response to half-sine pulse is like those shown in Figs. 8.18 and 8.19, obtained with the SAR model of Eq. (8.1a). Perhaps, the reduction of the amplitude results from a destructive inter-ference effect for certain pulse lengths. For the sawtooth pulse, since the pulse

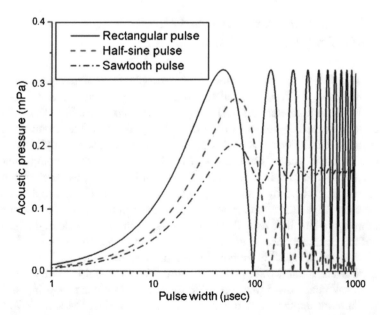

Fig. 8.22 Acoustic pressure, at the head center, of the fundamental mode as a function of pulse width for three different pulse shapes. The incident wave frequency is 2450 MHz.

amplitude increases over the pulse duration, even with a destructive interference effect, it does not completely annul the acoustic pressure amplitude.

These results suggest that microwave pulse-induced acoustic pressure waves in the head are a function of pulse widths and pulse shapes. However, for different SAR models and brain sizes, these results demonstrate that for equal average SAR, the induced pressures by various SAR patterns are comparable. Namely, for equal average SARs, the pressure generated by a surface localized SAR pattern is comparable to that generated by an SAR pattern that is peaked at the center.

8.7   Pressure Wave Measurements in Animal Heads

Physical measurements of acoustic pressure waves were made using hydrophone transducers in the brains of cats (4.5 kg, body mass), guinea pigs (700 g), and rats (475 g) under anesthesia. They involved exposures to pulsed 2450 and 5655 MHz microwaves [Olsen and Lin, 1983]. In addition, acoustic click-induced and pulsed microwave-induced brainstem evoked auditory potentials were acquired using microelectrodes that were buffered by an oscilloscope and summed using a signal averaging analyzer. The brainstem evoked auditory potentials aided to verify the perception of induced sound by the animal subjects. Figure 8.23 gives some details of the microwave exposure setups for the two experimental designs.

8.7.1   Pressure Sensing in Animal Head
and Frequency Analysis

Pressure waves were sensed by a small disk hydrophone element composed of lead zirconate-titanate (C-5500) material electroplated on each side. The disk elements were 3.18 mm in diameter and 0.51 mm thick. A coaxial cable (RG-174/U) was soldered to the metallic surfaces of the disk. The disk and cable connections were sealed using clear silicone rubber. The hydrophone signals were amplified and displayed on an oscilloscope. Calibration of the hydrophone was accomplished over the range of 50–200 kHz in a 76 × 33 × 37 cm tank of water using short bursts of acoustic energy emitted by a small spherical hydrophone. The hydrophone sensor was surgically implanted approximately 1.5 cm deep in the brain of an anesthetized animal through a burr hole drilled through the skull on the left side of the head near the top of the parietal bone. Dental acrylic cement was applied between the existing coaxial cable and the skull to immobilize the transducer.

At 2450 MHz, a microwave pulse generator (Epsco model PG5KB) with a peak power of 3 kW peak power was applied to the cats and guinea pigs using an Elmed Model 15 antenna applicator. In the rat experiments, an Elmed Model 3007 antenna applicator was used. The applicators were manually held against the heads of the animals. In addition, an open-ended WR-284 waveguide was used for rats. In this case, the rat's head was placed at the opening in the center of the waveguide. Pulse

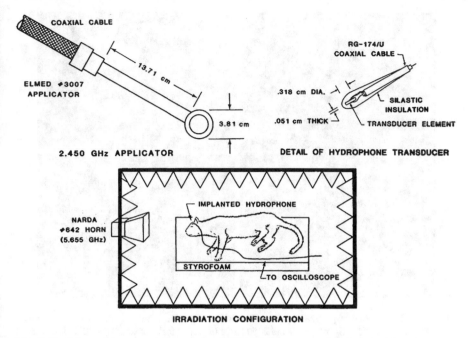

Fig. 8.23 Schematic representation of the 5655 MHz exposure configuration. Also shown are details of 2450 MHz applicator and the miniature hydrophone transducer

widths were 2.5μs for the cat and guinea pig exposures and 5–6μs for the rat experiments. The specific pulse repetition rates for the 2450 MHz energy were 2 Hz and 20 Hz for the evoked response and hydrophone, respectively.

At 5655 MHz, a microwave radar transmitter (AN/SPS-5D) produced 0.5μs wide pulses at a peak power of 200 kW. Pulse repetition rate for the radar transmitter was 2 Hz and 14 Hz for the evoked response and hydrophone measurements, respectively. Microwave energy was applied to the subjects via a standard gain horn (Narda Model 643) inside a $1.25 \times 1.25 \times 2.5$ m microwave anechoic chamber. The heads of the animals were located along the centerline of the horn. In most of these experiments, the animals under anesthesia were placed in the prone position facing the horn antenna.

Figures 8.24 and 8.25 give representative hydrophone output waveforms from a cat, guinea pig, and rat [Olsen and Lin 1983]. The hydrophone output signals recorded from cats and guinea pigs, and the accompanying spectrum of each recording show a rich harmonic content consisting of many modes of sound vibration. The hydrophone outputs are similar but with a complex envelop in all cases. The spectrum associated with each pressure recording shows a rich harmonic content with many modes of vibration. The spectral traces clearly indicate a more high-frequency response in the rat, which has a complex non-spherical head structure and smaller head sizes. Those recorded from the cat and guinea pig, and the accompanying spectrum of each recording show a rich harmonic content consisting of many modes of acoustic vibration. Although the waveforms are not identical, the records for rats

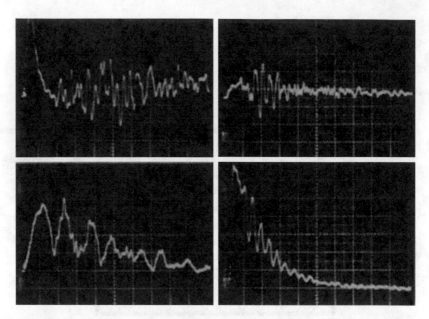

Fig. 8.24 Hydrophone responses to pulsed microwave exposure in the cat brain (left) and guinea pig brain (right) at 5655 MHz (upper) and 2450 MHz (lower). Horizontal scale is 20μs/div. Vertical scales are 20 μVdiv for the cat; 100 and 200μV/div for the guinea pig at 2450 and 5655 MHz, respectively

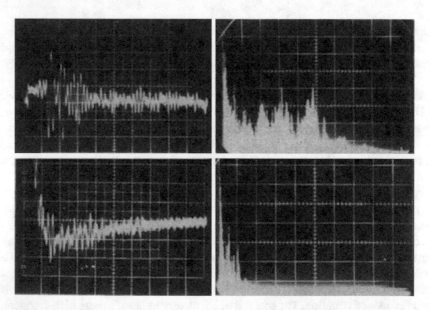

Fig. 8.25 Hydrophone response waveforms (left) and spectra (right) for rats exposed to pulsed 5655 MHz (upper) and 2450 MHz (lower) microwaves. The spectral trace vertical displays are 0.1 V/div, center frequency of 500 kHz, and 100 kHz/div. Vertical scales for the hydrophone waveforms are 100μV/div. Horizontal scale factor for the hydrophone waveforms are 10μs/div

showed a distinct frequency of vibration near 60 kHz, the computed fundamental mode for rat brain. Moreover, the second harmonics near 100 kHz are nearly identical to the 102 kHz predicted by the thermoelastic theory [Lin 1977a, b, c; 1978].

That the temporal recording and frequency spectrum or harmonic content of the various subjects and even from the same species were not always identical are noteworthy but not surprising. For example, the hydrophone output waveforms and their spectra for six rat experiments were all different but they all showed a distinct frequency of vibration near 60 kHz [Olsen and Lin 1983]. In addition to frequency modes that are attributable to the fundamental radial acoustic oscillation of the brain, higher order modes also are expected to be excited by the impinging microwave radiation, according to the theoretical predictions. Furthermore, the skull, as a spheroidal shell with unique anatomic structures, enclosing a brain of non-uniform dimensions, could also be a contributing source of the relatively complex frequency response components.

In one series of experiments, four-pulse bursts of 2450 MHz energy were applied such that the inter-pulse spacing of the burst was varied over a range of frequencies centered on the apparent "ringing" or resonance frequency of the cat brain observed in the single-pulse exposure. In so doing, a "tuning curve" for the fundamental mode of brain vibration was obtained [Olsen and Hammer, 1981; Olsen and Lin, 1981, 1983]. Some results of the "tuning-curve" experiments using the cat are shown in Fig. 8.26, where a peak in hydrophone outputs is observed at 39 kHz. Observation of the predominant mode in the cat brain vibrations induced at 2450 MHz, as seen in Fig. 8.24, plainly indicates a fundamental frequency near 40 kHz, and Fig. 8.26 shows a sharp frequency response of the brain to microwave-induced thermoelastic pressure wave.

In another experiment, the hydrophone sensed pressure waveforms are detected at several depths in the cat brain accessed through midline holes [Lin, et al., 1988]. Details are given in the following section. A frequency response is shown in Fig. 8.27. It is seen that a distinct fundamental component occurred at 39–40 kHz, the predicted fundamental mode of a 5-cm diameter brain sphere. The measured frequency response showed, in addition, several higher-order components as suggested by prior theoretical and experimental investigations [Lin, 1977a, b, c; Olsen and Lin, 1983]. Indeed, a sound wave near this frequency was expected based on experimental data obtained in the study of microwave pulse-induced cochlear microphonics [Chou et al., 1975; 1976]. Note that the thermoelastic theory prescribed a frequency response which is a function of the head size and speed of pressure wave propagation in brain tissue, independent of SAR pattern or microwave frequency. Thus, it is important to observe the comparable pressure frequency responses of the experimental studies and the predictions based on a spherical head with symmetric microwave power deposition that peaked at the center.

Fig. 8.26 Frequency response measured by a hydrophone transducer implanted in the cat's brain. "Tuning curve" results from cat brain implanted with hydrophone transducer. Four-pulse bursts of 2.450 GHz microwaves are used to expose the cat head by means of an Elmed #15 applicator. Inter-pulse spacings of each burst correspond to the abscissa frequency value

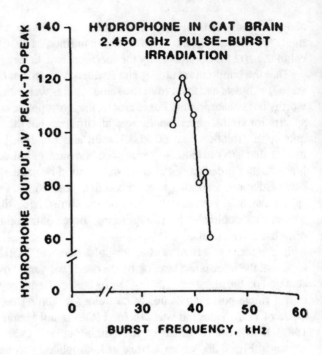

Fig. 8.27 Frequency spectrum of thermoelastic pressure wave measured at center of a cat's brain in response to pulsed microwave radiation. Horizontal scale is 49 kHz/div

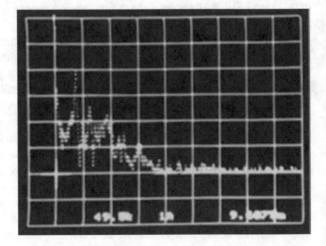

8.7.2   Comparison of Predicted and Measured Sound Frequency

The predicted fundamental frequency as a function of equivalent spherical head radius along with microwave-induced pressure wave frequencies of 39, 50, and 60 kHz measured in cats, guinea pigs, and rats are plotted in Fig. 8.28. It is seen that the sound frequency in various subjects differs according to their equivalent spherical head sizes, i.e., the smaller the head size, the higher the frequency. For rats, a distinct frequency near 60 kHz is shown. The average head radius for guinea pigs is about 1.5–2.5 cm. Figure 8.28 yields a range of 40–70 kHz for the corresponding fundamental sound frequency. The average head radius for cats is approximately 2.5–3.5 cm; the corresponding fundamental sound frequency is between 30 and 40 kHz. Note that these frequencies are close to the 50 kHz cochlear microphonics reported for guinea pigs [Chou et al., 1975] and 38–39 kHz acoustic oscillations reported for cats [Chou et al., 1976]. Human head sizes vary from 7 to 10 cm in radius for adults. From Fig. 8.28, it is seen that the predicted fundamental sound frequency ranges from 8 to 16 kHz. This is consistent with the known facts on human auditory response is in the range of a few Hz to 20 kHz. It also is in concordance with the observation that a necessary condition for auditory perception of microwaves is the ability to perceive auditory signals above 5–8 kHz [Frey, 1961; Guy et al., 1975a, b; Cain and Rissmann, 1978; Tyazhelov et al., 1979].

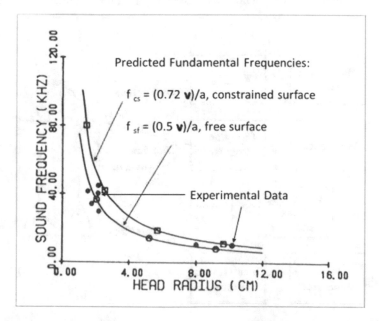

Fig. 8.28 Comparison of predicted and experimentally measured frequencies of microwave pulse-induced pressure waves in heads of experimental subjects for v = 1550 m/s

The measured sound frequency and resonance frequency of pressure waves compare favorably with those predicted by the thermoelastic theory based on homogeneous brain-tissue equivalent spheres. Note that the frequencies predicted by the constrained-surface formulation are higher than those calculated based on stress-free boundary conditions. Since most mammalian heads are neither entirely stress-free nor rigidly constrained, thus the actual fundamental sound frequency falls somewhere between those predicted under the two conditions.

8.7.3 Pressure Wave Propagation Measurement in Cat's Head

Measurements of the propagation of pressure waves have been made in the brains of cats exposed to pulsed microwaves [Lin, et al. 1988]. The experimental arrangement is illustrated in Fig. 8.29. Cats (4–5 kg body mass) were used in this experiment. Each animal was anesthetized using ketamine (25 mg/kg) injected intramuscularly and kept under the surgical level of anesthesia using intravenous sodium pentobarbital supplementation. Following skin incision and partial separation of the nuchal muscle, a matrix of small diameter holes (2 mm) was drilled through the dorsal portion of the skull. The locations of the 15 holes on the skull are shown in Fig. 8.30 (upper). Five holes (1–5) were drilled along the midline of the skull; in addition, ten holes, nos. 6–10 and 11–15, were drilled along two lines intersecting at the applicator. Figure 8.30 (lower) shows a midsagittal plane together with the locations of hydrophone and applicator during measurements. The animal

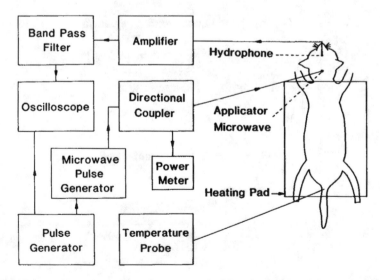

Fig. 8.29 Schematic representation of the anatomical location, microwave source, and hydrophone recording configuration

Fig. 8.30 Views of a cat head with implant locations from the top and along the mid-sagittal plane of a cat's skull showing antenna and hydrophone during measurements

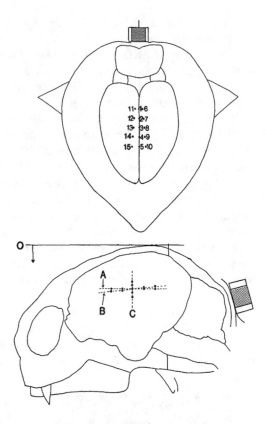

was placed in a stereotaxic head holder. By noting the stereotaxic coordinates of each location away from the center of applicator, a measure for the depth was obtained. The cat's body temperature was maintained at 37.5 ± 1.5 °C using a heating pad. At the end of each experiment, the animals were killed with an overdose of sodium pentobarbital.

Single pulses of microwave (2μs at 2450 MHz) were produced with a microwave generator (Epsco PH40k), which was controlled by an external pulse-forming circuits and instruments. Microwave pulses were applied to the surface of the head with a direct-contact Elmed antenna applicator (15 mm in diameter). The peak incident power of pulsed microwave was 15 kW.

A hydrophone transducer was used to detect the pressure wave. The cylindrical lead zirconate-titanate ceramic transducer element was enclosed in a water-proof Neoprene sheath (2×4 mm). It had a response sensitivity of -130 dB for the pertinent range of frequency (1–400 kHz). The directivity pattern was circular in the transverse plane at both 80 and 400 kHz. The hydrophone transducer was inserted stereotaxically through a matrix of holes drilled on the skull into the cat's brain and advanced precisely to the desired location with a micromanipulator. By noting the stereotaxic coordinates of each location away from the center of applicator, a measure for the depth was obtained.

Microwave-induced pressure waves detected by the hydrophone sensor were conditioned with a high-gain amplifier and a bandpass filter having cutoff frequencies at 1 kHz and 1 MHz. The first 100µs of response was displayed on an oscilloscope and photographed on film. In addition, a fast Fourier transform of the response was obtained with a digital oscilloscope. An example of the frequency response experiments in cats is shown in Fig. 8.27. A peak in hydrophone output is observed at 40 kHz along with some well-known higher frequency components.

Some of the hydrophone sensor output waveforms are shown in Fig. 8.31. Hydrophone responses detected in brain tissue at the same depth (27 mm) through five midline holes (separated by 1 cm) are shown in Fig. 8.31 (panel a). Sensor signals from along a straight line from the microwave antenna are displayed in Fig. 8.31 (panel b). Both figures clearly indicate time delays associated with pressure wave propagation. The hydrophone output signals obtained at a single location but different depths in brain tissue and at the same depth, but three different locations are given in Fig. 8.31 (panels c and d). The time delay between various traces in these figures is not as obvious since the measurement points are at approximately the same distance from the center of the antenna applicator and are located along the same wave front. These results confirm the propagating nature of microwave-induced acoustic pressure waves in the cat brain.

The speed of propagation of microwave-induced thermoelastic pressure wave in the cat brain may be derived from measured time delay and known distances traveled. As given in Table 8.4, the mean speed of propagation is 1522.77 m/s at body

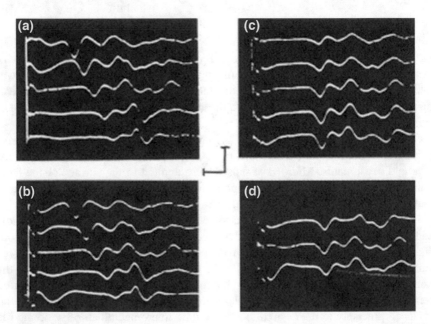

Fig. 8.31 Recordings from hydrophone implanted in cat brain. (**a**) Holes 1–5 at equal depth: 27 mm; see line **a** in Fig. 8.30; (**b**) Holes 1–5 at depth levels of 26, 26.5, 27, 27.5, and 28 mm, respectively; see line **b**; (**c**) Hole 3 at different depths: 23, 25, 27, 29, and 31 mm; see line **c**. (**d**) Holes 8, 3, and 13 at the same depth: 27 mm

Table 8.4 Measured speed of thermoelastic pressure wave propagation in live cat brain at 37.5 °C

Distance (cm)	Number of measurements (64 total)	Propagation speed (mean ± SD, m/s)
2.5–3.0	4	1463.75 ± 4.79
3.5–4.0	11	1532.78 ± 22.12
4.0–4.5	6	1554.92 ± 10.79
4.5–5.0	13	1498.03 ± 22.52
5.0–5.5	23	1531.33 ± 15.40
5.5–6.0	7	1531.09 ± 21.58
Average speed		1522.77 ± 28.45

temperature. This was based on an ensemble of 64 measurements made at six different sets of distances. This value is about the same as the ones used earlier for calculations, but slightly lower than those used for phantom modeling (1600 m/s at room temperature). The amplitude attenuation follows the typical exponential law and has an attenuation coefficient of 0.56 Np/m, which is higher than the 0.11 Np/m reported in the literature for ultrasound propagation in brain tissue [Dunn et al., 1969]. The speed of propagation and attenuation of sound pressure both are known to vary with frequency and temperature in general. Aside from giving a microwave-induced pressure wave propagating speed of 1523 m/s in the cat brain, the result lends support to the notion that a thermoelastic pressure wave is induced in the heads of humans and animals exposed to pulsed microwaves. It is important to note that propagating microwave pulse-induced pressure waves have been detected in the brain tissue of live cats.

8.8   Predicted and Measured Characteristics in Mammalian Heads

The acoustic pressure wave measurements described in the previous sections on various sized spherical head models and for cats, rats, and guinea pigs exposed to pulsed microwaves not only showed the existence of thermoelastic pressure waves, these experimental results verified a unique acoustic frequency response that corresponded to a geometry- or size-dependent mechanical resonance in the head as predicted by the thermoelastic theory of interaction. While these measurements advance the thermoelastic theory, its credibility as the primary mechanism of interaction was further boosted by corroborative observations from behavioral and physiological experiments.

These observations also include the measured cochlear microphonics in cats and guinea pigs, predicted fundamental acoustic frequency, and the well-documented requirements for human perception of pulsed microwaves: the ability to hear sonic energy above 5 to 8 kHz. Clearly, there is an agreement between observed and calculated sound frequency. Furthermore, the theory suggests that the frequency of sound perceived by a subject exposed to microwave pulses is the same regardless of the frequency of impinging microwaves and is independent of the pattern of absorbed energy or SAR. Reports have shown that the same cochlear microphonics are produced by 918 and 2450 MHz microwave radiation. Moreover, brainstem evoked electrical potential measurements made by moving a small contact applicator antenna around the head of a cat showed remarkable preservation in amplitude and temporal relations of the BER signals (Fig. 8.32; Also see Fig. 6.29).

8.8.1   Dependence of Response Amplitude on Microwave Pulse Width

Another body of support for the theory comes from the prediction of a sound pressure amplitude that initially increased with pulse width but soon reached a peak and then gradually oscillate toward a constant or to a lower value with further increase in pulse width for stress-free and constrained brain surfaces. Several laboratory studies have provided corroborating evidence from cats [Lin, 1980, 1981; Lin et al., 1982], guinea pigs [Chou and Guy, 1979], and humans [Tyazhelov et al., 1979].

Figure 8.33 shows a comparison of the calculated sound pressure amplitude and the variation of microwave-induced auditory brainstem response in cats with the width of impinging microwave pulses, while the peak power and pulse repetition rates are kept constant [Lin, 1980]. Similar relationship between relative loudness or sound pressure in guinea pigs and pulse width is given in Fig. 8.34 [Chou and Guy, 1979]. This is equivalent to the observation that increasing the pulse width decreases SAR required to elicit a threshold auditory brainstem response. The

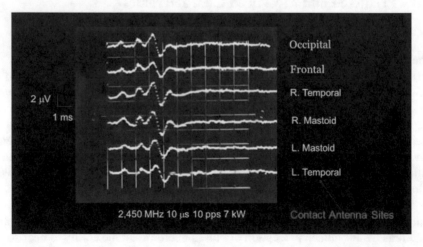

Fig. 8.32 Microwave pulse evoked brainstem potentials via contact antenna sited around a cat's head

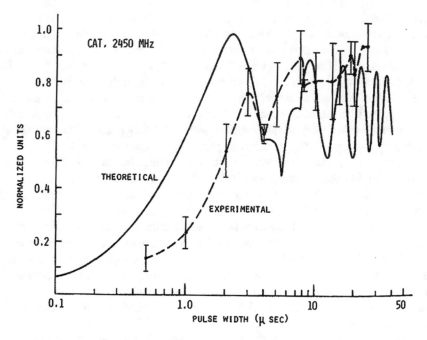

Fig. 8.33 Comparison of calculated sound pressure amplitude and variation of microwave-induced auditory brainstem response in cats with the width of impinging microwave pulses

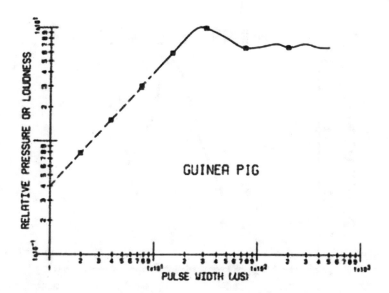

Fig. 8.34 Relative loudness or sound pressure in guinea pigs induced by microwaves as a function of pulse width

general characteristics of these curves are remarkably similar. The sound pressure, loudness, or brainstem response increased then decreased or oscillates as suggested by the theory. Furthermore, this non-monotonic dependence of the response on pulse width was also seen in some single auditory neurons of cats [Lebovitz, 1975; Lebovitz and Seaman, 1977], especially for longer pulse widths.

The loudness of sound perception as a function of pulse width for human subjects exposed to pulse-modulated 800 MHz microwave radiation was determined in 18 men and women with normal high-frequency auditory acuity [Tyazhelov et al., 1979; See also Lin, 1980]. Microwave pulses are delivered using an open-ended waveguide to the parietal area of a subject's head. The pulses are 5–150μs wide and the repetition frequency varied from 50 Hz to 20 kHz. The loudness for sensation at 800 MHz is shown in Fig. 8.35. As the width of rectangular pulses of constant peak power density is gradually increased from 5 to 150μs, a complex oscillatory loudness function is observed. The loudness increased as pulse width increased from 5μs and reached a peak at 50 μs, then diminished with further increase of pulse width from 70 to 100μs, and then increased again or oscillates with longer pulse widths. The nonlinear character of the experimental results corroborates previously described theoretical predictions (Fig. 8.19). These sound pressures were calculated using the SAR model of Eq. (8.1a) for human size stress-free and constrained surface head spheres at a comparable frequency (Figs. 8.17 and 8.19). The similarities between these curves are clearly visible. However, it has been shown [Yitzhak, et al., 2009] that the dependence of pressure amplitude on pulse width would vary if it is calculated using the SAR model of Eq. (8.1b). In the human experiment

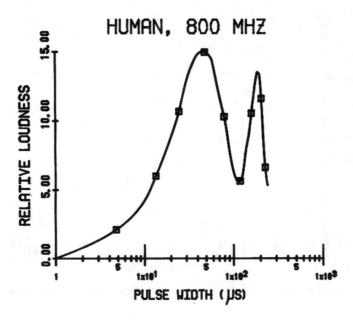

Fig. 8.35 The loudness of sound perception as a function of pulse width for human subjects exposed to 800 MHz microwave radiation

[Tyazhelov et al 1979], the first loudness peak occurs at 50µs which agrees with a calculation for the SAR model of Eq. (8.1b) shown as the solid curve in Fig. 8.21. The shapes of the loudness and pressure curves are consistent with data reported by Frey and Messenger [1973] and Guy et al. [1975a, b]. The difference between the pulse widths at which sound pressure or loudness reached a maximum likely stemmed from differences in actual and assumed head size, geometry, and structure. In general, sound pressure peaks at wider pulse widths for larger head sizes. Thus, an array of data substantiates the thermoelastic mechanism of microwave-induced auditory effect.

An observation from the human experiment suggested that the pitch (perceived frequency) of sound induced by microwave pulses of less than 50µs persisted as the subject's head was lowered into a saline-filled tank, while the loudness diminished roughly in proportion to the depth of immersion [Tyazhelov et al., 1979]. On complete immersion, auditory sensation disappeared. For pulse width longer than 50µs even partial immersion resulted in the loss of perception. This was interpreted as being at odds with the thermoelastic theory. However, there is an explanation that seems to fit the data [Lin 1990]. The theory suggests that the frequency of sound (pitch) evoked by microwave pulses is the same regardless of SAR or absorbed energy distribution and that the amplitude of sound (loudness) increases in proportion to SAR or the rate of energy absorption. Moreover, the amplitude of sound or loudness rises, peaks, and then falls or oscillates as the pulse width increases. Thus, as the subject's head was lowered into water, the absorbed energy and its distribution would have changed (that is decreased), but the acoustic property would have remained unaltered. Immersion would not alter the microwave-induced acoustic characteristic by an appreciable amount. Therefore, the perceptual quality of microwave-induced sound persists while the loudness diminished in proportion to the depth of immersion. The reason for the disappearance of auditory sensation or perception on complete immersion is that in this case, the impinging microwave pulse was mostly absorbed by water before reaching the head. The small fraction that did reach the head likely fell below the necessary SAR level needed to reach the threshold of perception. The observation that even partial immersion resulted in the loss of perception for pulse widths greater than 50µs could be resolved by recalling that 50µs was reported as being the optimal pulse width for the auditory perception by the subjects. Hence, microwave pulses of width greater than 50µs would not be as effective in eliciting an acoustic sensation and would be diminished when the head was lowered partially into water. Thus, these findings would strengthen rather than weaken the thermoelastic theory of microwave auditory effect.

Also, it was found in the human study that a beat frequency experiment that matched microwave (10µs, 8 kHz) pulse-induced sound to a phase-shifted 8 kHz sinusoidal sound input resulted in a loss (cancellation) of auditory perception [Tyazhelov et al., 1979]. However, cancellation also occurred when a 5 kHz train of pulses was properly phased with a 10µs sound signal. It is well-known that sensitivity to sound frequency or pitch varies for different subjects which may account for this finding. It is possible that the discrepancy was related to harmonic generation by the audio equipment as well.

The dependence of the microwave auditory effect on pulse repetition frequency (PRF) has been measured in the same experiment mentioned above. Data were obtained from two human subjects with different hearing acuities [Tyazhelov et al. 1979]. Their high-frequency auditory limits (HFAL) were 14 kHz and 17 kHz. The results of perception threshold measured as a function of the PRF are shown in Fig. 8.36 [Yitzhak, et al., 2009]. The theoretical calculations of microwave pulse-induced pressure were made under stress-free surface conditions using the SAR pattern of Eq. (8.1b) for 20-μs wide, 800 MHz rectangular pulses. The curves show the ratio of the acoustic pressure induced at the center of spherical head models at a PRF of 10 kHz to the corresponding pressure at other PRF, expressed in the relative dB scale. Also, the experimental data were normalized to their value at a PRF of 10 kHz. The calculated results show general agreements with the experimental data, especially regarding the feature of a broad threshold increase around a PRF of 7 kHz.

It should be mentioned that the theoretical solution is applicable to a host of physiological and psychophysical observations, as well as physical measurements made in phantom models and living animals. That it may be incomplete and thus require further extension or more meticulous computational approaches to account for certain additional experimental findings or satisfy the demands of scientific

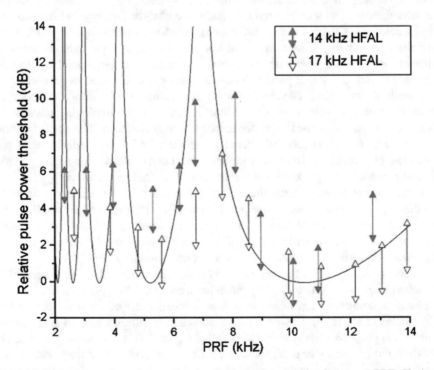

Fig. 8.36 Dependence of sound perception threshold on pulse repetition frequency (PRF). Vertical bars represent experimental data. The solid curves are calculated pressure induced in spherical head models by exposure to 20-μs wide rectangular microwave pulses

exactness would enhance the fundamental role of the thermoelastic theory in understanding the microwave auditory effect or phenomenon.

The mathematical development of the thermoelastic theory can describe most, if not all of the salient aspects of the microwave auditory effect occurring in heads of humans and animals exposed to pulsed microwaves. However, the theoretical calculations are based on assumptions made for mathematical feasibility and pragmatism. Specifically, the assumptions include a canonical model of head-equivalent homogeneous sphere filled only with brain tissue matter and exposure of the head model in isolation from the rest of the body. In reality, the detailed nature of the dependence of the microwave auditory effect on head structure, tissue composition, parameters of the impinging microwave radiation, and the precise amplitude and frequency of the induced thermoelastic pressure wave inside the head may be better determined by solving the governing differential equations by computer simulation or through numerical computation under more exact anatomical descriptions, which would indeed form the focus of the following or succeeding chapter.

References

Cain CA, Rissmann WJ (1978) Mammalian auditory responses to 3.0 GHz microwave pulses. IEEE Trans Biomed Eng BME-25(3):288–293

Carslaw HS, Jaeger JC (1959) Conduction of heat in solids, 2nd edn. Oxford University Press, Oxford

Chou C-K, Guy AW (1979) Microwave-induced auditory responses in guinea pigs: relationship of threshold and microwave-pulse duration. Radio Sci 14(6S):193–197

Chou C-K, Galambos R, Guy AW, Lovely RH (1975) Cochlear microphonics generated by microwave pulses. J Microwave Power 10(4):361–367

Chou CK, Guy AW, Galambos R (1976) Microwave induced cochlear microphonics in cats. J Microwave Power 11:171–173

Churchill RV (1958) Operational mathematics, 2nd edn. McGraw-Hill Book, New York

Dunn F, Edmonds PD, Fry WJ (1969) Absorption and dispersion of ultrasound in biological media. In: Schwan HP (ed) Biological engineering. McGraw, New York, pp 205–332

Frey AH (1961) Auditory system response to radio frequency energy. Aerospace Med 32:1140–1142

Frey AH, Messenger R Jr (1973) Human perception of illumination with pulsed ultra-high-frequency, electromagnetic energy. Science 181:356358

Guy AW, Lin JC, Chou CK (1975a) Electrophysiological effects of electromagnctic fields on animals. In: Fundamentals and applied aspects of nonionizing radiation. Plenum Press, pp 167–211

Guy AW, Chou CK, Lin JC, Christensen D (1975b) Microwave induced acoustic effects in mammalian auditory systems and physical materials. Ann NY Academ Sci 247:194–218

Jahnke E, Emde F (1945) Tables of functions. Dover Publications, New York

Johnson CC, Guy AW (1972) Nonionizing electromagnetic wave effects in biological materials and systems. Proc IEEE 60(6):692–718

Kritikos HN, Schwan HP (1975) The distribution of heating potential inside lossy spheres. IEEE Trans Biomed Eng BME-22(6):457–463

Lebovitz RM (1975) Detection of weak electromagnetic radiation by the mammalian vestibulocochlear apparatus. Ann NY Acad Sci 247:182–193

Lebovitz RM, Seaman RL (1977) Single auditory unit responses to weak, pulsed microwave radiation. Brain Res 126(2):370–375

Lin JC (1976a) Interaction of two cross-polarized electromagnetic waves with mammalian cranial structures. IEEE Trans Biomed Eng 23:371–375

Lin JC (1976b) Theoretical analysis of microwave-generated auditory effects in animals and man, biological effects of electromagnetic waves. BRH/DEW I:36–48

Lin JC (1977a) On microwave-induced hearing sensation. IEEE Trans Microwave Theory Tech 25:605–613

Lin JC (1977b) Further studies on the microwave auditory effects. IEEE Trans Microwave Theory Tech 25:936–941

Lin JC (1977c) Theoretical calculations of frequencies and threshold of microwave-induced auditory signals. Radio Sci Supplement, Biological Effects of Electromagnetic Waves 12/SS-1:237–252

Lin JC (1978) Microwave auditory effects and applications. CC Thomas, Springfield

Lin JC (1980) The microwave auditory phenomenon. Proc IEEE 68:67–73

Lin JC (1990) Auditory perception of pulsed microwave radiation. In: Gandhi OP (ed) Biological effects and medical applications of electromagnetic fields. Prentice-Hall, New York, pp 277–318

Lin JC, Wang ZW (2007) Hearing of microwave pulses by humans and animals: effects, mechanism, and thresholds. Health Phys 92(6):621–628

Lin JC, Guy AW, Kraft GH (1973) Microwave selection brain heating. J Microwave Power 8:275–286

Lin JC, Meltzer RJ, Redding FK (1982) Comparison of measured and predicted characteristics of microwave-induced sound. Radio Sci 17(5S):159S–163S

Love AEH (1927) The mathematical theory of elasticity. Cambridge University Press, New York

Nan H, Arbabian A (2017) Peak-power-limited frequency-domain microwave-induced thermoacoustic imaging for handheld diagnostic and screening tools. IEEE Trans Microwave Theory Tech 65(7):2607–2616

Olsen RG, Hammer WC (1981) Evidence for microwave-induced acoustical resonances in biological material. J Microwave Power 16:263–269

Olsen RG, Lin JC (1981) Microwave pulse-induced acoustic resonances in spherical head models. IEEE Trans Microwave Theory Tech MTT-29:1114–1117

Olsen RG, Lin JC (1983) Microwave-induced pressure waves in mammalian brains. IEEE Trans Biomed Eng BME-30(5):289–294

Shapiro R, Lutomirski RF, Yura HT (1971) Induced fields and heating within a cranial structure irradiated by an electromagnetic plane wave. IEEE Trans Microwave Theory Tech 19(2):187–196

Shibata T, Fujiwara O, Kato K, Azakami T (1986) "Calculation of thermal stress inside human head by pulsed microwave irradiation" (in Japanese). IEICE Trans Commun J69-B(10):1144–1146

Sokolnikoff IS (1956) Mathematical theory of elasticity. McGraw, New York

Stratton JA (1941) Electromagnetic theory. McGraw Hill, New York

Tyazhelov VV, Tigranian RE, Khizhniak EO, Akoev IG (1979) Some peculiarities of auditory sensations evoked by pulsed microwave fields. Radio Sci 14(6S):259–263

Uzunoglu NK, Polychronopoulos SI (1988) Microwave-induced auditory effect in a dielectric sphere. IEEE Trans Microwave Theory Tech 36:1418–1425

Yitzhak NM, Ruppin R, Hareuveny R (2009) Generalized model of the microwave auditory effect. Phys Med Biol 54:4037–4049

Chapter 9
Computer Simulation of Pressure Waves in Anatomic Models

The analytical expressions and mathematical calculations presented in Chap. 8 are important in predicting, prescribing, and quantifying the dependence of induced sound pressure characteristics on RF and microwave pulse characteristics in the spherical model of the head structures. However, detailed numerical computations using anatomically realistic head models and discrete formulation of the analytical expressions provide more precise descriptions of specific features and accurate information on the acoustic pressure waves generated inside the head and sound perceived by human observers. This chapter presents a computer simulation of the thermoelastic pressure waves generated in anatomical human heads exposed to pulsed RF and microwave radiation.

Anatomically realistic head models were first used by Watanabe et al. [2000] to numerically calculate the microwave-induced pressure waves in humans. The computational algorithm followed the steps of the analytical scheme and approach developed by Lin [1976, 1977a, b, 1978] for a mathematical solution. The paper applied the numerical method of finite-difference, time-domain (FDTD) formulation to sequentially obtain solutions for Maxwell's electromagnetic equations, biological heat transfer equation, and lastly, the thermoelastic equation of motion, which gives rise to the tissue expansions and acoustic pressure wave. Note that in the Watanabe case, the FDTD calculations of pressure waves generated by the thermoelastic mechanism were limited to 915 MHz plane waves.

Lin and Wang [2007, 2010] used an FDTD formulation and algorithms to computationally simulate the thermoelastic pressure waves generated in an anatomic head model exposed to localized 300- and 400-MHz sources – a birdcage magnetic resonance imaging (MRI) coil antenna, driven by a rectangular RF pulse. In addition, whole-body anatomic models for FDTD calculations of RF and microwave pulse-induced thermoelastic pressure waves were conducted by Yitzhak et al. [2014] for plane wave exposures ranging in RF and microwave frequencies from 40 MHz to 3 GHz. This report also included computer simulation results for anatomical head models.

J. C. Lin, *Auditory Effects of Microwave Radiation*,
https://doi.org/10.1007/978-3-030-64544-1_9

9.1   FDTD Formulation for Microwave Thermoelastic Pressure Waves

As mentioned, the hearing of microwave-pulse-induced sound involves a cascade of events. A minuscule but rapid rise in tissue temperature, resulting from the absorption of pulsed microwave energy, creates a thermoelastic expansion in the brain matter. This small theoretical temperature elevation can launch an acoustic wave of pressure that travels inside the head to the inner ear. There, it activates the sensory hair cells in the cochlea, which then relays the neural signal to the central auditory system for perception, via the same process involved in normal hearing. The following discussion begins with FDTD formulation for the solution of Maxwell's electromagnetic equations.

9.1.1   Maxwell's Electromagnetic Equations

The computer implementation in this multidisciplinary approach begins with discretizing Maxwell's electrodynamic equations (Eqs. 2.5 and 2.6). The FDTD computation of SAR distribution or microwave energy absorption inside the head or whole body is performed by directly modeling propagation of microwaves from a source of the impinging microwave radiation of interest into a volume of space containing the biological body. Details of the FDTD formulations along with some results from FDTD simulation of SAR distribution in anatomically realistic models are given in Chap. 5 for near field and plane wave scenarios.

9.1.2   Biological Heat Transfer Equation

The distribution of SAR at each FDTD cell as given by Eq. (5.17) becomes the source function for the biological heat transfer equation in the ensuing simulation of temperature elevation inside the biological body model [Bernardi et al., 1998, 2003; Lin and Bernardi, 2007]. Specifically, the biological heat transfer or Bioheat equation [Pennes, 1948] is given by

$$\rho C_p \frac{\partial T}{\partial t} = K \nabla^2 T + \rho SAR + A - B(T - T_b) \tag{9.1}$$

and the associated boundary condition is specified by.

$$-K \left(\frac{\partial T}{\partial n} \right)_s = H(T - T_a) \tag{9.2}$$

where ρ (kg/m³) is tissue mass density, $T = T$ (x, y, z, t) is the temperature (°C) at time t, C_p (J/kg- °C) is specific heat, K (W/m-°C) is thermal conductivity, A in W/kg is for metabolic heat production, B in W/m³ °C is a factor related to blood flow, T_b is blood temperature, T_a is the ambient temperature, n is the unit vector normal to the surfaces of the object, H is the convective heat transfer coefficient (W/m² °C), and SAR (W/kg) is the absorbed RF and microwave radiation - source for temperature elevation in the tissue.

The central difference scheme can be used to express the partial differentials for FDTD formulations of the Bioheat equation (9.1), such that

$$
T^{n+1}(i,j,k) = T^n(i,j,k) + \frac{\Delta t}{C_p(i,j,k)\rho(i,j,k)}\Big[\rho(i,j,k)SAR(i,j,k) + A(i,j,k)\Big]
$$

$$
+ \frac{\Delta t B(i,j,k)}{C_p(i,j,k)\rho(i,j,k)}\Big(T^n(i,j,k) - T_b\Big) + \frac{K(i,j,k)\Delta t}{C_p(i,j,k)\rho(i,j,k)}
$$

$$
\left[\left(\frac{T^n(i+1,j,k) + T^n(i-1,j,k) - 2T^n(i,j,k)}{\Delta x^2} + \frac{T^n(i,j+1,k) + T^n(i,j-1,k) - 2T^n(i,j,k)}{\Delta y^2}\right.\right.
$$
$$
\left.\left.+ \frac{T^n(i,j,k+1) + T^n(i,j,k-1) - 2T^n(i,j,k)}{\Delta z^2}\right)\right]
\tag{9.3}
$$

where (i, j, k) is the index for each grid, Δx, Δy, and Δz are the spatial steps in x, y, and z directions, respectively, and Δt is the incremental time step.

The temperature at the boundary of objects is calculated as follows:

$$
T^n\left(i_{min},j,k\right) = \frac{K(i,j,k)T^n\left(i_{min}+1,j,k\right)}{K(i,j,k) + H \cdot \Delta x} + \frac{H \cdot \Delta x \cdot T_a}{K(i,j,k) + H \cdot \Delta x}
\tag{9.4}
$$

or

$$
T^n\left(i_{max},j,k\right) = \frac{K(i,j,k)T^n\left(i_{max}-1,j,k\right)}{K(i,j,k) + H \cdot \Delta x} + \frac{H \cdot \Delta x \cdot T_a}{K(i,j,k) + H \cdot \Delta x}
\tag{9.5}
$$

Equations (9.4 and 9.5) are given only along the x-direction. Similar FDTD expressions can be obtained along the y- and z-directions.

The choice for the maximum size of the time step is derived from Von Neumann's condition to promote numerical stability. In this case, it is given by,

$$
\Delta t \leq \min_{m \in M}\left(\frac{2\rho_m C_{pm}\Delta}{12K_m + B_m\Delta^2}\right)
\tag{9.6}
$$

where M is the entire set of tissues modeled.

The convective heat-transfer coefficient H is typically set to 8.37 J/m²s°C with the ambient temperature set to 24 °C and blood temperature set to 37 °C for calculations under normal conditions. The initial temperature distribution inside the model is calculated first by setting SAR = 0, which is regarded as the temperature in the unexposed body model at thermal equilibrium. T (x, y, z, t) is the resulting distribution of tempera-ture (°C) at time t, inside the exposed body. The temperature elevation depends both on microwave pulse strength and width. For short microwave pulses, the temperature ele-vation is miniscule but finite. The tiny temperature elevation can cause the body tissue to expand and launch an acoustic pressure wave in the elastic tissue medium.

9.1.3   Thermoelastic Equation of Motion

Thus, as adopted in Chap. 8 for the analytical development, the last step in the mul-tidisciplinary approach is to derive an FDTD formulation for solving the microwave thermoelastic equation of motion for the acoustic pressure waves generated inside the anatomical head or body model.

One approach for this is to begin with the equivalent rectangular expressions of Eqs. (8.17) and (8.35). Instead, we will begin with Hooke's law for thermal stress, leading to a thermoelastic equation of motion [Landau and Lifshitz, 1986]. Note that pressure, p (x, y, z, t) is related to the normal component of stress tensor, σ_{ij} by

$$p(x,y,z,t) = \frac{1}{3} \sum_{i=x,y,z} \sigma_{ii} \tag{9.7}$$

In rectangular coordinates, the normal components of the stress tensor from Hooke's law for SAR- or microwave absorption-induced temperature, T(x, y, z, t) are given by

$$\frac{\partial \sigma_{xx}}{\partial t} = (\lambda + 2\mu)\frac{\partial v_x}{\partial x} + \lambda\frac{\partial v_y}{\partial y} + \lambda\frac{\partial v_z}{\partial z} - \alpha(3\lambda + 2\mu)\frac{\partial T}{\partial t} \tag{9.8}$$

$$\frac{\partial \sigma_{yy}}{\partial t} = (\lambda + 2\mu)\frac{\partial v_y}{\partial y} + \lambda\frac{\partial v_z}{\partial z} + \lambda\frac{\partial v_x}{\partial x} - \alpha(3\lambda + 2\mu)\frac{\partial T}{\partial t} \tag{9.9}$$

$$\frac{\partial \sigma_{zz}}{\partial t} = (\lambda + 2\mu)\frac{\partial v_z}{\partial z} + \lambda\frac{\partial v_x}{\partial x} + \lambda\frac{\partial v_y}{\partial y} - \alpha(3\lambda + 2\mu)\frac{\partial T}{\partial t} \tag{9.10}$$

and the perpendicular components are expressed as

$$\frac{\partial \sigma_{xy}}{\partial t} = \mu\left(\frac{\partial v_x}{\partial y} + \frac{\partial v_y}{\partial x}\right) \tag{9.11}$$

$$\frac{\partial \sigma_{yz}}{\partial t} = \mu\left(\frac{\partial v_y}{\partial z} + \frac{\partial v_z}{\partial y}\right) \tag{9.12}$$

$$\frac{\partial \sigma_{zx}}{\partial t} = \mu \left(\frac{\partial v_z}{\partial x} + \frac{\partial v_x}{\partial z} \right) \tag{9.13}$$

where σ_{ij} are the components of the stress tensor, v_i are the components of the particle velocity, λ and μ are the Lame's elastic constants, and α is the coefficient of linear thermal expansion. These six equations are then solved together with the three Newton equations of motion,

$$\rho \frac{\partial v_x}{\partial t} = \frac{\partial \sigma_{xx}}{\partial x} + \frac{\partial \sigma_{xy}}{\partial y} + \frac{\partial \sigma_{zx}}{\partial z} \tag{9.14}$$

$$\rho \frac{\partial v_y}{\partial t} = \frac{\partial \sigma_{yy}}{\partial y} + \frac{\partial \sigma_{yz}}{\partial z} + \frac{\partial \sigma_{xy}}{\partial x} \tag{9.15}$$

$$\rho \frac{\partial v_z}{\partial t} = \frac{\partial \sigma_{zz}}{\partial z} + \frac{\partial \sigma_{zx}}{\partial x} + \frac{\partial \sigma_{yz}}{\partial y} \tag{9.16}$$

where ρ is the tissue mass density and σ_{ij} are the rectangular components of the stress tensor. The two sets of Eqs. (9.8, 9.9, 9.10, 9.11, 9.12, and 9.13) and (9.14, 9.15, and 9.16) can be discretized using the Yee scheme to complete the FDTD formulation and solve numerically for microwave-pulse-generated pressure waves.

The following are the explicit three-dimensional FDTD equations for computer simulation of microwave-pulse-generated thermoelastic acoustic pressure waves in biological bodies. Since the stress tensor is a symmetric tensor and consists of two components, it can be subdivided into a normal pressure part which represents the forces per unit area normal to given planes and a transverse shear component representing forces per unit area parallel to their planes of reference.

The normal pressure components are:

$$\sigma_{xx}^n \left(i + \frac{1}{2} j + \frac{1}{2}, k + \frac{1}{2} \right) = \sigma_{xx}^{n-1}\left(\left(i + \frac{1}{2}, j + \frac{1}{2}, k + \frac{1}{2} \right) \right) + \frac{(\lambda + 2\mu)\Delta t}{\Delta x}\left[v_x^{n-1/2}\left(i + 1, j + \frac{1}{2}, k + \frac{1}{2} \right) - v_x^{n-1/2}\left(i, j + \frac{1}{2}, k + \frac{1}{2} \right) \right]$$

$$+ \frac{\lambda \Delta t}{\Delta y}\left[v_y^{n-1/2}\left(i + \frac{1}{2}, j + 1, k + \frac{1}{2} \right) - v_y^{n-1/2}\left(i + \frac{1}{2}, j, k + \frac{1}{2} \right) \right]$$

$$+ \frac{\lambda \Delta t}{\Delta z}\left[v_z^{n-1/2}\left(i + \frac{1}{2}, j + \frac{1}{2}, k + 1 \right) - v_z^{n-1/2}(i + \frac{1}{2}, j + \frac{1}{2}, k) \right] - \alpha(3\lambda + 2\mu)\Delta t \frac{\partial T\left(i + \frac{1}{2}, j + \frac{1}{2}, k + \frac{1}{2} \right)}{\partial t}$$

$$\tag{9.17}$$

$$\sigma_{yy}^n \left(i + \frac{1}{2}, j + \frac{1}{2}, k + \frac{1}{2} \right) = \sigma_{yy}^{n-1}\left(i + \frac{1}{2}, j + \frac{1}{2}, k + \frac{1}{2} \right) + \frac{(\lambda + 2\mu)\Delta t}{\Delta y}\left[v_y^{n-1/2}\left(i + \frac{1}{2}, j + 1, k + \frac{1}{2} \right) - v_y^{n-1/2}\left(i + \frac{1}{2}, j, k + \frac{1}{2} \right) \right]$$

$$+ \frac{\lambda \Delta t}{\Delta z}\left[v_z^{n-1/2}\left((i + \frac{1}{2}, j + \frac{1}{2}, k + 1 \right) - v_z^{n-1/2}\left(i + \frac{1}{2}, j + \frac{1}{2}, k \right) \right]$$

$$+ \frac{\lambda \Delta t}{\Delta x}\left[v_x^{n-1/2}\left(i + 1, j + \frac{1}{2}, k + \frac{1}{2} \right) - v_x^{n-1/2}\left(i, j + \frac{1}{2}, k + \frac{1}{2} \right) \right] - \alpha(3\lambda + 2\mu)\Delta t \frac{\partial T\left(i + \frac{1}{2}, j + \frac{1}{2}, k + \frac{1}{2} \right)}{\partial t}$$

$$\tag{9.18}$$

$$\sigma_{zz}^{n}\left(i+\frac{1}{2},j+\frac{1}{2},k+\frac{1}{2}\right)=\sigma_{zz}^{n-1}\left((i+\frac{1}{2},j+\frac{1}{2},k+\frac{1}{2})\right)+\frac{(\lambda+2\mu)\Delta t}{\Delta z}\left[v_{z}^{n-1/2}\left(i+\frac{1}{2},j+\frac{1}{2},k+1\right)-v_{z}^{n-1/2}\left(i+\frac{1}{2},j+\frac{1}{2},k\right)\right]$$

$$+\frac{\lambda\Delta t}{\Delta x}\left[v_{x}^{n-1/2}\left(i+1,j+\frac{1}{2},k+\frac{1}{2}\right)-v_{x}^{n-1/2}\left(i,j+\frac{1}{2},k+\frac{1}{2}\right)\right]$$

$$+\frac{\lambda\Delta t}{\Delta y}\left[v_{y}^{n-1/2}\left(i+\frac{1}{2},j+1,k+\frac{1}{2}\right)-v_{y}^{n-1/2}\left(i+\frac{1}{2},j,k+\frac{1}{2}\right)\right]-\alpha(3\lambda+2\mu)\Delta t\frac{\partial T\left(i+\frac{1}{2},j+\frac{1}{2},k+\frac{1}{2}\right)}{\partial t}$$

$$(9.19)$$

and the transverse shear components are:

$$\sigma_{xy}^{n}\left(i,j,k+\frac{1}{2}\right)=\sigma_{xy}^{n-1}\left(i,j,k+\frac{1}{2}\right)+\frac{\mu\Delta t}{\Delta x}\left[v_{y}^{n-1/2}\left(i+\frac{1}{2},j,k+\frac{1}{2}\right)-v_{y}^{n-1/2}\left(i-\frac{1}{2},j,k+\frac{1}{2}\right)\right]$$

$$+\frac{\mu\Delta t}{\Delta y}\left[v_{x}^{n-1/2}\left(i,j+\frac{1}{2},k+\frac{1}{2}\right)-v_{x}^{n-1/2}\left(i,j-\frac{1}{2},k+\frac{1}{2}\right)\right]$$

$$(9.20)$$

$$\sigma_{yz}^{n}\left(i+\frac{1}{2},j,k\right)=\sigma_{yz}^{n-1}\left(i+\frac{1}{2},j,k\right)+\frac{\mu\Delta t}{\Delta y}\left[v_{z}^{n-1/2}\left(i+\frac{1}{2},j+\frac{1}{2},k\right)-v_{z}^{n-1/2}\left(i,j-\frac{1}{2},k\right)\right]$$

$$+\frac{\mu\Delta t}{\Delta z}\left[v_{y}^{n-1/2}\left(i+\frac{1}{2},j,k+\frac{1}{2}\right)-v_{y}^{n-1/2}\left(i+\frac{1}{2},j,k-\frac{1}{2}\right)\right]$$

$$(9.21)$$

$$\sigma_{zx}^{n}\left(i,j+\frac{1}{2},k\right)=\sigma_{zx}^{n-1}\left(i,j+\frac{1}{2},k\right)+\frac{\mu\Delta t}{\Delta z}\left[v_{x}^{n-1/2}\left(i,j+\frac{1}{2},k+\frac{1}{2}\right)-v_{x}^{n-1/2}\left(i,j+\frac{1}{2},k-\frac{1}{2}\right)\right]$$

$$+\frac{\mu\Delta t}{\Delta x}\left[v_{z}^{n-1/2}\left(i+\frac{1}{2},j+\frac{1}{2},k\right)-v_{z}^{n-1/2}\left(i-\frac{1}{2},j+\frac{1}{2},k\right)\right]$$

$$(9.22)$$

$$v_{x}^{n+1/2}\left(i,j+\frac{1}{2},k+\frac{1}{2}\right)=v_{x}^{n-1/2}\left(i,j+\frac{1}{2},k+\frac{1}{2}\right)+\frac{\Delta t}{\rho\Delta x}\left[\sigma_{xx}^{n}\left(i+\frac{1}{2},j+\frac{1}{2},k+\frac{1}{2}\right)-\sigma_{xx}^{n}\left(i-\frac{1}{2},j+\frac{1}{2},k+\frac{1}{2}\right)\right]$$

$$+\frac{\Delta t}{\rho\Delta y}\left[\sigma_{xy}^{n}\left(i,j+1,k+\frac{1}{2}\right)-\sigma_{xy}^{n}\left(i,j,k+\frac{1}{2}\right)\right]+\frac{\Delta t}{\rho\Delta z}\left[\sigma_{zx}^{n}\left(i,j+\frac{1}{2},k+1\right)-\sigma_{zx}^{n}\left(i,j+\frac{1}{2},k\right)\right]$$

$$(9.23)$$

$$v_{y}^{n+1/2}\left(i+\frac{1}{2},j,k+\frac{1}{2}\right)=v_{y}^{n-1/2}\left(i+\frac{1}{2},j,k+\frac{1}{2}\right)+\frac{\Delta t}{\rho\Delta y}\left[\sigma_{yy}^{n}\left(i+\frac{1}{2},j+\frac{1}{2},k+\frac{1}{2}\right)-\sigma_{yy}^{n}\left(i+\frac{1}{2},j-\frac{1}{2},k+\frac{1}{2}\right)\right]$$

$$+\frac{\Delta t}{\rho\Delta z}\left[\sigma_{yz}^{n}\left(i+\frac{1}{2},j,k+1\right)-\sigma_{yz}^{n}\left(i+\frac{1}{2},j,k\right)\right]+\frac{\Delta t}{\rho\Delta x}\left[\sigma_{xy}^{n}\left(i+1,j,k+\frac{1}{2}\right)-\sigma_{xy}^{n}\left(i,j,k+\frac{1}{2}\right)\right]$$

$$(9.24)$$

$$v_z^{n+1/2}\left(i+\frac{1}{2},j+\frac{1}{2},k\right) = v_z^{n-1/2}\left(i+\frac{1}{2},j+\frac{1}{2},k\right) + \frac{\Delta t}{\rho\Delta z}\left[\sigma_{zz}^n\left(i+\frac{1}{2},j+\frac{1}{2},k+\frac{1}{2}\right) - \sigma_{zz}^n\left(i+\frac{1}{2},j+\frac{1}{2},k-\frac{1}{2}\right)\right]$$

$$+\frac{\Delta t}{\rho\Delta x}\left[\sigma_{zx}^n\left(i+1,j+\frac{1}{2},k\right) - \sigma_{zx}^n\left(i,j+\frac{1}{2},k\right)\right] + \frac{\Delta t}{\rho\Delta y}\left[\sigma_{yz}^n\left(i+\frac{1}{2},j+1,k\right) - \sigma_{yz}^n\left(i+\frac{1}{2},j,k\right)\right]$$

$$(9.25)$$

As noted earlier, the pulsed microwave-induced thermoelastic pressure is obtained from the stress tensor by the simple relationship given by Eq. 9.7; specifically,

$$p(x,y,z,t) = \frac{1}{3}\sum_{i=x,y,z} \sigma_{ii} \qquad (9.7)$$

9.1.4 Modulated Rectangular Pulse Functions

Since the sharp corners of rectangular pulses in the time domain give rise to extremely high-frequency components that cannot be efficiently computed by the FDTD method, rectangular pulses may be modulated by an error function to reduce the spurious high-frequency components to facilitate computation. A modulated pulse function with an angle parameter, b is introduced for the computations, such that,

$$\frac{\partial}{\partial t}T(x,y,z,t) = \frac{SAR(x,y,z)}{2Cp}\left\{\left[erf(bt)\right] - erf\left[b(t-t_0)\right]\right\} \qquad (9.26)$$

where C_p is specific heat, t_0 is the pulse duration, and $erf(bt)$ is the error function. The angle parameter may be conveniently set to 6×10^5 /s. Other investigators have chosen $b = 4 \times 10^5$ /s for the angle parameter [Watanabe et al., 2000].

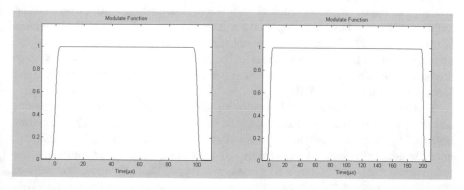

Fig. 9.1 Modulated pulse functions with the angle parameter set to 6×10^5 /s for pulse width of 100μs and 200μs to render slightly less high-frequency content

Examples of the pulse functions with $b = 6 \times 10^5$ /s for pulse width $t_0 = 100\mu s$ and $200\mu s$ are shown in Fig. 9.1. Note the slightly rounded corners of each pulse function. The introduction of a modulated function is not necessary for analytical calculations since the sharp corners of rectangular pulses do not present any mathematical issue.

9.2   Sound Pressure Waves Induced by RF Pulses in MRI Systems

Advantages of magnetic resonance imaging (MRI) have made it the radiological modality of choice for many diagnostic medical procedures. The clinical successes have also heightened the desire for increased spatial resolution from MRI systems. This demand has prompted the exploration of higher strength static magnetic fields, and the associated use of higher RF spectra and increasing levels of RF power. Pulsed RF magnetic fields are used to elicit magnetic resonance signals from tissues. In a typical MRI procedure, the patient is exposed to numerous pulses of RF radiation. For the common 1.5 tesla (1.5 T) clinical MRI scanner, the associated RF frequency for proton imaging at 1.5 T is about 64 MHz, although increasingly, 3.0 T (128 MHz) MRI scanners are installed in radiological imaging centers. At present, several experimental ultra-high-field strength and ultra-fast MRI systems operate at 7.0 T, 9.4 T, or higher, with a corresponding 300-MHz, 400 MHz, or higher RF frequencies for proton imaging. Clearly, MRI systems employ strong RF fields whose frequency depends on the static magnetic field strength.

Aside from some acute potential risks such as those from aneurysm clips, certain implanted or attached metal objects, etc., adverse health effects have not been associated with clinical MRI imaging. While overheating of tissues in cases with varying body morphologies has been identified [Chakeres et al., 2003; Chakeres and de Vocht, 2005; de Vocht et al., 2007; Glover et al., 2007; Houpt et al., 2005; Shellock and Crues, 2004; Wang et al., 2007, 2008; Wang and Lin, 2012; FDA, 2018], with appropriate attention to patient protocols, the safety issue may be mitigated or realistically managed in most clinical situations. Interestingly, RF-pulse-induced auditory responses have been reported for human subjects during MRI procedures [Roschmann, 1991]. Indeed, a theoretical study via computer simulation has demonstrated that acoustic pressure waves in the human head induced by MRI system are two–three fold above minimum sound perception pressure in the cochlea of the inner ear [Lin and Wang, 2010].

This section discusses FDTD computer simulation studies of the frequency, intensity, and pattern of thermoelastic sound pressure waves generated by RF pulses absorbed by the head of human subjects in the MRI scanning system. As mentioned, acoustic pressure waves have been shown to arise from a rapid (~μs) but miniscule temperature rise (~10^{-6} °C) in head tissues, which generates a thermoelastic wave of pressure and propagates to the cochlea inside the head for sound perception [Lin

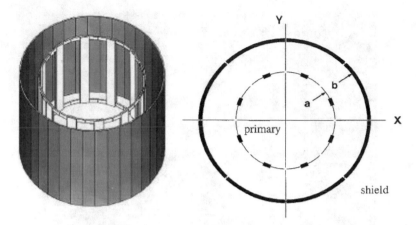

Fig. 9.2 Schematic diagram of a birdcage antenna for near field exposure: a for coil radius; b for shield radius

and Wang, 2007; 2010]. The computer simulations are performed by sequentially implementing the FDTD solutions to Maxwell's electrodynamic equations, and the biological heat transfer and thermoelastic expansion equations.

A high-pass birdcage MRI coil antenna of conventional design is adopted [Jin, 1999; Ibrahim et al., 2000]. Its dimensions are 28 cm in diameter (distance between legs or rungs on opposite sides of the coil) and 22 cm in length (Fig. 9.2). The coil consisted of 16 rungs with a 1 cm rung width, equipped with a cylindrical shield with a diameter and length of 36 cm and 28 cm, respectively. Current sources are placed at the midpoint of each end-ring element. Each current source has a sinusoidal waveform with a 22.5° phase shift between elements, producing a circularly polarized field inside the coil antenna. The birdcage coils are popular because they can produce more homogeneous static magnetic (B_1) field over a large volume within the coil for the current generation of clinical imagers. The human head models examined include homogeneous brain spheres and anatomical image-based data sets obtained from the "VH Project" from the U.S. National Library of Medicine.

9.2.1 Homogeneous Spherical Head Model

9.2.1.1 Pressure Wave and Spectrum

Computed thermoelastic pressures and power spectra for a homogeneous spherical model of adult-size head (18-cm diameter) inside 1.5 T (64 MHz) and 7.0 T (300 MHz) MRI coil antennas for 100μs and 200μs pulses are shown in Fig. 9.3 [Lin and Wang, 2007; 2010]. A similar set of computed thermoelastic pressures and power spectra for a homogeneous spherical model of child-size head (10-cm diameter) are given in Fig. 9.4. The results show obvious variations of the pressure wave for 100- and

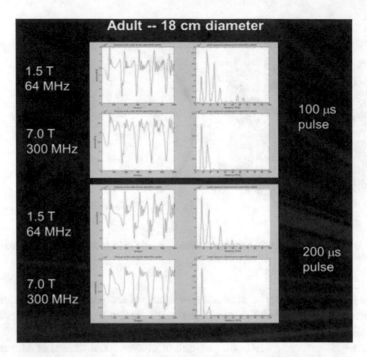

Fig. 9.3 Computed thermoelastic pressures and power spectra for a spherical model of adult-size head (18-cm diameter) inside 1.5 T and 7.0 T MRI imaging coils for 100μs and 200μs pulses

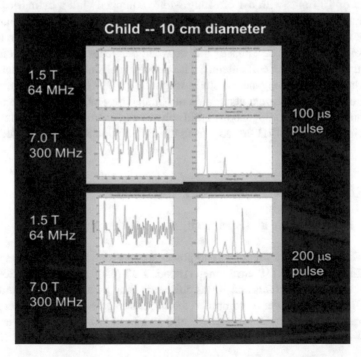

Fig. 9.4 Computed thermoelastic pressures and power spectra for a spherical model of child-size head (10-cm diameter) inside 1.5 T and 7.0 T MRI imaging coils for 100μs and 200μs pulses

200-μs pulses, both in the spherical adult human and child head models. It should be emphasized that the thermoelastic pressures oscillate as a function of time both during and after the pulse application and reach a peak shortly after the cessation of the 100- or 200-μs pulse. Further details of the computation, tissue properties, and results are given in sections that follow.

A complex wave of pressure oscillating as functions of time with different characters can be seen. Initially, a negative expansion pressure begins with zero intensity, then grows to a peak value at about 40-μs, and is followed by oscillation to a positive peak at about 70-μs, which rises to even higher peaks, after the end of the pulse. The power spectra of the pressure waves showed that the spectral amplitudes for the 200-μs pulses are greater than that of the 100-μs pulses. However, a fundamental frequency component at about 8 kHz inside the adult-head sphere, followed by a series of higher harmonics are observed both for 64 MHz (1.5 T) and 300 MHz (7.0 T). The fundamental frequency component at about 14.5 kHz inside the child-head sphere along with a series of even higher harmonics are associated with the smaller sphere. This means that the fundamental frequency of the thermoelastic pressure wave is not a function of MRI field strength, but depends on the dimensions of the head sphere exposed to the birdcage MRI coil antenna, as predicted by the microwave thermoelastic theory discussed above and elsewhere [Lin 1977a, b, c; Lin and Wang 2010]. Moreover, while the spectral power at 7 T (300 MHz) was much higher than that at 1.5 T (64 MHz), similarities in the spectral content indicate that the spectral power peaks correspond to the resonant frequencies of pressure waves reverberating inside the spherical head model.

9.2.1.2 SAR Distributions

An FDTD study of the characteristics of SAR induced by birdcage coil antennas in canonical models such as the homogeneous brain sphere may facilitate a better understanding of some of the observations in addition to its inherent interest of itself. However, SAR distributions in anatomical head models for different frequencies, as shall be seen later, tend to be asymmetric, nonuniform, and irregular. While they may be underscored by the fact that the head is asymmetric and heterogeneous with complex permittivity properties. Unique aspects of the physical interaction of RF with the biological structure's geometry and tissue composition also have influences on the SAR behavior.

For FDTD simulation, a homogeneous spherical head model is inserted in the center of the computational domain for it to coincide with the center of the birdcage coil antenna. Figure 9.5 gives the results of FDTD-computed SAR distributions from absorption of continuous sinusoidal 1.5 T (64 MHz) and 7.0 T (300 MHz) RF fields in an adult- and child-size head sphere. Note that SAR distributions are the same as those for CW and pulsed exposures. In this case, the birdcage coil antenna is oriented with its length along the z-direction of a rectangular coordinate system. Thus, the xy, yz, and zx planes become the corresponding axial, coronal, and

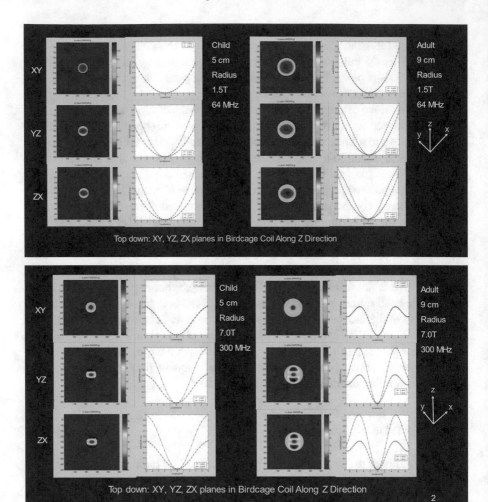

Fig. 9.5 FDTD-computed SAR distributions from absorption of continuous sinusoidal RF energy in spherical head models inside a birdcage coil antenna of MRI scanner

sagittal planes of reference in human anatomy. In Fig. 9.5, localized SAR variations shown in line graphs and 2-D images are displayed in relative units.

The SAR distributions are nearly symmetric in any one direction about the center of the spheres but not equal along all directions. There are significant differences in SAR distributions even in a symmetric brain sphere for various head sizes at different frequencies. For a child's head of 10-cm in diameter or a lower 64 MHz frequency (or when the ratio of head size to wavelength is small), the SAR has a value of zero at the center and rises steadily to a maximum toward the edge of the head sphere due to induced circular flow of eddy currents. Magnitudes of the circulating eddy currents are directly proportional to the distance from the sphere's center. At

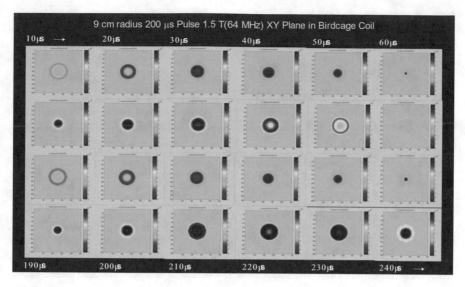

Fig. 9.6 Two-dimensional views of computed pressure wave intensities in relative units over the central axial plane in an adult spherical head exposed to a 200-μs pulse at 64 MHz (1.5 T)

the higher frequency (300 MHz), geometric resonance adds a standing-wave pattern to the flow of eddy currents, producing multiple SAR peaks in varying distributions inside the homogeneous adult-size spherical head along different directions and planes. However, SAR at the center remains at a minimum. To some extent, the same behavior is observable for a child-size head at 64 MHz. It is interesting to note from the coronal and sagittal planes of the adult head sphere, at 300 MHz, there are two mirrored SAR resonance patterns with maxima located above and below the center of the sphere where SAR is minimum.

9.2.1.3   Thermoelastic Pressure Waves

Using the SAR distributions from above as inputs and a modulated rectangular RF pulse applied singly to a birdcage coil, thermoelastic pressure waves inside homogeneous spherical head models are computed by FDTD simulation. Figures 9.6 and 9.7 show 2-D spatial views of computed pressure wave intensities in relative units over the central axial plane in an adult spherical head exposed to a 200-μs pulse at 64 MHz (1.5 T) and 300 MHz (7.0 T), as a function of time, in 10-μs steps for the first 240μs. A high-SAR region at the periphery of the adult head sphere, as seen from the SAR distribution in Fig. 9.5 for 64 MHz, initiates an expanding negative pressure that begins with zero intensity and travels toward the center, grows to a peak negative value and reversing itself, and rises to a peak positive pressure then falls again before going into resonant oscillation, or a reverberation mode as the pressure waves travel back and forth between the edge of sphere and its center.

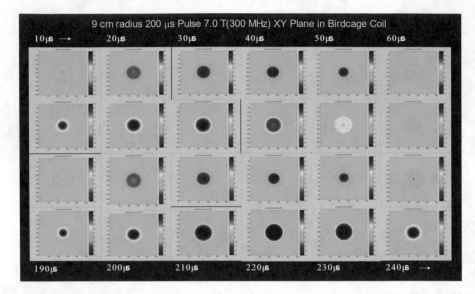

Fig. 9.7 Two-dimensional views of computed pressure wave intensities in relative units over the central axial plane in an adult spherical head exposed to a 200-μs pulse at 300 MHz (7.0 T)

Figure 9.7 shows similar general variations of the pressure wave for a 200-μs pulse in the adult spherical model at 300 MHz (7.0 T). However, the more homogeneous SAR distribution in this case, instead of allowing the pressure wave to behave as a traveling wave, enables the initial negative pressure wave to go into oscillation sooner. Thus, it provides a different space–time relationship of the pressure wave as it reverberates inside the head sphere. Thus, acoustic pressure waves are indeed generated in homogeneous head spheres by RF pulses from both 64 and 300 MHz birdcage coils. The induced thermoelastic pressure waves oscillate as a function of time during and after the pulse application and reach a peak shortly after the cessation of the 200-μs pulse (see Fig. 9.3 for time-domain signals). Moreover, the birdcage MRI-induced sound-pressure waves are complex functions of head size, pulse width and strength, MRI frequency, and SAR distribution pattern.

9.2.2   Anatomical Head Model

The anatomical human head model with a 3-mm spatial resolution is derived from segmented images obtained from the "VH Project" of the National Library of Medicine, which is a 3-D digital image library representing an adult human male and female (http://www.nlm.nih.gov/research/visible/visible_human.html). The segmentation procedure had been carried out for the male model by the Air Force Research Laboratory, Brooks City, Texas, USA [Mason et al. 2000]. The

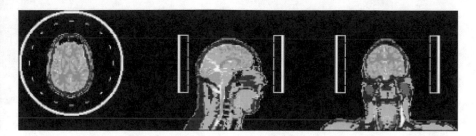

Fig. 9.8 The anatomical head model embedded in MRI birdcage coil antenna. Views of coil and head model geometry on the axial, sagittal, and coronal planes passing through the coil center

Table 9.1 Dielectric permittivity and conductivity of tissues for the anatomical head model

Tissue material	64 MHz		300 MHz		400 MHz		Density
	σ (S/m)	ε_r	σ (S/m)	ε_r	σ (S/m)	ε_r	ρ (kg/m³)
Blood	1.206	86.506	1.316	65.689	1.350	64.181	1058
Blood vessel	0.429	68.687	0.537	48.357	0.561	47.003	1040
Body fluid	1.503	69.127	1.518	69.017	1.529	69.000	1010
Bone.(cancellous)	0.161	30.888	0.215	23.181	0.234	22.442	1920
Bone.(cortical)	0.059	16.690	0.082	13.446	0.091	13.147	1990
Bone marrow	0.021	7.215	0.027	5.7609	0.029	5.6728	1040
Cartilage	0.452	62.959	0.552	46.808	0.586	45.467	1097
Cerebellum	0.719	116.520	0.972	59.819	1.030	55.991	1038
Cerebral spinal fluid (CSF)	2.066	97.354	2.224	72.786	2.251	70.997	1007
Eye (aqueous Humor)	1.503	69.127	1.517	69.017	1.529	69.000	1009
Eye (cornea)	1.000	87.447	1.150	61.431	1.193	59.275	1076
Eye (lens)	0.586	60.569	0.647	48.972	0.668	48.150	1053
Eye (retina)	0.883	75.347	0.975	58.934	1.000	57.672	1026
Eye (sclera/wall)	0.883	75.347	0.975	58.934	1.000	57.672	1026
Fat	0.035	6.509	0.040	5.6354	0.041	5.580	916
Glands	0.778	73.979	0.851	62.472	0.877	61.548	1050
Gray Matter	0.511	97.538	0.691	60.090	0.738	57.432	1038
Ligaments	0.474	59.524	0.537	48.001	0.560	47.289	1220
Lymph	0.778	73.979	0.851	62.472	0.877	61.548	1040
Mucous Membrane	0.488	76.797	0.630	51.959	0.669	49.895	1040
Muscle	0.688	72.274	0.770	58.229	0.796	57.127	1047
Nerve (spine)	0.312	55.109	0.418	36.951	0.447	35.406	1038
Skin/dermis	0.435	92.290	0.640	49.902	0.688	46.783	1125
Tooth	0.059	16.690	0.082	13.446	0.091	13.147	2160
White matter	0.291	0.291	0.413	43.821	0.445	42.070	1038

final-segmented model, made freely available to the scientific community, is comprised of 586 × 340 × 1878 voxels, each with a resolution of 1 × 1 × 1 mm³, and is segmented in about 40 different tissue types (ftp://starview.brooks.af.mil/EMF/dosimetry_models/). The human head model used in this case had approximate dimensions of 137 × 182 × 180 mm, and consisted of 25 tissue types, including grey matter, white matter, cerebral spinal fluid (CSF), muscle, fat, bone, tooth, etc. (Fig. 9.8). Their relative permittivity and conductivity values are listed in Table 9.1, where a four-parameter Cole–Cole interpolation technique [Cole and Cole, 1941] was used to determine values for the dielectric properties of the tissues at 1.5 T, 7.0 T, and 9.4 T (64, 300, and 400 MHz).

The elastic properties for soft tissues are characterized by the two Lame's elasticity constants: $\lambda = 2.24$ GPa and $\mu = 1.052 \times 10^{-6}$ GPa, and the coefficient of linear

Table 9.2 Thermal properties of biological tissues for the head model

Tissue material	Specific heat	Blood perfusion coefficient	Thermal conductivity
	J/(kg- °C)	J/(m³-s- °C)	J/(m-s- °C)
Blood	3900	0	0.49
Blood vessel	3553	9000	0.46
Body fluid	4155	0	0.62
Bone (cancellous)	1300	3300	0.40
Bone (cortical)	1300	3400	0.40
Bone marrow	2700	32,000	0.22
Cartilage	3500	9000	0.47
Cerebellum	3700	40,000	0.57
CSF	4200	0	0.62
Eye (cornea)	4200	0	0.58
Eye (lens)	3000	0	0.40
Eye (retina)	3680	35,000	0.57
Eye (sclera/wall)	4200	0	0.58
Eye (aqueous humor)	4200	0	0.60
Fat	2500	1700	0.25
Glands	3600	360,000	0.53
Gray matter	3700	40,000	0.57
Ligaments	2802	4830	0.31
Lymph	3686	31,800	0.49
Mucous membrane	3300	9000	0.43
Muscle	3600	2700	0.50
Nerve (spine)	3500	40,000	0.46
Skin/dermis	3500	9100	0.42
Tooth	1340	0	0.50
White matter	3600	40,000	0.50

thermal expansion, $\alpha = 4.1 \times 10^{-5}$ per °C. For bone and tooth tissue, the values are given by $\lambda = 6.923$ GPa, $\mu = 4.615$ GPa, and $\alpha = 1.06 \times 10^{-5}$ (1/°C) [Lin, 1978].

The thermal properties of tissues for the head model have been compiled from different sources ([Bernardi et al., 1998, 2003]; Lin and Bernardi [2007]). Some available numerical values are given in Table 9.2. Note that for brief (~100µs) RF pulses, the short durations are insufficient for significant conductive or convective heat transfer to contribute to tissue temperature rise. In this case for a specific heat of C_p, the time rate of rise in temperature is related directly and linearly to SAR, i.e.,

$$Cp\frac{\Delta T}{\Delta t} = SAR \qquad (9.27)$$

Accordingly, using the properties listed in Table 9.2, a single, 100-µs RF pulse would induce a rapid but miniscule temperature elevation of about 10^{-6} °C in brain tissues. This small theoretical temperature elevation is undetectable by any currently available temperature sensors and cannot be felt as a thermal sensation or heat. Nevertheless, it can launch an acoustic wave of pressure that propagates inside the head to the inner ear for perception by humans.

9.2.2.1   Computed SAR Distributions

The anatomical head model is embedded both in the center of the MRI birdcage coil antenna (Fig. 9.8) and in the computational domain, which consists of a $150 \times 150 \times 150$ grid of $3 \times 3 \times 3$ mm voxel Yee cells [Yee, 1966]. A Berenger PML

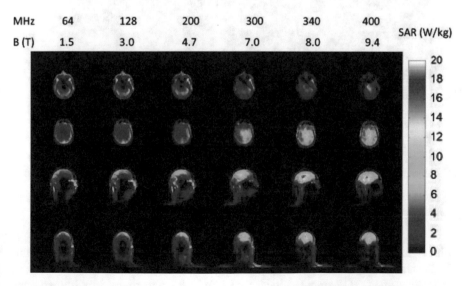

Fig. 9.9 SAR distributions for a single voxel averaging mass in anatomic head model, shown in axial (2 views), coronal, and sagittal planes normalized to an average whole-head SAR of 3 W/kg

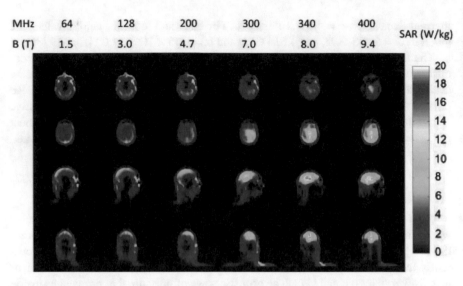

Fig. 9.10 SARs for 1-g averaging mass shown in the coronal, sagittal, and two axial planes in normalized values for an average whole-head SAR of 3 W/kg

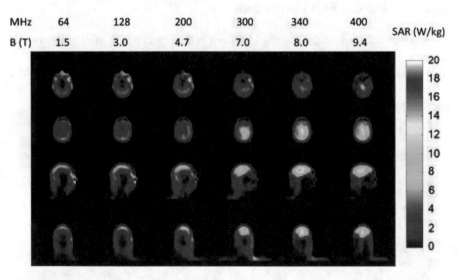

Fig. 9.11 SARs for 10-g averaging mass shown in the coronal, sagittal, and two axial planes in normalized values for an average whole-head SAR of 3 W/kg

with eight layers is implemented as the electromagnetic absorption boundary in the FDTD algorithm [Berenger, 1996].

Results of FDTD-computed local SARs from RF absorption in a human anatomical head model situated inside a birdcage coil antenna of MRI scanners are given in Figs. 9.9, 9.10 and 9.11. The image displayed is taken from an axial plane passing through the eyes (top), an axial plane passing through the center of the coil and brain (second row), and sagittal and coronal planes passing through the center of the coil and brain (third and bottom rows). The coil operating frequencies are 64 MHz (1.5 T), 128 MHz (3.0 T), 200 MHz (4.7 t), 300 MHz (7.0 T), 340 MHz (8.0 T), and 400 MHz (9.4 T). Note that the SARs for a single voxel (Yee cell), and 1-g and 10-g averaging mass shown in the coronal, sagittal, and two axial planes are normalized values for an average whole-head SAR of 3 W/kg.

Localized SAR distributions are asymmetric, nonuniform, and variable; they underscore the fact that the head is asymmetric and heterogeneous with complex permittivity properties. Although there are some differences for 1 voxel, 1-g or 10-g averaging mass and for different frequencies, these differences are more noticeable at higher frequency or T values than lower ones such as 64 MHz (1.5 T) and 128 MHz (3.0 T), where the SAR is relatively more uniformly distributed. It is also noteworthy that for each of the frequencies, computed SAR values decrease significantly as the averaging mass increases. It clearly shows that a larger averaging mass may not provide as accurate a representation of RF induced SARs in biological tissues, as reported previously [Cavagnaro and Lin, 2019].

The peak SAR in the head occurs at different locations for different frequencies in general. For example, the peak SARs are found in subcutaneous fat and muscle tissues at the lower frequencies, whereas they are associated with grey or white matter in the brain at higher frequencies (see Table 9.3). Note that the 1-g (SAR_{1g}) and 10-g (SAR_{10g}) regions are calculated by a standard method whereby SARs in a region of tissue-containing voxels (Yee cells) surrounding a central cell are averaged as the region is expanded by one cell at a time until a specified mass of tissue (1 g or 10 g) is reached.

Moreover, at lower frequencies, the computed SARs show local peak absorptions in the superior-peripheral regions such as the scalp and the top of the head. SAR decreases with distance toward the center and becomes nearly zero at the middle of the head. With increasing RF frequency, the SAR results not only show higher local absorptions in the peripheral regions, but also peak SARs are found in the

Table 9.3 Peak SARs for birdcage coils calculated for 1 voxel (SAR_{1cell}), 1-g (SAR_{1g}), and 10-g (SAR_{10g}) regions of specified mass of tissue

RF frequency	Magnetic field	Peak SAR_{1cell} (1 Voxel)		Peak SAR_{1g} (1 g)		Peak SAR_{10g} (10 g)	
MHz	Tesla (T)	W/kg	Tissue type	W/kg	Tissue type	W/kg	Tissue type
64	1.5	48.19	Mucous membrane	16.97	Muscle	10.95	Fat
128	3.0	47.46	Mucous membrane	16.21	Muscle	10.70	Ligament
200	4.7	41.73	Mucous membrane	17.35	Muscle	11.82	Muscle
300	7.0	39.21	CSF	14.93	White matter	11.66	Grey matter
340	8.0	48.51	CSF	21.48	White matter	15.03	Grey matter
400	9.4	43.61	CSF	20.54	White matter	13.41	Grey matter

central regions of the brain, especially at 300 MHz (7.0 T) or 400 MHz (9.4 T). It is interesting to note the region of high SAR in the upper portion of the head migrating toward its center as the frequency increases. It is most evident in the coronal and sagittal slices at higher frequencies, but not so below the center of the head model – a consequence of the upper portion of the head with its spheroidal geometry and homogeneity, and the asymmetry and heterogeneity of tissues in the lower portion of the head inside the birdcage coil antenna. As noted in the spherical head model, at these frequencies an eddy current is induced and circulates inside the head models. The magnitude of the circulating eddy current is directly proportional to the distance from its center of circulation and the current density is constrained by the cross-sections of different tissues through which current flows (SAR is proportional to the square of current density).

9.2.2.2   Temperature Computation

An equilibrium temperature distribution was first computed with SAR = 0 W/kg throughout the anatomical head model embedded both in the center of the birdcage coil antenna and in the FDTD computational domain mentioned above. The ambient temperature is set at 24 °C and the temperature of blood is a constant at 37 °C. It is assumed that the rate of blood perfusion is independent of time and temperature. The temperature elevations induced by the SAR patterns of Figs. 9.9, 9.10 and 9.11, after 30 min of exposure with a whole-head averaged SAR of 3.0 W/kg are given in Figs. 9.12, 9.13 and 9.14. The MRI system's operating frequencies are 64 MHz

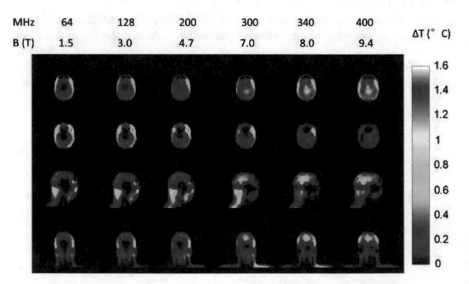

Fig. 9.12 Temperature elevations induced by the SAR$_{1c}$ patterns of Fig. 9.9 after 30 min of exposure

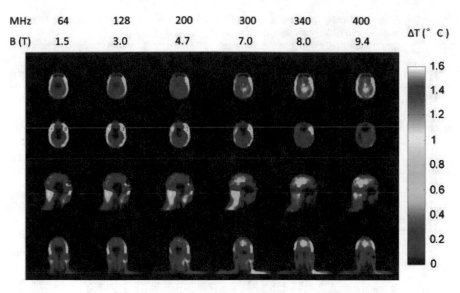

Fig. 9.13 Temperature elevations induced by the SAR_{1g} patterns of Fig. 9.10 after 30 min of exposure

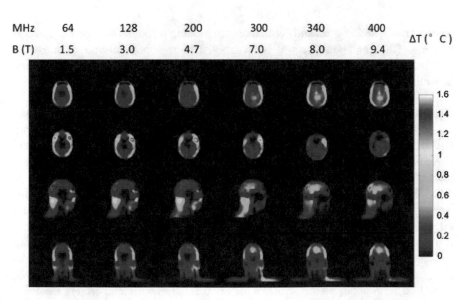

Fig. 9.14 Temperature elevations induced by the SAR_{10g} patterns of Fig. 9.11 after 30 min of exposure

(1.5 T), 128 MHz (3.0 T), 200 MHz (4.7 t), 300 MHz (7.0 T), 340 MHz (8.0 T), and 400 MHz (9.4 T). The image slices are taken from an axial plane passing through the eyes (top), an axial plane passing through the center of the coil and brain

Table 9.4 Calculated peak temperature in head tissues and temperature elevations for birdcage coil antennas at 30 min of RF exposure at head-average SAR of 3 W/kg

RF frequency	Magnetic field	Temperature (1 cell or voxel)		Temperature (1 g)		Temperature (10 g)	
		Max	ΔT	Max	ΔT	Max	ΔT
MHz	Tesla (T)	°C Tissue	°C Tissue	°C Tissue	°C Tissue	°C Tissue	°C Tissue
64	1.5	38.28 Muscle	1.65 Ligament	38.27 Ligament	1.64 Ligament	38.20 Ligament	1.59 Fat
128	3.0	38.40 Muscle	1.75 Ligament	38.39 Ligament	1.74 Ligament	38.31 Ligament	1.66 Fat
200	4.7	38.58 Muscle	1.89 Muscle	38.55 Ligament	1.89 Ligament	38.46 Ligament	1.76 Ligament
300	7.0	38.16 Muscle	1.48 Muscle	38.16 Muscle	1.57 Fat	38.16 Muscle	1.55 Fat
340	8.0	38.48 Grey matter	1.21 Grey matter	38.40 Ligament	1.14 Ligament	38.22 Grey matter	0.96 Ligament
400	9.4	38.41 Grey matter	1.12 Grey matter	38.36 Ligament	1.10 Ligament	38.18 Ligament	1.01 Cortical bone

(second row), and sagittal and coronal planes passing through the center of the coil and brain (third and bottom rows). Noticeably, the temperature elevations inside the anatomical head are affected by its structural asymmetry and tissue heterogeneity, as well as the operating coil's RF frequency.

At the head-average SAR of 3 W/kg currently allowed by the U.S. Food and Drug Administration [FDA, 2018), the highest rise in temperature, reaching a value of 1.89 °C, is found in peripheral tissues such as muscle for frequencies from 64 MHz to 300 MHz (Table 9.4). At higher frequencies, the maximum temperature elevation is around 1 °C (the limit of FDA allowable elevation under normal imaging operation) for tissues that are more centrally located such as grey matter at 400 MHz. Also note that the simulated temperatures and temperature elevations decrease as the tissue averaging mass increase from 1 voxel or Yee cell to 1 g and 10 g, as expected, at each frequency and in general – a strong indication that there is a spatial correlation between averaging mass for SAR and temperature elevation.

Recent results on the correlation of SAR with induced tissue temperature elevation, and the dependence on the mass of averaging tissue and exposure duration, show that, in general, SAR provides a better correlation with temperature elevation for exposure durations shorter than 2 min (short durations) at most frequencies examined [Cavagnaro and Lin, 2019]. For example, a mass of 1 g is optimal, but the correlation coefficient remains above 0.9 at 2 min for a 2-g mass. For longer exposures, the maximum correlation coefficient is reduced, and the correlation favors a larger averaging mass. At steady state (30 min), the correlation of temperature increase with SAR is maximum for a mass of 5–9 g at frequencies below 6000 MHz.

Aside from exposure duration, other major factors include heterogeneity of tissues, variation in perfusion rate between different tissues, and thermal conductivity distribution in regions of relatively low SAR.

It should be mentioned that the simulation of temperature distribution and elevation described above, and results shown are illustrative of FDTD computation of temperatures involving the biological heat transfer equation. The numerical data presented are specific to the head-average SAR at 3 W/kg and an exposure duration of 30 min, which in fact is close to steady-state exposure. However, in the multidisciplinary FDTD simulation for microwave thermoelastic pressure waves produced by a miniscule (~10^{-6} °C), but rapid (~μs) rise of temperature in the head as a result of the absorption of pulsed RF or microwave energy, the bioheat equation and elastic equation of motion are involved as an integral part of FDTD simulation cascade along with the applicable biological parameters, as discussed in the next section.

9.2.3   Thermoelastic Pressure Waves in Anatomical Head Model

The approach taken in this section is to employ the FDTD-computed SAR distributions directly as input to the FDTD thermoelastic pressure algorithm as a two-step process. Modulated rectangular RF pulses are applied singly with a pulse width of 100 or 200μs. The rectangular pulses are modulated by an error function to reduce spurious high-frequency components for more efficient computation by the FDTD algorithm (see Sect. 9.1.4 and Eq. 9.26). A stress-free boundary condition is imposed at the surface of the VH head model. The anatomical head model is embedded both in the center of the birdcage MRI coil and in the computational domain, same as previously described and consists of a 150 × 150 × 150 grid of 3 × 3 × 3 mm voxels [Lin and Wang, 2007; 2010]. The results shown below are normalized to an input current of 100 mA to the birdcage MRI coils. Values of SAR are proportional to the square of coil currents.

The FDTD-simulated SAR distributions in the anatomical head model inside high-pass, birdcage MRI coils operating at 64 MHz (1.5 T), 300 MHz (7.0 T), and

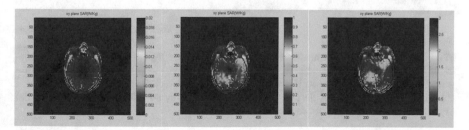

Fig. 9.15 Computer-simulated SAR distributions for exposures to 64 MHz (1.5 T), 300 MHz (7.0 T), and 400 MHz (9.4 T) birdcage coil antennas

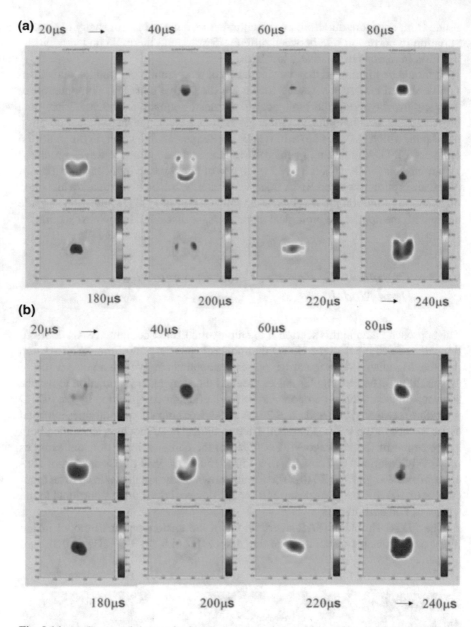

Fig. 9.16 (**a**) Computed thermoelastic pressure waves induced by a 100μs pulse in the VH anatomic head model exposed to 64 MHz (1.5 T) birdcage coil antenna. (**b**) Thermoelastic pressure waves induced by a 100μs pulse in the head model exposed to 128 MHz (3.0 T) birdcage coil antenna. (**c**) Computed thermoelastic pressure waves induced by a 100μs pulse in the head model exposed to 300 MHz (7.0 T) birdcage coil antenna

(c)

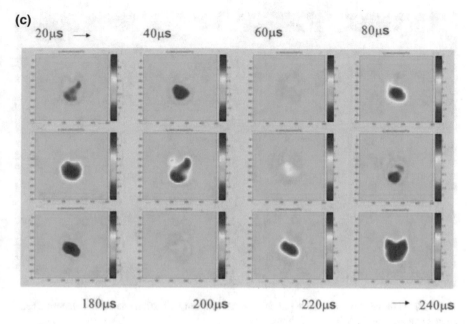

Fig. 9.16 (continued)

400 MHz (9.4 T) are shown in Fig. 9.15 for an axial plane at the level of the eyes. As expected, the computed SARs at 64 MHz are variable and displayed local peak absorptions in the peripheral regions such as the area of the eyes and nose. The highest SARs appeared in the skin and muscle (0.0024 W/kg for 100 mA), but the SAR values are much lower throughout the rest of the brain. For the higher fields at 7.0 and 9.4 T (300 and 400 MHz), in addition to the peripheral region, the SARs are higher in several regions deep in the brain. Peak SARs in the right-temporal and central-occipital regions are 2.24 and 6.82 W/kg for 300 and 400 MHz, respectively. The appearance of multiple SAR peaks inside the brain demonstrates a geometric resonance absorption phenomenon with standing-wave patterns inside the head at the higher RF frequencies. Note that, in all cases, the SAR was low near the center of the anatomical head model.

The FDTD-computed thermoelastic pressure waves inside the head at 64 MHz (1.5 T), 300 MHz (7.0 T), and 400 MHZ (9.4 T) are illustrated in Fig. 9.16 for 100µs pulses. The sequences of computed cross-section images shown in 20µs intervals show a complex pattern and propagation phenomenon of induced thermoelastic pressure waves in the axial plane. An expanding negative pressure wave, initiated at the location where the SAR is highest, launches an acoustic pressure wave front that propagates away from that location (periphery in Fig. 9.16). For 64 MHz (1.5 T), the expanding negative pressure waves first propagate toward the center of the head. When it reached the center, the pressure wave collapses, reverses itself, and starts to increase as a positive pressure. The reverberating pressure wave then begins

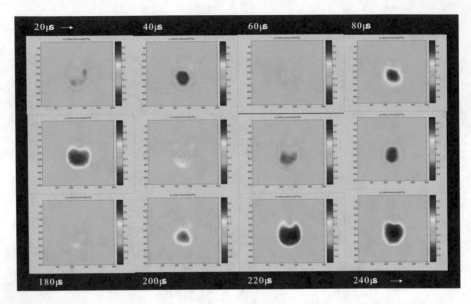

Fig. 9.17 Computed thermoelastic pressure waves induced by a 200μs pulse in a anatomic head model exposed to 300 MHz (7.0 T) birdcage coil antenna

propagating out toward the peripheral region. Upon reaching the head surface, the pressure again reverses in intensity and direction, then repeats itself in a reverberation mode of expansive and compressive pressure waves. However, for 300 MHz (7.0 T) and 400 MHZ (9.4 T), the "less" homogeneous SAR distributions with multiple SAR peaks deep inside brain tissues, instead of a traveling pressure wave, the initial negative pressure wave quickly spreads out and goes into oscillation and reverberation mode inside the anatomical head.

At 100μs, the driving energy for pressure generation from the RF pulse ceases, and the pressure variations thereafter are the result of reverberation from resonant acoustic pressure waves. Moreover, results for 200μs pulses (Fig. 9.17) indicate a pulse width effect on the pressure wave. In the first 100μs, images of the pressure distributions at 20, 40, 60, 80, and 100μs are comparable for 100μs and 200μs pulses, suggesting similar effects from the two different RF pulses. However, images of the pressure distribution for the 100-μs and 200-μs pulses became quite different at 120μs. For 100μs pulses, the driving energy from the RF pulse ceases and the pressure variation is influenced by resonances from the reverberating pressure wave. For 200μs pulses, energy from the RF pulse continues to affect pressure generation. Therefore, the peak pressure intensity for the 200μs cases is generally higher than for 100μs pulses. This observation is clearly illustrated by the series of temporal variations of the thermoelastic pressure waves at the center of the same axial planes

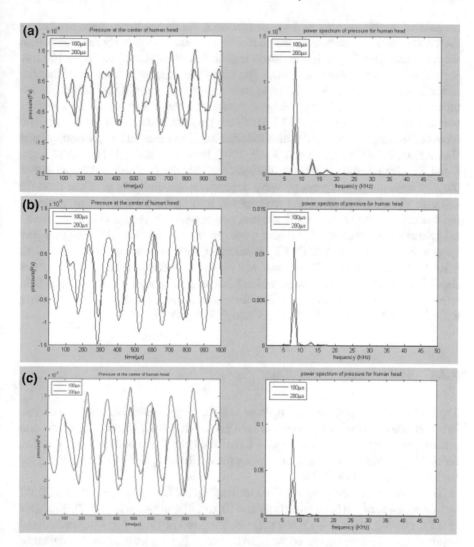

Fig. 9.18 Time variations and corresponding power spectra of the thermoelastic pressure waves computed in the center of axial planes of an anatomic head model at (a) 64 MHz (1.5 T); (b) 300 MHz (7.0 T); and (c) 400 MHz (9.4 T) for 100μs and 200μs RF pulses

of the head (Fig. 9.18). The corresponding power spectra at 64 MHz (1.5 T), 300 MHz (7.0 T), and 400 MHz (9.4 T) are also given for the first 1000μs.

The series of time-domain oscillating acoustic pressure waves in Fig. 9.18 for 100μs and 200μs RF pulses show the pressure intensities rise and fall as a function of time both during the pulse and after the pulse had ended. Note that in each case the pressure starts at zero then increases to a negative peak and is followed by oscillations that can rise to even higher peaks after the pulse ends. The pressure distributions are different for 64 MHz (1.5 T), 300 MHz (7.0 T), and 400 MHz (9.4 T) because of the differences in SAR distributions inside the head. Specifically, the SAR and pressure distributions inside the human head at 400 MHz are higher than the result at 300 MHz, which in turn is higher than 64 MHz for the same 100 mA normalizing current. Also, the power spectra of the pressure waves show that the spectral power for the 200-μs pulses is greater than that of the 100μs pulses. On the other hand, a fundamental frequency component at about 8 kHz inside the head is obtained for 64 MHz (1.5 T), 300 MHz (7.0 T), and 400 MHz (9.4 T). An acoustic resonance phenomenon for RF pulse-induced sound in the head has previously been shown [Lin, 1978, 1990; Lin and Wang, 2007; Roschmann, 1991] (see also Chaps. 6 and 8). Like microwave pulses, this indicates that the fundamental frequency of the thermoelastic sound pressure wave depends on the dimensions of the head (in this case VH model about 19 cm in diameter) loading the MRI coil. Similarities in the spectral content show that the spectral power peaks correspond to the mechanical resonance frequency of pressure waves inside the head model, however, the spectral power for 400 MHz (9.4 T) is much higher than that for 300 MHz (7.0 T) MRI birdcage antennas. Thus, the MRI-induced thermoelastic pressure wave depends on head dimension, pulse width, SAR, and RF frequency or magnetic field strength, which also confirms previous observations made in the case of microwave radar [Lin, 1978; 1980; 1990; Lin and Wang, 2007; 2010].

9.2.4   MRI Safety Guidelines

The FDA deems diagnostic MRI systems as having significant risk when the operating conditions involve static magnetic field greater than 8 T for adults, children, and infants aged older than 1 month and 4 T for infants aged 1 month or less; likewise, for SAR_{10g} equal to or greater than 4 W/kg for whole-body average, or 3.2 W/kg for the head average [FDA, 2014].

The birdcage coil currents required to induce peak SARs of 4, 8, 10, and 20 W/kg in anatomic human head models from a 100μs pulse at 64 MHz (1.5 T), 300 MHz (7.0 T), and 400 MHz (9.4 T), respectively, are given in Table 9.5. As expected, the required coil currents increased with SAR, which is proportional to the square of coil current. However, decreasing current amplitudes of 4080, 489, and 258 mA are required from 1.5, 7.0, and 9.4 T coils to produce the same local SAR of 4 W/kg. It

Table 9.5 Birdcage MRI coil currents required to induce peak SARs of 4, 8, 10, and 20 W/kg from a 100μs pulse in anatomic human head models

		64 MHz (1.5 T)	300 MHz (7.0 T)	400 MHz (9.4 T)
Pulse width (μs)	Head SAR(W/kg)	Coil current (mA)		
100	4	4080	489	258
100	8	5770	692	365
100	10	6451	773	408
100	20	9123	1093	577

Table 9.6 Peak thermoelastic pressure wave intensity induced in the anatomic head model by MRI systems

Threshold pressure (μPa)	Peak SAR (W/kg)	64 MHz (1.5 T)	300 MHz (7.0 T)	400 MHz (9.4 T)
		Acoustic pressure in (μPa) and (dB) for 100μs pulse		
20	4	29 (3.23)[a]	28 (2.92)[a]	24 (1.58)[a]
20	8	57 (9.10)	57 (9.10)	48 (7.60)
20	10	71 (11.0)	71 (11.0)	59 (9.40)
20	20	142 (17.0)	141 (17.0)	119 (15.5)

[a]Sound pressure level (dB) above the threshold pressure of 20μPa for auditory perception at cochlea

can be seen that the coil current required to induce a given SAR at 64 MHz is nearly 10 times at 300 MHz and more than 100 times at 400 MHz; reaffirming the trend of enhanced RF absorption at higher RF frequencies and stronger magnetic field strength. Note that the existing IEC/FDA guides for MRI scanners limit the local SAR_{10g} in the body – including the head and trunk – to 10 W/kg for patients with normal thermoregulatory function [IEC, 2010].

Table 9.6 lists the computed peak thermoelastic pressure intensity in the anatomic human head for a 100μs pulse at the frequencies of 64, 300, and 400 MHz as well as for SARs of 4, 8, 10, and 20 W/kg. It is interesting to note that the peak acoustic pressures generated in the anatomic head model are essentially the same at 64, 300, and 400 MHz (1.5, 7.0, and 9.4 T) for a given SAR, as they should be since the induced acoustic pressure is proportional to SAR. More significantly at 4 W/kg, the pressures generated in the anatomic head are comparable to the threshold pressure of 20μPa for sound perception by humans at the cochlea [Corso, 1963; Lin, 1978]. However, results in Table 9.6 indicate that the peak acoustic pressure in the brain tracks SAR; the auditory threshold at 8 W/kg doubles that of 4 W/kg. At SARs of 10 and 20 W/kg, the acoustic pressure in the brain is more than 3 to 6 times the auditory threshold. In other words, the sound pressure levels would be about 10–11 dB and 16–17 dB above the threshold of perception at the cochlea, but below the discomfort threshold.

In summary, the pressure distributions are different at different MRI frequencies and magnetic field strength because of the differences in SAR distributions inside the head. The SAR and thermoelastic pressure distributions inside the human head at 400 MHz is higher than that at 300 MHz, which in turn is higher than 64 MHz for the same coil current. Also, the power spectra of the pressure waves showed that the spectral intensity for wider pulses is greater than that of the narrower pulses. In contrast, a fundamental frequency component at about 8 kHz inside the head model is observed for all RF frequencies or magnetic field strengths. This means that the fundamental frequency of the thermoelastic sound pressure wave depends on the dimensions of the head loading the MRI coil antenna. This acoustic resonance phenomenon for the anatomic head model confirms previous prediction and behavior of RF and microwave pulse-induced sound. While the spectral power at 9.4 T (400 MHz) is much higher than that at 7.0 T (300 MHz), similarities in the spectral content indicate that the spectral power peaks correspond to the acoustic resonant frequencies of pressure waves inside the head. Thus, the MRI-induced thermoelastic pressure wave depends on head dimension, pulse width, SAR, and MRI frequency or magnetic field strength.

Current IEC/FDA guides limit the SAR in the body including the head and trunk which is 10 W/kg for patients with normal thermoregulatory function. The corresponding MRI safety limits for the extremities is 20 W/kg. There is also a partial body average SAR that varies from 2 to 10 W/kg [FDA, 2018]. The results above showed that current amplitudes of 6452, 773, and 408 mA are required from 1.5, 7.0, and 9.4 T coils to produce a peak local SAR of 10 W/kg inside the head. The associated thermoelastic pressure generated in the head varied from 59 to 71 μPa from these MRI coils for the same 10 W/kg SAR, which is a surprise for 9.4 T, since the induced thermoelastic pressure is proportional to SAR. The threshold pressure for auditory perception is 20 μPa at the cochlea. The thermoelastic sound pressure data for SARs of 4 to 20 W/kg range from 24 to 142 μPa. It is noteworthy that the sound pressures generated by MRI pulses in the head model are 1.6 to 17 dB above the threshold pressure of 20 μPa for sound perception by humans at the cochlea. While the MRI pulse-induced sound pressure waves in the head are above the auditory threshold for perception, they are well below the discomfort threshold for the airborne sound of about 110–120 dB.

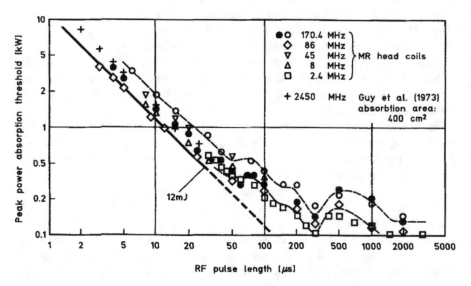

Fig. 9.19 RF auditory thresholds in peak power versus the RF pulse width for different MRI head coils for 2.4 MHz to 170 MHz (4 T). Note agreement with peak power density thresholds measurements at 2450 MHz when assuming an effective head absorption area of 400 cm²

9.2.5 Human Subjects Inside MRI RF Head Coils

A similar observation was made in an investigation of human subjects inside MRI RF coils [Roschmann 1991]. The threshold for auditory perception of pulsed RF energy absorption in the human head was studied on six subjects in MRI head coils for 2.4 MHz–170MHz (4 T) and pulse widths of 3–100µs. The RF energy thresholds were reported at 16 ±4 mJ per pulse for mostly saddle and quadrature head coils. The peak absorption for threshold results was adjusted to the peak power density of the reported data for microwave antenna measurements by assuming an effective cross-section of absorption of 400 cm² for the human head. After the correction, it showed that most of the thresholds observed with MRI coils for 2.4–170 MHz fall into the range of data reported for frequencies between 425 and 3000 MHz. Thus, at peak power levels currently available in clinical MRI systems, there is no evidence known for detrimental health effects arising from RF-evoked sound pressure levels in the human head rising to the discomfort threshold level.

The RF auditory effect threshold measurements for applied MRI peak power for humans are shown in Fig. 9.19 as a function of the RF pulse width or length [Roschmann 1991]. Aside from agreements found with peak power density thresholds reported for microwave measurements at 2450 MHz [Guy et al., 1973, 1975], measured thresholds fall monotonically as pulse widths increased from 3µs to about 50µs. The peak power threshold begins to fluctuate with pulse-width until reaching a minimum at 300µs. It then reverses trajectory to achieve a minor maximum at 500–600µs and falls again to another minimum for pulses wider than 1000µs when the auditory threshold of RF pulses continues to oscillate around peak power levels of typically 150 ±50 W for the head coils. The results from this study confirm theoretical predictions from the thermoelastic expansion model [Lin, 1977a, b].

The confirmation becomes much more evident by noting the experimental data in Fig. 9.19 are presented as the subject's threshold to RF-induced auditory sound. Note that the hearing threshold is the inverse of perceived loudness or sound pressure. Therefore, the inverted graphic results would show a loudness or pressure amplitude curve steadily rising from a low value as pulse widths increased from 3µs to about 50µs. The amplitude starts to fluctuate with pulse-width until reaching a maximum at 300µs. It then reverses course to go down to a lower value at 500–600µs and rises again to another maximum for pulses wider than 1000µs when the sound amplitude induced by RF pulses continues to oscillate around some average values. The amplitude variation curve by inversion of Fig. 9.19 in comparison with Figs. 8.16, 8.17 and 8.19 reveals remarkable similarity to perceived loudness or sound pressure characteristics predicted by the thermoelastic theory. Thus, the human MRI results furnish direct evidence in support of the theory of RF and microwave-pulse-induced auditory effects.

As a mechanism of interaction, the report falsely assumed that the RF pulse auditory phenomenon is a secondary cause associated with primary RF to thermal energy conversion in body tissues [Roschmann 1991]. As previously described, a single, 100-µs RF pulse would induce a rapid but miniscule temperature elevation of circa 10^{-6} °C in brain tissues. This small theoretical temperature elevation is undetectable by any currently available sensors and cannot be felt as a thermal sensation or heat. Thus, the temperature elevation in thermoelastic-expansion-produced pressure wave is merely a theoretical construct to facilitate our understanding, not to be taken literally. The process can simply be labeled as RF and microwave elastic or microwave acoustic conversion in body tissues, without the use of the thermo prefix. Nevertheless, the induced elastic expansion can launch an acoustic wave of pressure that propagates inside the head to the inner ear, where it becomes a nerve signal and is eventually perceived as sound by the cerebral cortex in humans.

9.3   Induced Pressure in Human Heads from Far-Field Exposure

The FDTD algorithm was used by Watanabe et al. [2000] to simulate the sound pressure waves generated by microwave-pulse-induced thermoelastic contraction and expansion in anatomic head models. The exposure scenario consists of a vertically polarized plane-wave 915 MHz microwave pulse impinging from the back of the head. The incident power density is 10 W/m^2. The FDTD simulation was

Table 9.7 Dielectric properties of head tissues at 915 MHz

Tissue	ε_r	σ [S/m]
Bone	16.6	0.244
Brain	45.7	0.772
Cartilage	42.6	0.789
Muscle	55.9	0.975
Eye ball	68.9	1.64
Fat	11.3	0.110
Skin	46.0	0.850
Lens	35.8	0.489

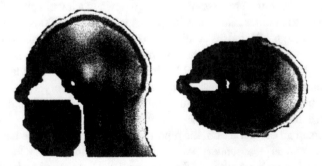

Fig. 9.20 SAR distribution in VH male exposed to 915-MHz plane wave incident on the back of the head

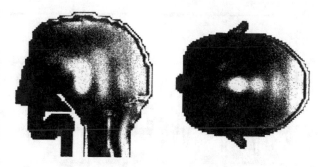

Fig. 9.21 SAR distribution in Japanese male exposed to 915-MHz plane wave incident on the back of the head

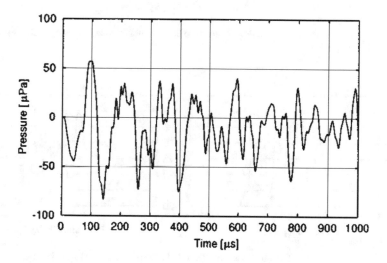

Fig. 9.22 Pressure wave at the cochlea of VH head model exposed to a single 20-μs pulse of 915 MHz at 10 W/m2

performed similarly as above in two successive steps by first computing SAR distribution and followed by using SAR as the driving force to solve the thermoelastic equation of motion for induced acoustic pressure waves. Numerical values of the dielectric properties of the set of head tissues included for the computation are given in Table 9.7.

Figures 9.20 and 9.21 are the sagittal and axial cross-sectional views of SAR distributions computed for 915 MHz plane waves impinging from the back of the head of two head models. The highest SARs are located close to the back of the head toward the incident microwaves. However, smaller SAR peaks are seen near

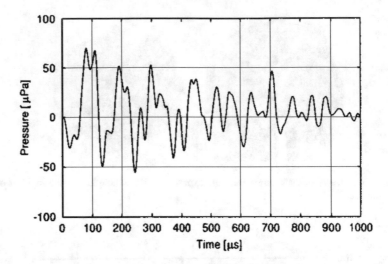

Fig. 9.23 Pressure wave at the cochlea of Japanese head model exposed to a single 20-μs pulse of 915 MHz at 10 W/m^2

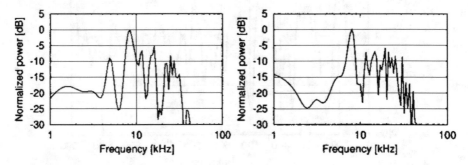

Fig. 9.24 Power spectra of the microwave-induced elastic pressure waves at the cochlea in: (**a**) VH model. (**b**) Japanese model

the middle of the adult brain. The results are in good agreement with the calculations for an adult-size brain sphere given in Chap. 5. The peak SAR$_{1g}$ for 10 W/m^2 incident from the back is 0.30 and 0.34 W/kg for the VH and Japanese models, respectively.

Computed results indicate in this case pressure wave fronts are initiated in the back of the head, where the SAR is the highest, and travels toward the center of the head and then gets into a reverberation mode due to an acoustic resonance phenomenon. Figures 9.22 and 9.23 show computed pressure waves at the cochlea of the anatomical head models generated by a 20-μs 915 MHz plane wave pulse. The associated sound power spectra shown in Fig. 9.24 suggest the pressure waves (in log scale) have dominant or fundamental frequency components around 7–9 kHz,

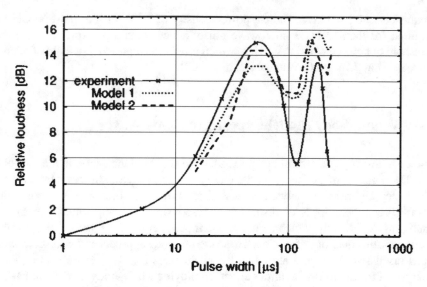

Fig. 9.25 Dependence of relative loudness or sound pressure on microwave pulse width. Sources: Solid curve [Lin, 1980, 1990]; experimental data [Tyazhelov et al., 1979]; FDTD-simulated peak sound pressures at the cochlea in the two anatomic head models [Watanabe et al., 2000]

which corresponds to the resonant frequencies of sound pressure wave inside the head. (There is a minor peak at 50 kHz for the VH model.) The waveforms are complex and somewhat different than those of the spherical models of Chap. 8. While they are like those shown in Fig. 9.18 for anatomical head models exposed in MRI coils, the pressure waves are more complicated. The variations are likely related to differences in SAR distribution attributable to the differences in RF and microwave sources, complex anatomy, and heterogeneity of the head models.

One of the distinctive features of the thermoelastic theory described in Chap. 8 is the prediction of perceived loudness of microwave auditory effect or induced sound pressure's dependence on pulse width, an observation that was demonstrated in human experiments [Tyazhelov et al., 1979; Lin, 1980, 1990]. Fig. 9.25 shows that the FDTD-simulated peak sound pressures at the cochlea in the two anatomic head models display similar characteristics [Watanabe et al., 2000]. The peak sound pressure or the loudness increases as the microwave pulse width increases from 5 to 50μs, weakens with further increase of pulse width from 70 to 100μs, and then increases again with longer pulse width, etc., as it cycles through the constructive and destructive interference phases of reverberation.

The peak values of acoustic pressure computed at the cochlea and inside the anatomical head vary from 83 and 270μPa for the VH model and 70 and 590μPa for the Japanese model, respectively, in response to a 20-μs pulse with the incident microwave power density of 10 W/m^2 at 915 MHz. The highest pressures tend to appear near the center of the head (see also Figs. 9.15 and 9.16). The threshold for bone conduction hearing is about 20 mPa or 60 dB (Re: 20 μPa) for sound frequencies of 5–10 kHz. Thus, a power density of 3 kW/m^2 per pulse would be required for the 915-MHz microwave pulse-induced sound pressure to reach the threshold of bone conduction hearing, which is near the reported threshold of microwave hearing [Lin, 1980; Watanabe et al., 2000].

9.4   Whole-Body Model Exposed to Plane Waves

For a whole-body anatomical human model exposed to plane waves in the far field, FDTD simulation of the thermoelastic theory for microwave auditory effect involves a two-step algorithmic analysis, as described above. The first step is to solve the Maxwell equations for SAR distributions responsible for temperature rise, big or small, inside the body, which serves as the input energy in the second step for the thermoelastic equations of motion. These are the same approaches for the computations discussed above and are taken by Yitzhak et al. [2014] to implement their analysis of the microwave auditory effect, employing whole-body models for FDTD calculation of induced thermoelastic pressure waves. The study covered the RF and microwave frequency range between 40 MHz and 3 GHz. In addition to whole bodies, it also includes computations for an anatomical head model. Specifically, the VH models used are a male body, with a height of 1.875 m, resolution of $5 \times 5 \times 5$ mm^3, and 39 different tissue types; a female body, with a height of 1.725 m, resolution of $5 \times 5 \times 5$ mm^3 and 32 different tissue types; a male head and shoulders, with a height of 0.348 m, resolution of $2 \times 2 \times 2$ mm^3, and 26 different tissue types.

Many FDTD computations have been performed for SAR distributions using anatomical human body models (including models for head and shoulders) and are widely published (see Chap. 5). The acoustic pressure waves generated inside a human body and at the cochlea of the inner ear by exposure to RF or microwave pulses was studied through computer simulation using the FDTD algorithm for plane waves for male, female, and head and shoulder models [Yitzhak et al., 2014]. The power density of the incident 70-μs wide 915 MHz pulse is 10 W/m^2. The value $b = 4 \times 10^5$ /s is chosen for the pulse angle parameter of the modulated rectangular pulse (Eq. 9.26). A stress-free boundary condition is used for the top portion of the head and a PML layer of 10 cells is applied to the bottom portion of the head model.

The RF and microwave frequency dependence of the acoustic pressure computed throughout the body or specifically at the cochlea in the three-body models is shown in Fig. 9.26 [Yitzhak et al. 2014]. The features of the pressure amplitude variations are similar for all three models; a peak occurs at about 200 MHz. However, the pressure amplitude for the male head model has the highest value and the pressure in the female model is lower for frequencies higher than 200 MHz. In general, the variations in acoustic pressure amplitude follow the microwave or electromagnetic (EM) frequency dependence of average SAR in the head, which is similar for all three models (Fig. 9.27).

The results of RF and microwave frequency dependence of acoustic pressure at the cochlea of the male body model for different polarizations and directions of

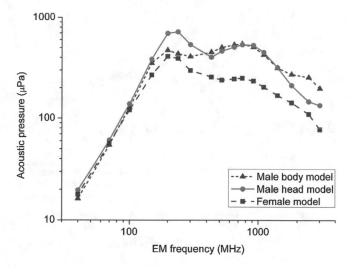

Fig. 9.26 Dependence of the pressure amplitude at the cochlea for the three anatomical human models exposed from the backside with a horizontally polarized 70μs wide RF or microwave pulses at 10 W/m²

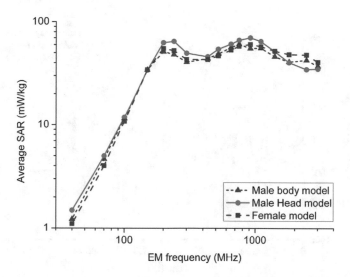

Fig. 9.27 Electromagnetic frequency dependence of average SAR in the head for the three anatomical human models exposed from the backside with a horizontally polarized 70μs wide RF or microwave pulses at 10 W/m²

incidence are shown in Fig. 9.28. For plane waves of vertical polarization, there are two distinct acoustic pressure peaks at about 60 MHz and 200 MHz, which clearly reflect the whole-body resonance SAR peak and that in the head (Fig. 9.27). Otherwise, the variations are similar for front and back incidences, but they are

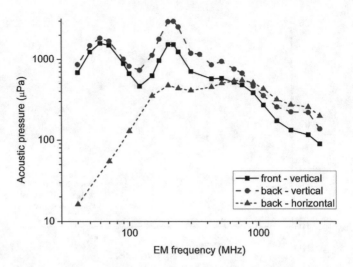

Fig. 9.28 Dependence of the pressure amplitude at the cochlea for male anatomical models exposed to 70μs wide 10 W/m² RF or microwave pulses of different polarization and incidence direction

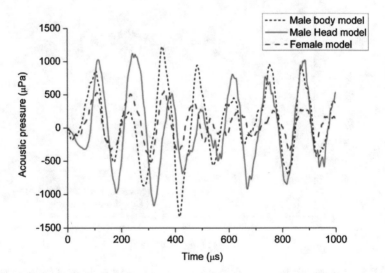

Fig. 9.29 Computed pressure waves at the cochlea of three anatomical models for horizontally polarized plane waves incident from the backside (915 MHz, 70-μs pulse)

substantially different from horizontal polarization. In the horizontal polarization case, a minor peak or a flat range of acoustic pressure is reached, due to the frequency dependence of the SAR on head size. Moreover, the pressure amplitudes are much more reduced at lower RF frequencies.

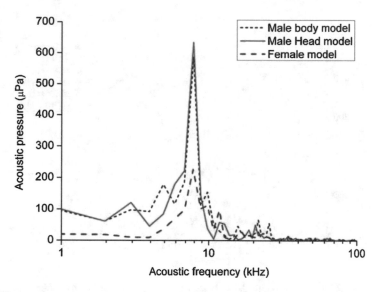

Fig. 9.30 Spectrum of pressure waves at the cochlea of three anatomical models for horizontally polarized plane waves incident from the backside (915 MHz, 70-µs pulse)

The FDTD-simulated acoustic pressure waves generated by microwave-induced thermoelastic expansion at the cochlea in the three models are shown in Fig. 9.29 for a plane wave, 70-µs wide 915-MHz pulse incident from the back of the body with a horizontal polarization [Yitzhak et al. 2014]. Figure 9.30 provides the corresponding power spectra. The general features of the acoustic response for the head model are consistent with the results given in Sects. 9.2 and 9.3 for isolated head exposures. In fact, the microwave-induced pressure waves inside the head showed essentially the same characteristics for all three models. A minor exception is the female model has a smaller pressure amplitude due to associated lower SARs. It is worthy of note that the microwave thermoelastic pressure wave is initiated and launched at the spot where SAR is highest and propagates away from it to the rest of the body. In all VH models, the highest acoustic pressure amplitude occurs at a frequency of about 8 kHz, which represents the fundamental resonance frequency of sound generation in the human head [Lin, 1977a, b; Watanabe et al., 2000; Lin and Wang, 2010]. Moreover, they further confirm microwave pulse induced acoustic resonance inside the head is independent of microwave frequency, pulse width, polarization, and direction of incidence.

The analytical solutions described in Sect. 8.5 demonstrated a distinctive pulse-width dependence of the induced pressure amplitude inside the spherical head model. The predicted characteristics from the calculations [Lin, 1977a, b, c] were confirmed by experimental data from human subjects [Tyazhelov et al., 1979; Lin, 1980], and FDTD simulations [Watanabe et al. 2000]. It is interesting to note that a Fourier decomposition of the pressure [Yitzhak et al., 2014] indicated that its

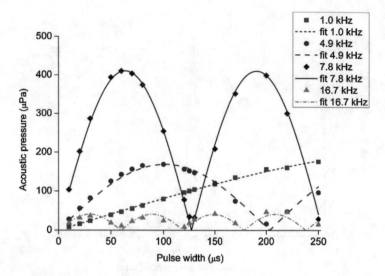

Fig. 9.31 Microwave pulse width dependence of the amplitude of acoustic frequency components. The curves were obtained by fitting Eq. (9.28) for the male body model irradiated from the front by a vertically polarized 915 MHz plane wave

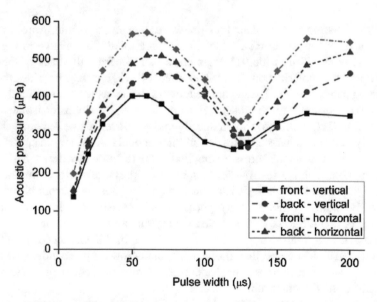

Fig. 9.32 Microwave pulse width dependence of the pressure amplitude at the cochlea for the male model irradiated by a 915 MHz plane wave for different incidence directions and polarizations

component curves at various acoustic angular frequencies ω_m can be fitted to the functional form of Eq. (9.28),

$$P(t) = P_0 \left| \frac{\sin(\omega_m t)}{2} \right| \tag{9.28}$$

where t is the pulse width and the amplitude factor P_0 is independent of the pulse width. The pulse width dependence of the amplitudes of selected acoustic frequency components obtained by Fourier decomposition of the pressure is given in Fig. 9.31. It shows various acoustic frequencies clearly follow the simple sinusoidal relation (Eq. 9.28), suggesting that there is an effective resonant cavity for each acoustic frequency in the complex head structure. However, the power spectra of the microwave thermoelastic pressure waves at the cochlea are dominated by the acoustic head cavity resonating at about 8 kHz. Figure 9.32 shows the pulse width dependence of the pressure amplitude at the cochlea due to a 915-MHz pulse for different polarization and direction of exposure. Indeed, the finding of a characteristic oscillatory dependence on the pulse width (from 10 to 250μs in this case) holds for homogeneous spherical and anatomical head models, for plane wave and near zone exposures, and it retains its applicability even in complicated biological experiments involving different human subjects.

In this study, the FDTD computations for the male model exposed to plane waves with vertical polarization from the back by 10- and 20-μs wide 2450 MHz microwave pulses at 10 W/m² produced sound pressures of 88.7 and 144μPa [Yitzhak et al., 2014]. By applying the Zwislocki [1957] and Thurlow and Bowman [1957] pulse width and threshold level (dB) adjustment to the data, and recognizing that the computed sound pressure values are obtained with an incident power density of 10 W/m², the power density required to induce the auditory effect is estimated to be about 106 kW/m² for a 10-μs pulse and about 49.6 kW/m² for a 20-μs pulse. The corresponding threshold levels for the microwave-induced auditory response measured for human subjects with normal hearing acuity by Guy et al. [1973, 1975] are 40 and 21.5 kW/m², respectively, for near-zone exposure to a horn antenna.

In summary, as expected, the dependence of the amplitude of the thermoelastic pressure waves on RF and microwave frequency is correlated with SAR in the head of whole-body models. Also, the dependence of the pressure wave amplitude on the pulse width derived with whole-body models is consistent with those obtained from detached head models. However, for plane wave whole-body exposures, a frequency-dependent geometrical body resonance may occur for vertically polarized incidence. Thus, for some frequencies, SAR inside the head may not be as high as in other parts of the body, which renders the frequency of incident microwaves less pertinent in predictions of frequency dependence of microwave auditory effect or induced sound pressure in the head.

References

Berenger P (1996) Perfectly matching layer for the FDTD solution of wave-structure interaction problems. IEEE Trans Antenn Propag 44:110–117

Bernardi P, Cavagnaro M, Pisa S, Piuzzi E (1998) SAR distribution and temperature increase in an anatomical model of the human eye exposed to the field radiated by the user antenna in a wireless LAN. IEEE Trans Microw Theory Tech 46:2074–2082

Bernardi P, Cavagnaro M, Pisa S, Piuzzi E (2003) Specific absorption rate and temperature elevation in a subject exposed in the far-field of radio-frequency sources operating in the 10-900-MHz range. IEEE Trans Biomed Eng 50:295

Cavagnaro M, Lin JC (2019) Importance of exposure duration and metrics on correlation between RF energy absorption and temperature increase in a human model. IEEE Trans Biomed Eng 66(8):2253–2258

Chakeres DW, de Vocht F (2005) Static magnetic field effects on human subjects related to magnetic resonance imaging systems. Prog Biophys Mol Biol 87:255–265

Chakeres DW, Kangarlu A, Boudoulas H, Young DC (2003) Effect of static magnetic field exposure of up to 8 Tesla on sequential human vital sign measurements. J Magn Reson Imaging 18:346–352

Cole KS, Cole RH (1941) Dispersion and absorption in dielectrics: alternating current characteristics. J Chem Phys 9:341–351

Corso JF (1963) Bone-conduction thresholds for sonic and ultrasonic frequencies. J Acoust Soc Am 35:1738–1743

de Vocht F, Stevens T, Glover P, Sunderland A, Gowland P, Kromhout H (2007) Cognitive effects of head-movements in stray fields generated by a 7 Tesla whole-body MRI magnet. Bioelectromagnetics 28:247–255

FDA (2014) Guidance for magnetic resonance diagnostic devices – criteria for significant risk investigations of magnetic resonance diagnostic devices. Issued June 20, 2014. U.S. Food and Drug Administration. Rockville, MD, USA

FDA, MRI (Magnetic Resonance Imaging). August 29, 2018. U.S. Food and Drug Administration. Rockville, MD, USA, 2018. https://www.fda.gov/radiation-emitting-products/medical-imaging/mri-magnetic-resonance-imaging

Glover PM, Cavin I, Qian W, Bowtell R, Gowland PA (2007) Magnetic-field-induced vertigo: a theoretical and experimental investigation. Bioelectromagnetics 28:349–361

Guy AW, Taylor EM, Ashleman B, Lin JC (1973) Microwave interaction with the auditory systems of humans and cats. In: Proceedings of the IEEE INTernational microwave symposium Boulder, June 1973, pp 321–323

Guy AW, Chou CK, Lin JC, Christensen D (1975) Microwave induced acoustic effects in mammalian auditory systems and physical materials. Ann NY Acad Sci 247:194–218

Houpt TA, Pittman DW, Riccardi C, Cassell JA, Lockwood DR, Barranco JM, Kwon B, Smith JC (2005) Behavioral effects on rats of high strength magnetic fields generated by a resistive electromagnet. Physiol Behav 86:379–389

Ibrahim TS, Lee R, Baertlein BA, Yu Y, P-ML R (2000) Computational analysis of the high pass birdcage resonator: finite difference time domain simulations for high-field MRI. Magn Reson Imaging 18:835–843

IEC (International Electrotechnical Commission) (2010) International standard, medical equipment IEC 60601-2-33: particular requirements for the safety of magnetic resonance equipment, 3rd edn. IEC, Geneva

Jin JM (1999) Electromagnetic analysis and design in magnetic resonance imaging. CRC Press, Boca Raton

Landau LD, Lifshitz EM (1986) Theory of elasticity. Pergamon Press, Oxford

Lin JC (1976) Theoretical analysis of microwave-generated auditory effects in animals and man, biological effects of electromagnetic waves. BRH/DEW I:36–48

Lin JC (1977a) On microwave-induced hearing sensation. IEEE Trans Microw Theory Tech 25:605–613

Lin JC (1977b) Further studies on the microwave auditory effects. IEEE Trans Microw Theory Tech 25:936–941

Lin JC (1977c) Theoretical calculations of frequencies and threshold of microwave-induced auditory signals. Radio Sci 12(SS-1):237–252

Lin JC (1978) Microwave auditory effects and applications. Springfield, Charles C. Thomas

Lin JC (1980) The microwave auditory phenomenon. Proc IEEE 68(1):67–73

Lin JC (1990) Auditory perception of pulsed microwave radiation. In: Gandhi OP (ed) Biological effects and medical applications of electromagnetic fields. Prentice-Hall, New York. Chapter 12, pp 277–318

Lin JC, Bernardi P (2007) Computer methods for predicting field intensity and temperature change. In: Barnes F, Greenebaum B (eds) Bioengineering and biophysical aspects of electromagnetic fields. CRC Press, Boca Raton, pp 293–380

Lin JC, Wang ZW (2005) SAR and temperature distributions in canonical head models exposed to near- and far-field electromagnetic radiation at different frequencies. Electromagn Biol Med 24:405–421

Lin JC, Wang ZW (2007) Hearing of microwave pulses by humans and animals: effects, mechanism, and thresholds. Health Phys 92:621–628

Lin JC, Wang ZW (2010) Acoustic pressure waves induced in human heads by RF pulses from high-field MRI scanners. Health Phys 98(4):603–613

Liu W, Collins CM, Smith MB (2005) Calculations of B1 distribution, specific energy absorption rate, and intrinsic signal-to-noise ratio for a body-size birdcage coil loaded with different human subjects at 64 and 128 MHz. Appl Magn Reson 29:5–18

Love AEH (1927) A mathematical theory of elasticity. Cambridge University Press, New York

Mason PA et al (2000) In Klauenberg BJ, Miklavcic D (eds) Radio frequency radiation dosimetry. Brook Air Force Laboratory, Texas, pp 141–157

Pennes HH (1948) Analysis of tissue and arterial blood temperatures in resting forearm. J Appl Physiol 1:93–122

Roschmann P (1991) Human auditory system response to pulsed radiofrequency energy in RF coils for magnetic resonance at 2.4 to 170 MHz. Magn Reson Med 21:197–215

Shellock FG, Crues JV (2004) MR procedures: biologic effects, safety, and patient care. Radiology 232:635–652

Thurlow WR, Bowman R (1957) Threshold for thermal noise as a function of duration and interruption rate. J Acoust Soc Am 29(2):281–283

Tyazhelov VV, Tigranian RE, Khizhniak EO, Akoev IG (1979) Some peculiarities of auditory sensations evoked by pulsed microwave fields. Radio Sci 14(6S):259–263

Wang ZW, Lin JC (2012) Partial-body SAR calculations in Magnetic Resonance Image (MRI) scanning systems. IEEE Antenn Propag Mag 54(2):230–237

Wang Z, Lin JC, Mao W, Liu W, Smith MB, Collins CM (2007) SAR and temperature: simulations and comparison to regulatory limits for MRI. J Magn Reson Imaging 26(2):437–441

Wang Z, Lin JC, Vaughan JT, Collins CM (2008) Consideration of physiological response in numerical models of temperature during MRI of the human head. J Magn Reson Imaging 28(5):1303–1308

Watanabe Y, Tanaka T, Taki M, Watanabe SI (2000) FDTD analysis of microwave hearing effect. IEEE Trans Microw Theory Tech 48:2126–2132

Yee KS (1966) Numerical solution of initial boundary value problems involving maxwell's equations in isotropic media. IEEE Trans Antennas Propag 14(3):302–307

Yitzhak NM, Ruppin R, Hareuveny R (2014) Numerical simulation of pressure waves in the cochlea induced by a microwave pulse. Bioelectromagnetics 35:491–496

Zwislocki J (1957) In search of the bone-conduction threshold in a free sound field. J Acoust Soc Am 29:795–804

Chapter 10
Applied Aspects and Applications

The microwave auditory effect or the hearing of microwave pulse-induced sound involves a cascade of events. Minuscule but rapid (~μs) rise in tissue temperature (~10^{-6} °C), resulting from the absorption of pulsed microwave energy, creates a thermoelastic expansion of brain matter. This small theoretical temperature elevation is undetectable by any currently available temperature sensors, and at threshold levels, it cannot be felt as a thermal sensation or heat. Nevertheless, it can launch an acoustic wave of pressure that travels inside the head to the inner ear. There, it activates the nerve cells in the cochlea, and then relays the neural signal to the central auditory system for perception, via the same process involved for normal hearing. Thus, the discovery of microwave thermoelastic or thermoacoustic pressure wave generation in biological tissues by deposition of short microwave pulses in tissue came about as the result of an intense effort in search of a mechanism to help understand a newly observed auditory response to microwave radiation [Lin, 1976a, b; 1977a, b; Lin, 1978]. Moreover, recognition of the propagation nature of the acoustic wave of pressure in biological tissues soon prompted the exploration of its potential for applications in biological and medical imaging [Lin, 1982; Olsen, 1982; Lin and Chan, 1983; Olsen and Lin, 1983; Chan et al., 1984].

Preceding chapters have described in detail the physiology, behavior, biophysics, and the thermoelastic mechanism of transduction for pulsed microwave-induced auditory effects in humans and animals. This chapter explores where and how this microwave-induced thermoelastic wave phenomenon may be applied for practical purposes in life science, medicine, and other realms of human endeavors. A few interesting diagnostic imaging applications of the microwave thermoelastic pressure wave interaction and some applied aspects of the microwave auditory effects are described. They include a summary of early investigations of microwave thermoacoustic tomography and imaging and some applied aspects of the microwave auditory effects such as reported personnel attacks at some diplomatic missions and potential startle responses associated with pilot disorientation.

© Springer Nature Switzerland AG 2021
J. C. Lin, *Auditory Effects of Microwave Radiation*,
https://doi.org/10.1007/978-3-030-64544-1_10

10.1   Microwave Thermoacoustic Tomography and Imaging

The first 2D microwave thermoacoustic image was obtained from a phantom model of the human hand [Olsen and Lin, 1983; Lin, 2005; Manohar and Razansky, 2016]. The phantom hand model consisted of a surgical glove filled with homogeneous muscle-equivalent materials. The 5.66-GHz microwave pulses from an open-ended waveguide antenna provided the irradiation of the phantom hand model inside a water-filled tank. Induced thermoacoustic pressure waves were detected with a 20 × 20 detector array. Acoustic signals from the array were processed to yield the projection image shown in Fig. 10.1, in which the hand model can be clearly identi-fied. A year earlier, simple one-dimensional (1D) and two-dimensional (2D) projec-tion images of objects were obtained by detecting the thermoacoustic pressure waves following exposure of the simple objects with microwave pulses [Olsen, 1982; Lin and Chan, 1983]. Shortly thereafter, a complete design of a 2.45-GHz microwave thermoelastic tissue imaging system was published [Lin and Chan, 1984]. Some thermoacoustic A-scans from interfaces of a variety of materials using pulses from a 2.45 GHz microwave source were reported later [Nasoni et al., 1984]. In the latter case, samples placed in the microwave-oven type cavity included a three-layer tissue-mimicking phantom made of a lipid material and layers simulat-ing the microwave properties of bone and muscle.

Since the generation and detection of thermoelastic pressure waves depends on pulsed microwave, dielectric permittivity, and elastic properties of biological tissues, microwave thermoelastic imaging possesses the characteristic features of a "dual-modality" imaging system. In this dual-modality imaging system, the distinctive attributes of the high contrast offered by dielectric permittivity-based microwave propagation and absorption [Lin, 1985, 1986] and the well-known fine spatial reso-lution provided by ultrasonic acoustics in tissue media are being explored to provide an imaging modality for nonionizing and noninvasive imaging of tissue properties. This section describes the research being conducted in developing microwave ther-moelastic tomography and imaging for medical diagnosis, especially for early detec-tion of breast cancer. The account builds upon the biophysics of thermoelastic wave

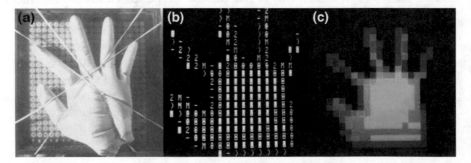

Fig. 10.1 Microwave thermoacoustic image of a phantom model of the human hand: (**a**) Phantom hand hydrophone transducer array; (**b**) Original digital image; (**c**) Color display

generation and propagation in biological tissues discussed in previous chapters. It includes the experimentation with prototype microwave thermoelastic tomographic imaging systems [Olsen and Lin, 1983; Lin and Chan, 1983, 1984; Chan et al., 1984; Chan and Lin, 1988] and the reconstruction of tomographic images using filtered-back projection algorithms [Su and Lin, 1985; 1990; 1991]. Note that the quality of reconstructed images showed significant improvements using algorithms that incorporate a rebinning methodology and by taking into account the diffraction and refraction effects of wave propagation [Su and Chen, 1992].

Here are some quantitative illustrations to help with understanding the contrast enhancement of microwave imaging, and the resolution advantage of ultrasonic imaging in combination would make microwave thermoelastic imaging of biological tissues a potentially valuable *dual modality* for diagnostic imaging. For example, the wavelengths in muscle equivalent tissues for microwaves are 17.5 mm at 2450 MHz and 95.2 mm at 433 MHz [Lin, 1985]. For ultrasonic energy, the wavelength is a mere 0.5 mm at 3 MHz and 1.5 mm at 1 MHz. Thus, the potential gain in spatial resolution is tremendous for tissue imaging compared to employing microwave radiation alone as typically done in microwave tissue imaging modalities.

It is noted that the term microwave thermoelastic tomography or imaging was introduced to recognize the extensive scientific research being conducted to understand the mechanism for microwave auditory effect and the development of an innovative technique based on the science of microwave-pulse-induced thermoelastic pressure waves to form images of biological objects in the 1980s. After the initial work garnered increasingly greater attention around the year 2000, more investigators began to use the term microwave-induced thermoacoustic tomography or microwave-induced thermoacoustic imaging. As the title of this section suggests, this chapter has opted to embrace the use of "microwave thermoacoustic tomography (MTT)" or "microwave thermoacoustic tomographic (MTT) imaging" in referring to this dual modality imaging system and technology.

In short, as an emerging medical diagnostic imaging technology with great application prospects, MTT is still facing many challenges in translating it to routine clinical applications. Clearly, multidisciplinary cooperative efforts could offer a promising and needed alternative methodology to X-ray mammography that depends on the effects of ionizing radiation for image formation.

10.1.1   *Microwave Thermoacoustic Imaging*

The modality is based on the observation that sonic and ultrasonic waves can be generated in biological tissues by deposition of short high-power microwave pulses. Absorption of the pulsed microwave energy causes a rapid rise ($\sim\mu s$) in tissue temperature ($\sim 10^{-6}$ °C). A thermoelastic pressure wave is launched in response that propagates away in all directions, and it may be detected with an acoustic sensor or ultrasonic transducer array. The detected acoustic waves are processed or reconstructed to form 2D images of the tissue media.

A prototype MTT system was designed to promote further experimentation and for model testing [Lin and Chan, 1984; Chan and Lin, 1988]. Cylindrical phantom models with dielectric properties of normal and malignant breast tissues are immersed in a tank of water; microwave pulses are applied at whose surface through a 2450-MHz open-waveguide antenna applicator (Fig. 10.2). Microwave-induced thermoelastic pressure waves are detected by a 20 × 20 piezoelectric transducer array (ITC, lead zirconate titanate), located at the bottom of the tank. These transducers have a free field voltage sensitivity of −220 dB (re: 1 V/μPa) at 70 kHz. For each data point, a waveform was captured on the oscilloscope screen and stored on disk (Fig. 10.3). The received signals were then amplified, filtered, and converted

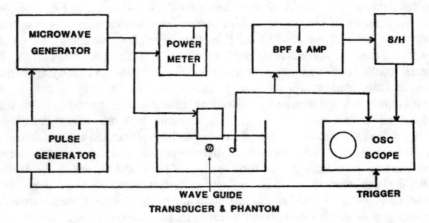

Fig. 10.2 Functional representation of MTT imaging system showing the microwave source, phantom model in a water tank, and thermoelastic signal recording configuration of a single transducer or transducer array

Fig. 10.3 Experimentally obtained 2450 MHz microwave pulse (10μs) induced thermoelastic pressure signal from a soft tissue medium (1 V vertical; 49.55μs horizontal per div)

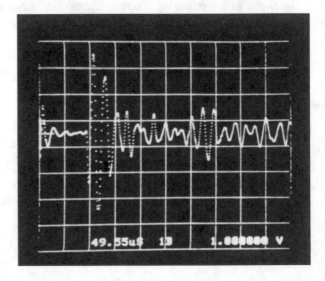

Fig. 10.4 A time-of-flight thermoelastic image obtained by the microwave pulse–hydrophone transducer array arrangement, with no object present in the field

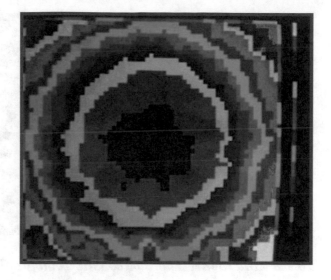

into digital form, offline, using a 2-MHz Nicolet oscilloscope. The maximum peak of each waveform was then read off the oscilloscope and key into the computer. For each image, 400 data points were acquired. This resulted in 400 image-pixel values (produced by a 0.9 × 0.9 cm hydrophone acoustic transducer). Figure 10.4 illustrates a time-of-flight thermoelastic image obtained by the microwave pulse–hydrophone transducer array arrangement, with no object present in the field. It represents the receiving characteristics of the transducer array. The rings seen in the image correspond to different propagation delay times, the shortest of which is seen in the middle and longest at the corners.

A projection-type thermoacoustic map of the phantom produced interactively is given in Fig. 10.5, where the processing techniques of subtraction, normalization, convolution filtering, grey scale transformation, and thresholding were used to enhance the image [Chan and Lin, 1988]. Briefly, the phantom consists of two glass test tubes 0.9 cm in diameter, filled with tissue equivalent materials, and they were placed 1.0 cm apart. The equivalent dielectric materials were an ethylene glycol in the saline mixture and a semisolid mixture consisting of polyethylene powder, saline, and a gelling agent. It can be seen from Fig. 10.5 that the two test objects can be clearly differentiated. Note that the ratio between the amplitude of the acoustic signals from the phantoms was 6.75 to 3.51 or nearly 2-to-1, which approximates the difference in dielectric permittivity between malignant and normal mammary tissues. It also demonstrates the above-mentioned prediction that malignant breast tissues would generate significantly higher amplitude of thermoelastic pressure waves – leading to a better signal-to-noise ratio, which should render greater discriminations between malignant and normal breast tissues. Preliminary investigation of the spatial resolution indicated that the system could resolve objects that are of a one-unit transducer size, separated by the one-transducer unit. This translates into about 1 cm in the above studies [Lin, 2005]. It suggests that decreasing the size of the transducer may lead to better spatial resolution.

Fig. 10.5 Microwave
thermoelastic image of a
phantom consisting of two
glass test tubes 0.9 cm in
diameter filled with tissue
equivalent materials and
placed 1.0 cm apart

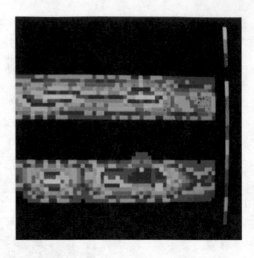

10.1.2   Microwave Thermoacoustic Tomography

Tomography involves the formation of internal images of an object through which
beams from an incident energy source have passed through, and the measurement of
the transmitted beams from many different directions or angles within a plane
through the body. From these measurements, an image of the cross-section of the
body is mathematically reconstructed using a computer algorithm and then formatted
as an image displayed on a screen. A tomographic system typically consists of three
major components: a signal generation component consisting of a radiant energy
source, a signal detector component, and a computer-based system that manages the
procedure and conditions the collected data, reconstructs the image, and manipulates
the image for display on a viewer. Clearly, the source–detector pair is the salient
component of any tomographic system, which provides the physical apparatus to
acquire the raw information for image formation. Thus, MTT as a dual modality
imaging system is unique in that the source is pulse-microwave energy and the
detectors are acoustic pressure sensors.

A prototype MTT system (Fig. 10.6) is essentially based on the same design
previously described and shown above (Fig. 10.2), with the addition of a gantry
platform for rotating the phantom [Su and Lin, 1990; 1991]. The pulsed microwave
energy is provided by a microwave pulse generator (Epsco PH40K) with a 2450-
MHz plug in. The duration and the peak power of the microwave pulse were
externally controllable from 0.4 to 25µs and up to 40 kW, respectively. For many of
the experiments, a 2-µs pulse width and a peak power of 30 kW were used. The
projections were obtained by rotating the phantom through calibrated $\pi/16$ positions
in a specially fabricated Delrin gantry-phantom platform. The same 20×20 element
piezoelectric transducer array, in a water tank, was used for data collection. The data
used to reconstruct the images were pre-processed to reduce noise in the signals by
applying a smoothing filter, which would also reduce the sharpness of the images.
Furthermore, a thresholding technique was applied for better object differentiation.

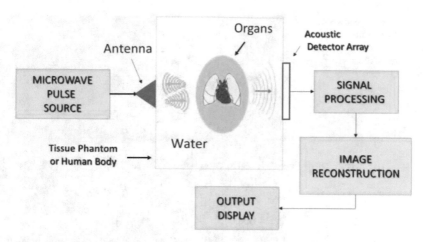

Fig. 10.6 A schematic diagram of the MTT (microwave thermoacoustic tomographic) imaging system

The investigation of microwave thermoelastic tomography was initiated by first conducting a series of computer simulations of microwave thermoelastic tomographic imaging of tissue phantoms consisting of one, two, or three circular cylinders. Both parallel and fan beams, along with a filtered back-projection algorithm, were used in conjunction with the Shepp–Logan filter for the reconstruction [Su and Lin, 1985; 1990; 1991]. In addition, both the transmission and emission cases were simulated for the tomographic reconstruction, using the same phantoms. As mentioned previously, tissues and organs with different dielectric, thermal, and elastic properties, upon absorption of pulsed microwave energy would emit acoustic waves at different strengths and frequencies. Results show that in shape, size, and location, the reconstructed images correspond well with the original phantoms. The grey or color levels of the reconstructed images agree well with the attenuation constants in the transmission case. In the case of emission tomography, however, they were proportional to the ratio of the emitted signal strength and the attenuation constant. It was found by experimenting with the phantoms for varying numbers of projections – measurement from different directions – that 16 projections are the minimum number required to yield a reasonably accurate reconstruction. Figure 10.7 shows the reconstructed cross-sectional images using 16 projections assuming perfect data and absence of noise.

Fan beam-reconstructed tomographic images from phantoms with one, two, and three muscle-equivalent cylinders are given in Fig. 10.8 The left figure is for a cylinder, 1.5 cm in diameter, located at the middle of the reconstruction region. The center image represents two cylinders, 1.2 and 2.0 cm in diameter, respectively, separated 1 cm apart. The image on the right is for three 1.5-cm cylinders, separated by 1 or 2 cm. Clearly, the phantom locations are well represented in the tomographic images, especially for an object at the middle of the reconstruction zone. Also, the objects with separation distances of 1 cm are easily resolvable. The MTT images

Fig. 10.7 Reconstructed cross-sectional images of two phantom objects using 16 projections assuming perfect data and absence of noise

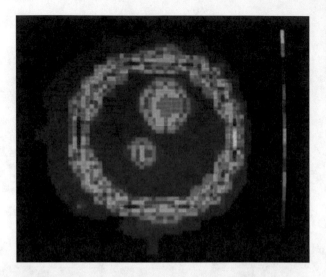

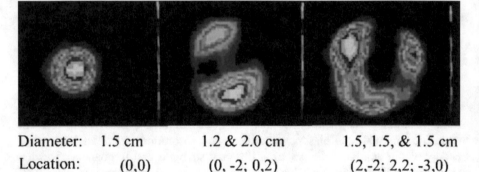

Diameter:	1.5 cm	1.2 & 2.0 cm	1.5, 1.5, & 1.5 cm
Location:	(0,0)	(0, -2; 0,2)	(2,-2; 2,2; -3,0)

Fig. 10.8 Reconstructed fan beam tomographic images of phantoms with one, two, and three muscle-equivalent cylinders in a 25 × 30 × 50 cm water tank. (Microwave energy: 2450 MHz source, 3 cm contact applicator, 2μs pulse, 30 kW peak power.) Location refers to 2D planar rectangular coordinates centered at (0, 0)

consistently displayed relative shape, size, location, and object property. For three objects, the images are degraded by the phenomena of wave diffraction and refraction, and the umbrae of object in the field. Moreover, the images are distorted when objects in the phantoms are off-center. Some of these difficulties may be minimized by choosing a different set of microwave frequencies and pulse widths. They may also be alleviated by using a microwave antenna that can provide a more parallel beam for the projections, or a finer array of smaller ultrasonic detectors. Furthermore, computer algorithm-based software corrections are practical approaches to mitigate some of these problems as well. Nevertheless, the results show that biological materials with high, moderate, and low water contents can be differentiated from reconstructed MTT images, and thus, demonstrating its feasibility as a tomographic imaging system for biological tissues.

The matter of tissue specificity and spatial resolution of MTT images have been investigated further utilizing phantom models consisting of a water tank within which are placed several test tubes of varying sizes filled with solutions of different contents [Su and Lin, 1987]. Specifically, they include four sizes of test tubes, 1.2, 1.5, 1.6, and 2.0 cm in diameter and are filled with water, glycerol, or glycol to simulate body fluid, muscle, or fat. The test tubes were held in stable positions using semicircular grooves machined on a plastic holder. The holder was designed in such a way that the separation between two test tubes was maintained in fixed increments, namely, 2.0, 2.5, 3.0, 3.5, 4.0, and 4.5 cm. The pulsed microwave energy (2µs, 2450 MHz, 20 kW peak power) was generated by an Epsco PH40k source under the control of an external pulse generator. Microwave energy was applied to the simulated tissue models through a rectangular waveguide (7.2 × 3.4 cm). A quarter-wavelength dielectric plate was used to maintain the required impedance matching condition between air and water. A double stub tuner was employed to further reduce the reflection coefficient. A thin absorbing layer of foam plate was attached to the output flange to reduce the multipath effects due to the reflection of excited acoustic waves. The influence of reflecting water tank boundaries was eliminated by appropriate time gating to simplify the acoustic signal analysis.

A spherical hydrophone acoustic transducer (0.7 cm in diameter) was used in the experiments. The barium titanate piezoelectric element of the transducer has a response of 81.3 Pa/mV for the pertinent range of frequencies (80–200 kHz). The acoustic transducer was placed outside the radiation beam of the waveguide and perpendicularly to the propagation direction of microwaves (Fig. 10.2). Measurements were made of the acoustic responses from a single tube and from multiple test tubes separated by known distances. Typical transducer responses from a single tube filled with glycol are shown in Fig. 10.9. Note the characteristic dependence of some acoustic wave peaks on the edges of the test tubes and the magnitude of variation with space and/or time. Based on these acoustic wave characteristics, a pattern extraction algorithm was developed to analyze the wave.

The thermoelastic pressure waves for different filling materials and different sizes of cylindrical tubes are shown in Figs. 10.10 and 10.11. It can be seen from Fig. 10.10 that the pressure wave magnitude depends on the material inside the test tube; the higher the differential microwave absorption rate and thermoelastic conversion efficiency of the medium, the greater the generated pressure, which translates into enhanced tissue image contrast. As expected, the response from water is the lowest but note the effect of glass tube is very small. Figure 10.11 shows that the larger the tube size, the greater the magnitude of pressure peaks.

The correlation between imaged and physical distances of separation between the phantom models is shown in Fig. 10.12. Estimated distance based on images is in close agreement with the actual distance. Moreover, two cylindrical test objects (diameter = 1.6 cm) are imaged with a spatial resolution of 9 mm. When the separation is less than 9 mm, the interference of responses from adjacent pairs of objects becomes sufficiently severe to cause uncertainty. In summary, these experimental results show that the peaks of microwave pulse-induced thermoelastic pressure waves depend both on the size and composition of test objects, which is

Fig. 10.9 Acoustic
transducer recordings after
successively shifting
transducer to the right
0.5 cm. P_1 and P_2
correspond to the two sides
of a test tube filled with
glycol

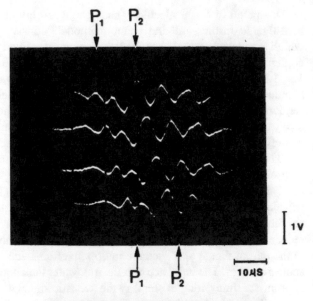

Fig. 10.10 Thermoelastic
responses for 2.0 cm
diameter tubes filled with
different solutions: (**a**)
water, (**b**) 0.9% saline,
(**c**) glycerol, (**d**) diethylene
glycol, and (**e**) glycol

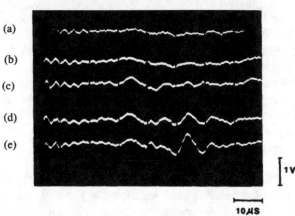

useful in identifying objects of unknown composition with enhanced contrast.
Furthermore, for this prototype MTT system, two phantom objects are detectable
with a spatial resolution of 9 mm, without invoking common signal conditioning
routines or sophisticated digital image processing techniques.

Another set of attributes of interest to imaging applications are the thermoelastic
pressure wave frequency and intensity generated by microwave pulse as functions
of tissue mass or volume. The amplitude of microwave-pulse-induced thermoelastic
pressure in a spherical model of tissue mass with free and constrained surfaces is
given in Sect. 8.2 and 8.3, respectively. Typically, the pressure amplitude is directly
related to microwave power; it is therefore controllable and will not be discussed
further here.

Fig. 10.11 Thermoelastic responses for cylindrical tubes filled with glycol with different tube diameters: (a) 2.0 cm, (b) 1.6 cm, (c) 1.5 cm, and (d) 1.2 cm

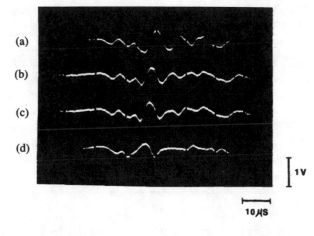

Fig. 10.12 Correlation of MTT-estimated distance and actual distance. The numbers 2 and 3 denotes number of data sets having the same value

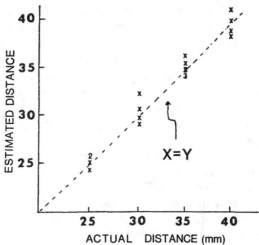

Some calculated acoustic frequencies as a function of spherical tissue mass show that the microwave-induced acoustic frequencies are higher for a constrained tissue surface compared to those for a free surface (see Figs. 8.2 and 8.3). For tissue spheres ranging in radii from 1 to 50 mm, the fundamental acoustic frequencies vary from 16 to 780 kHz for free surfaces and 22 kHz to 1.11 MHz for constrained surfaces such as bones and skulls [Lin, 2005]. Since most tissue boundaries or organ surfaces are not rigidly constrained inside the body, microwave pulse-generated thermoelastic pressure wave inside the human body have acoustic frequencies or wavelengths closer to numbers represented by the free surfaces.

The formula for calculating the frequency of microwave-pulse-generated thermoelastic pressure waves by a spherical tissue model with a stress-free surface is given in Eq. (8.27) or

$$f_m = \frac{mc}{2a}$$

(10.1)

Table 10.1 Calculated acoustic harmonic frequencies of thermoelastic pressure waves in spherical tissue models with stress-free surfaces

Sphere radius (mm)	Acoustic frequency (MHz)		
	Fundamental (f_1)	3rd Harmonic (f_3)	5th Harmonic (f_5)
1	0.78	2.33	3.88
3	0.26	0.78	1.29
5	0.16	0.47	0.78
10	0.08	0.23	0.39
20	0.052	0.12	0.19
30	0.026	0.078	0.13
40	0.019	0.058	0.097
50	0.016	0.047	0.078

where f_m is the frequency of acoustic vibrations inside the spherical tissue model. There is an infinite number of modes of acoustic vibration of a spherical tissue mass exposed to pulse-modulated microwave radiation for m = 1, 2, 3 …. Note that the acoustic pressure frequency is only a function of a, the model size and c, the speed of pressure wave propagation, which depends on the elastic properties of tissue or organ. Table 10.1 gives a list of calculated acoustic frequencies as a function of radius of spherical tissue models for a speed of thermoelastic pressure waves at 1550 m/s. It lists the fundamental acoustic frequency, and third and fifth harmonics for a free surface. For tissue spheres ranging in radii from 1 to 50 mm, the fundamental acoustic frequencies vary from 16 to 780 kHz for free surfaces. Note that frequencies of the 3rd and 5th harmonic cover the range from 47 kHz to 2.33 MHz and 78 kHz to 3.88 MHz for tissue spheres with free surfaces whose radii are 1 to 50 mm. These results indicate that a minimum bandwidth of acoustic sensors required to appropriately detect the tissue responses is between 16 kHz and 3.88 MHz. A better choice is an acoustic sensor that can cover up to the tenth harmonics or about 7.75 MHz. The wavelength associated with these frequencies is down in the submillimeter range. Thus, the theoretical spatial image resolution afforded by these microwave pulse-induced acoustic signals would certainly exceed the 9-mm limit based on simple cylindrical phantom experiments by a significant margin in soft tissues.

10.1.3   Further Investigations in MTT

The prototype MTT system and its use in successfully demonstrating the feasibility of MTT as a nonionizing and noninvasive imaging modality for anatomical structures, with the possibility of yielding unique information on tissue characteristics with enhanced spatial resolution, has been encouraging and promising. Thus, around the year 2000, the potential of MTT as an innovative imaging method started to attract considerable research interest. Greater efforts toward developing MTT for near real-time detection and imaging of mammary tumors began to emerge especially for early breast cancer detection [Kruger et al., 1999, 2000; Wang et al., 1999; Ku and Wang, 2000a, b; 2001; Xu et al., 2001; Nie et al., 2008; Ji et al., 2012]. These efforts have reaffirmed that an MTT imaging system that combines micro-wave energy absorption and ultrasonic energy emission is a unique nonionizing and noninvasive imaging modality. Most of the investigations were targeting MTT as a modality that may complement mammography for early detection of breast cancer. In addition, there have been some studies aimed toward a more exact reconstruction of tissue properties [Ku and Wang, 2000a, b; Xu and Wang, 2002]. Note that although it is an approximation, the often-used filtered back projection algorithms have given reconstructed images that agree well with the original specimens [Su and Lin, 1990; 1991]. Furthermore, enhanced algorithms that incorporate a rebinning methodology and consider the diffraction and refraction effects of wave propagation can render significant improvements in the quality of reconstructed images [Su and Chen, 1992].

Nevertheless, the task is complex due to the intrinsic heterogeneity of normal breast tissues, which are composed of fat, connective tissue, glandular tissue, muscle, milk ducts, and blood vasculatures. Moreover, the tissue compositions may differ across patients of various age groups. Subjects in different stages of menstruation, pregnancy, or lactation have different compositions. Normal breast tissues have different water or fat content. Furthermore, malignant breast tissues often are composed of small parts of malignant cells infiltrating within a large number of normal cells, which obviously renders imaging of mammary tumors and early detection of breast cancers a challenge.

Some promising results were obtained in a series of studies using pulses of 434 MHz microwave energy. The group experimented with several variations of MTT instrument design. The MTT systems were specifically targeted for human breast imaging [Kruger et al, 1999; 2000], and consist of a hemispherical, 300-mm-diameter bowl in which 64 piezo-ceramic transducers were attached in a spiral pattern. The bowl was mounted on a cylindrical shaft and could be rotated 360° about its axis using a stepper motor. The entire device was immersed in deionized water contained in a cylindrical tank. For human breast imaging, the tank itself was placed beneath an examination table, in which a circular hole was cut and mated to the top of the MTT imaging tank. Initially, a helical antenna was installed at the base of the transducer array. A semi-rigid coaxial cable, which passed through the center of the shaft, was used to feed microwave energy to the helical antenna. This antenna arrangement produced circularly polarized microwaves, and a slowly varying spatial intensity pattern within the imaging volume. Microwave pulses (0.5-μs wide; 25 kW peak power) were delivered at a nominal repetition rate of 4000 Hz. An updated version of the system replaced the helical antenna with wave-guide antennas [Kruger and Kiser, 2001; Kruger et al., 2002; 2003]. The wave-guide antennas produced linear polarization and had averted the strong, thermoelastic signals generated at the surface of the helical antenna. These unwanted, thermoelastic signals overlapped in time with the much weaker thermo-elastic signals from the microwave absorbing tissues. A schematic diagram and a photograph of the latter system are shown in Figs. 10.13 and 10.14. Note that this system has eight waveguide antennas and 128 ultrasonic transducers arrayed on the hemispherical surface.

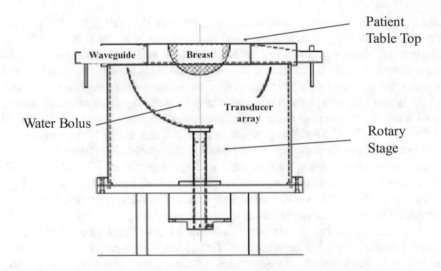

Fig. 10.13 A schematic diagram of 434 MHz microwave thermoacoustic tomography system

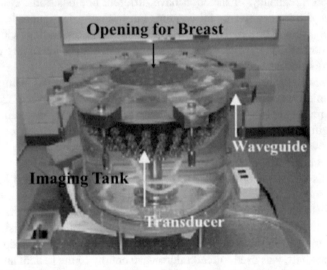

Fig. 10.14 A photograph of 434 MHz microwave thermoacoustic tomography system

A series of coronal sections and one sagittal section of the MTT images from a patient's breast in the region of the cancerous mass are shown in Fig. 10.15. The arrows in these images indicate a large, lobulated mass. An increase in contrast of approximately twofold was seen in the region of this large mass [Kruger et al., 2000]. The X-ray mammogram (M in Fig. 10.15) from the same patient indicated a large, poorly defined mass in the upper right breast. The transverse ultrasonographic scan (U) showed the same mass (indicated by calipers). Examples of cysts, fibroadenomas, and ductal carcinomas in MTT images are given in Fig. 10.16 [Kruger and Kiser, 2001; Kruger et al., 2002]. In Fig. 10.16a, the cysts can be readily visualized as dark regions surrounded by brighter glandular tissue, with higher microwave

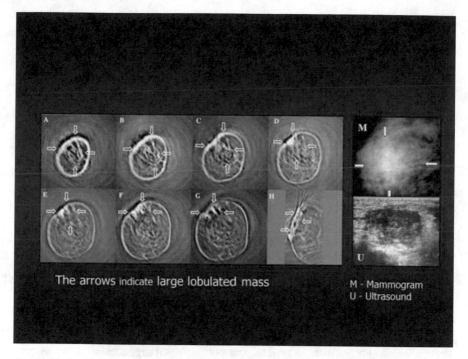

The arrows indicate large lobulated mass

M - Mammogram
U - Ultrasound

Fig. 10.15 A series of coronal sections and one sagittal section of the MTT images from a patient's breast in the region of the cancerous mass

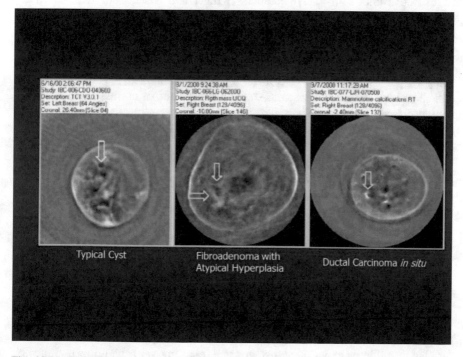

Typical Cyst

Fibroadenoma with Atypical Hyperplasia

Ductal Carcinoma *in situ*

Fig. 10.16 Example of 434 MHz MTT images: (**a**) cysts, (**b**) fibroadenomas, and (**c**) ductal carcinomas

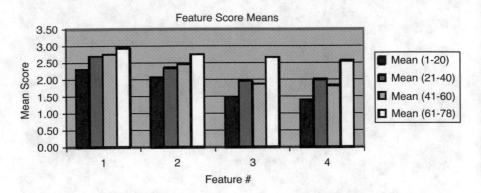

Fig. 10.17 MTT image quality grades for 4 features: areolar complex, skin margin, glandular tissue, and Cooper's ligament, average for patient's number 1–20, 21–40, 41–60, and 61–78

absorption. Fibroadenomas absorb microwave energy less and are represented by a weaker contrast enhancement in MTT images. Figure 10.16c illustrates the appearance of ductal carcinoma in situ. Thus, the fibrocystic disease was well seen – cysts appearing as areas of low microwave absorption. Fibroadenomas did not demonstrate contrast enhancement in general, cancer displayed higher microwave absorption than surrounding tissues.

The potential clinical utility of using MTT to image the breast was assessed in a retrospective pilot study of 78 patients [Kruger and Kiser, 2001; Kruger et al., 2002]. The patient volunteers were classified into three age groups (<40, 40–50, >50 years). The study population was further segregated into normal and suspicious groups, based on the previous X-ray mammography and ultrasound. MTT image quality was evaluated qualitatively by consensus of two trained mammographers using a 4-point scale (1 = not visualized; 2 = poorly visualized; 3 = adequately visualized; and 4 = excellently visualized). The appearance of normal anatomy, cysts, benign disease, and cancer is noted. Figure 10.17 gives results of image quality evaluation for four features (areolar complex, skin margin, glandular tissue, and Cooper's ligament), averaged for patient's number 1–20, 21–40, 41–60, and 61–78. Two trends are evident. In general, image evaluation scores improved throughout the course of the study. Also, the scores of the four features followed the pattern: areolar complex > skin margin > glandular tissue > Cooper's ligament. The mean scores of all four features were averaged for patients with different density of breast. Figure 10.18 shows that the denser the breast the better the MTT images are scored, in contrast to X-ray mammography.

10.1.4   Recent Developments in MTT

As an emerging biomedical imaging modality, MTT has gained substantial momentum but is still in its experimental stages and it has not yet been utilized clinically, although many studies have demonstrated its potential for clinical applications.

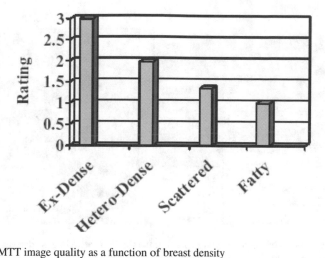

Fig. 10.18 MTT image quality as a function of breast density

Much of the progress in the interim are well documented in a review article [Cui et al., 2017]. Several research groups are working toward reaching the goal of translating MTT into a routine procedure in diagnostic radiology. Recent developments in MTT have proceeded along the direction of improvements through computer simulation, practical system design to fulfill its premise of a low-cost, nonionizing, noninvasive imaging modality with high tissue contrast and high spatial resolution, and extension to novel application scenarios beyond the initial focus on breast tumor detection [Aliroteh and Arbabian, 2018; Wen et al., 2017; Zhao et al., 2013].

   In general, computer simulation is a flexible method to facilitate system design and optimization. Using a setup like the prototype MTT system (see Fig. 10.2), Wang, et al., [2012] developed a numerical MTT imaging model for breast cancer imaging application for FDTD simulations of the entire MTT operation (Fig. 10.19). The simulated hardware systems include the feeding antenna, matching mechanism, liquid medium (fluid environment), 3-D breast model, and acoustic transducer. A rectangular waveguide is used as the transmitting antenna. The waveguide antenna is driven by a 1-μs-wide square pulse-modulated 2800 MHz microwave generator at a peak power of 1 W. Thermoelastic pressures were simulated for detection by a planar (49 × 49 sensor) acoustic transducer array with 0.5-mm spacing. It is interesting to note that the tissue phantoms showed unique thermoelastic response due to differences in microwave energy absorption as functions of microwave frequency, which may be useful for tissue identification purposes. The simulation provided quantitative relationships between the input microwave peak power and the resulting specific absorption rate as well as the output acoustic pressure signal. Moreover, it was noted that the acoustic pressures of all tissues exhibited discernible increase from 2.3 to 6 GHz and decrease from 6 to 12 GHz. Also, the reconstructed MTT images were found to be consistent with the models used.

   An integrated computer simulation based on finite integration time-domain (FITD) and pseudo-spectral time-domain (PSTD) methodologies was developed by

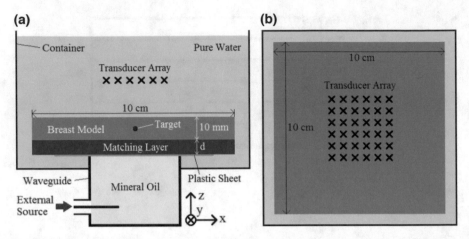

Fig. 10.19 Schematics of a numerical MTT model: (**a**) side view with longer transverse dimension of waveguide in the x-direction.; (**b**) top view

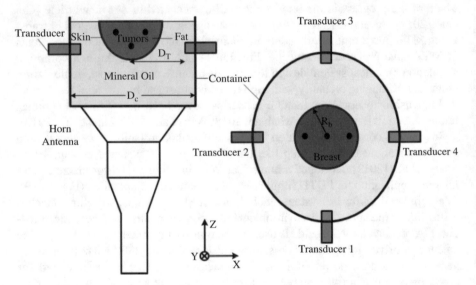

Fig. 10.20 Schematic illustration of system model for MTT imaging of breast tumors: (a) side view; (b) top view

Song et al. [2013]. The simulated system for MTT imaging consisted of a wave-guide horn antenna, fluid coupling medium, 3-D hemispherical breast phantom, and signal collection by acoustic transducers for image reconstruction. A 41-mm radius, 3-component breast phantom model with skin, tumor, and fatty tissue is embedded in a 109.5 mm diameter imaging domain (Fig. 10.20). One advantage of the PSTD method is that it minimizes the dispersion error associated with the FDTD approximation of spatial derivatives. Authors reported that the simulation result matches well with their experimental result.

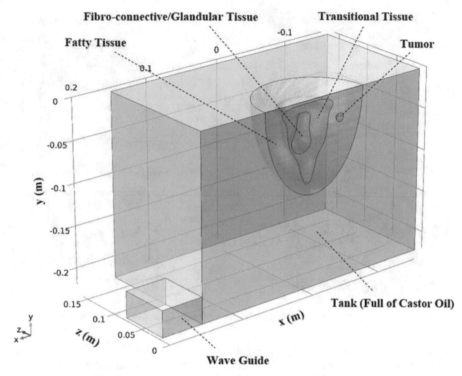

Fig. 10.21 Three-dimensional view of waveguide, tank, breast tissue, and tumor in a computer model of MTT imaging

In another computer experiment (Fig. 10.21), using a commercial computational software [Soltani et al., 2019], a three-dimensional numerical simulation of an anatomically realistic breast phantom consisting of a tumor and three different tissue types are placed in a tank containing fluid medium and is irradiated by a 1-ms wide, 2450-MHz pulsed microwave from a rectangular waveguide. Note that while immersing a biological tissue sample in castor oil (a mineral oil) may be pragmatic in this case, the use of a water-base fluid medium would be more appropriate and necessary for biomedical applications. Nevertheless, the results showed a slight temperature rise (2.47×10^{-3} °C) from absorption of 1-ms wide, 1.0 kW microwave pulse in a 1-cm diameter tumor located in fatty breast tissue. The small temperature rise in the tumor can produce a pressure variation of 0.58 kPa in the tumor. This pressure produced an acoustic signal, detectable with an array of transducers to allow MTT image reconstruction.

The prototypical systems mentioned above commonly employ rectangular waveguide or waveguide horn antennas and microwave pulses that are µs to ms wide for irradiation of tissue targets in the near field. A different scheme for MTT imaging system for breast tumors using ultrashort (2–10 ns) microwave pulses with fast rise time (800 ps) and high peak power was developed and tested [Lou, et al., 2012; Fu et al. 2014]. This enlarged field of view MTT or nsMTT imaging system (the

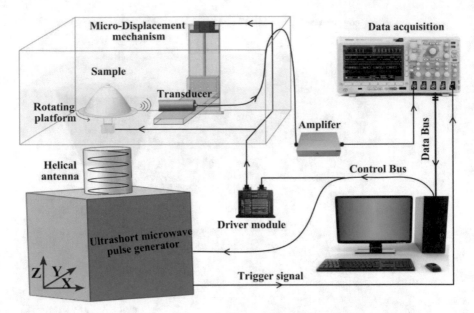

Fig. 10.22 A systems representation of an ultrashort pulse MTT or nsMTT imaging system

suggested descriptive name of this scheme) requires about 2% of the microwave energy density compared to the common MTT technique. A helical antenna with a center frequency of 434 MHz is used to deliver a short high-voltage pulse triggered microwave radiation. A line-focus ultrasonic transducer with a center frequency of 5 MHz (80% bandwidth, −6 dB), focal length of 50 mm, and a transducer element diameter of 8 mm is immersed in mineral oil to detect the thermoelastic signals (see Fig. 10.22). Mineral oil is used for acoustic coupling. Computer simulated and experimental results demonstrate that the nsMTT system developed can achieve 61 mm imaging depth, 230μm spatial resolution, and 12×12 cm^2 microwave radiation area, sufficiently large to permit imaging of the entire breast volume. A breast tumor phantom can be localized to within an accuracy of 1 mm.

Recently, coherent frequency-domain MTT imaging was investigated as an alternative to the short-pulse-based approaches mentioned above, but with reduced peak power requirement [Nan and Arbabian, 2017; Nan et al., 2015]. Using a 2100-MHz microwave source providing 120 W of peak power, it demonstrated that the proposed frequency-domain MTT method can obtain sufficient soft-tissue contrast and resolution with a pork belly sample of layered structure, while showing a 27-dB signal-to-noise ratio improvement over the common pulse method. The reduced peak-power requirement would open future possibilities of employing a solid-state microwave source to enable portable MTT imaging for various diagnostic applications.

Another recent innovation is the development of a compact, portable, and low-cost MTT system that is proposed for mobile imaging applications. In support of this effort, a handheld dipole antenna with an aperture of 6 cm and a mass of 230 g has been developed, which is coupled with a linear acoustic transducer array [Huang et al., 2019]. Note that the handheld dipole antenna-induced thermoelastic pressure is about two times higher than that produced by an aperture waveguide antenna. Aside from ambulance and bedside uses, a portable MTT system could be valuable for imaging many specific tissue types at constrained organ sites.

10.1.5   Other MTT Application Domains

While breast tumor and its early detection have been the prime motivation of many MTT investigations, other potential applications of MTT imaging in the medical diagnosis domains are beginning to attract research attention. The advantages of high resolution and high contrast along with low cost and portability of MTT systems are being explored for imaging of soft tissues such as the kidney, prostate, and subcutaneous blood vasculature as well as imaging of hard tissues such as the knee and finger joints [Cao et al., 2010; Patch et al., 2016; Aliroteh and Arbabian, 2018].

The feasibility of MTT imaging of renal calculi was explored in a project based on microwave absorption differences among the calcium oxalate calculus, uric acid calculus, and normal kidney tissue [Cao et al., 2010]. A schematic of the experimental setup is shown in Fig. 10.23. The 1200 MHz microwave pulses (0.5 μs and 300 kW peak power) irradiated the kidney sample in a mineral oil-filled tank through a rectangular waveguide horn antenna with a cross-section of 12.7 cm × 6.3 cm. A multi-element linear transducer array captured the thermoelastic pressure signals using circular scanning rotations. The reconstructed images indicate that the calculi embedded in the swine kidney are clearly imaged with good contrast and resolution in the three orthogonal MTT thermoacoustic images shown in (Fig. 10.24).

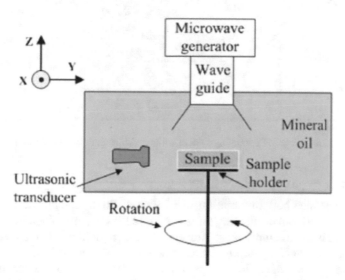

Fig. 10.23 Schematic of the experimental arrangement of 1200-MHz MTT imaging system

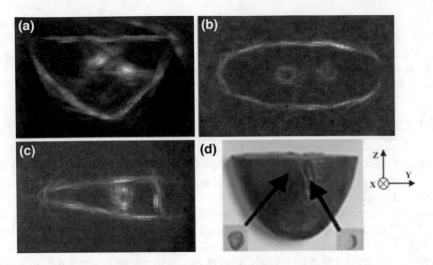

Fig. 10.24 MTT images of renal calculi in swine kidney in orthogonal views: (**a**) Coronal plane. (**b**) Sagittal plane. (**c**) Axial plane. (**d**) Photograph of sample

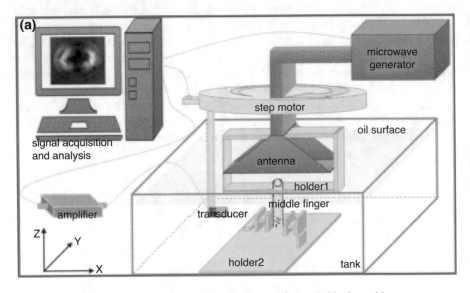

Fig. 10.25 A 3000-MHz MTT imaging system with hand/fingers holder in position

The potential of MTT to image the anatomical structures of the extremities has been demonstrated both in rabbit knees [Chi et al., 2016] and human finger joints [Chi et al., 2019]. Schematic diagrams for the MTT system is given in Fig. 10.25; the same for both structures except for the jigs used to hold the rabbit knee or positioning the hand and fingers, which are immersed in a mineral oil-filled tank. Pulsed microwaves (3000 MHz, 70 kW peak power, 750 ns pulse) are fed through a

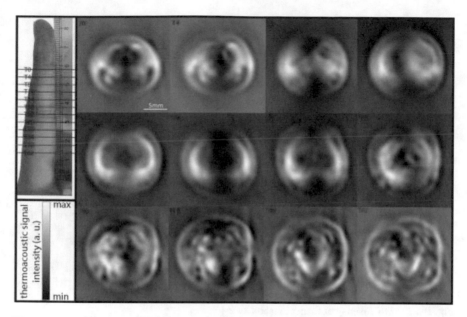

Fig. 10.26 MTT images at different scanning positions along the axis of the left middle finger. The spatial interval between two adjacent scans is 4 mm. [Chi, et al. 2019]

waveguide horn antenna (aperture size: 114 mm × 144 mm). The acoustic pressure wave is detected using a 5 MHz cylindrically focused ultrasound transducer with an active aperture diameter of 10 mm and a focal length of 53 mm. For data acquisition, the transducer was rotated with a radius of 7 mm at 2° intervals to 180 positions to cover the 360° around the finger. Note that in principle, half as many positions would provide the same reconstructed images due to redundancy. The middle and index fingers from healthy volunteers are tested in vivo. Axial slices showing the structure along the length of the left middle finger of one subject are presented in Fig. 10.26. Various details of intra- and extra-articular tissues are identifiable [Chi, et al., 2019].

Adult male New Zealand rabbits (2 kg) were employed in a knee imaging study [Chi et al., 2016]. The rabbit is placed in its holder after intravenously injection of pentobarbital (30 mg/kg) through the ear. The holder and rabbit are then positioned under the horn antenna of the MTT system for irradiation, signal acquisition, and image reconstruction (Fig. 10.27). After the experiment, the imaged knee joint was dissected and photographed. Identical procedures were performed for four rabbits. Comparative analyses between MTT images and anatomical pictures of the knee joint showed great similarities between the reconstructed images and the anatomical pictures in the shape and size of various knee joint tissues. MTT images of the ligament, fat pad, and other joint tissues are clear and can render details of the rabbit knee joint with good resolution. Thus, MTT may offer a new tool for noninvasive detection of joint diseases such as osteoarthritis, or perhaps orthopedic as well.

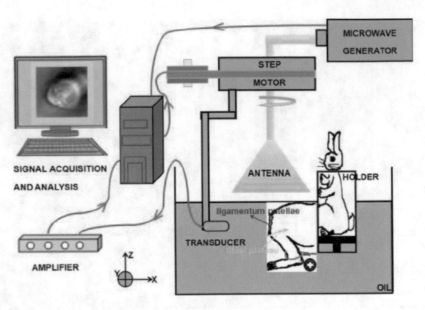

Fig. 10.27 MTT imaging of tissues in rabbit knee joint

An experiment to investigate the feasibility of using MTT to image brain structures was initiated with the preserved head of a 7-month-old Rhesus monkey [Xu and Wang, 2006; Jin et al., 2008], and the study was followed up later with a phantom-filled adult human skull [Yan, et al., 2019]. The experimental setup is like that shown in Figs. 10.2 and 10.17, but with microwave antenna transmitting 0.5 µs wide, 3000 MHz pulse at 20 kW peak power at the bottom of the mineral-oil-filled tank. The signal detectors were a cylindrically focused ultrasonic transducer with an active-element aperture of ultrasonic transducers. The center frequency of the ultrasonic transducers is 1 MHz with a bandwidth of 0.8 MHz. Figure 10.28b shows an MTT image of an axial cross-section at the nose level (Fig. 10.28a). The corpus callosum, 3 cm from the head surface, is distinctly visible from the image.

For a phantom-filled human head, a 3000 MHz microwave pulse of 1.2µs at a peak power of 60 kW was delivered through the waveguide horn antenna. The thermoacoustic pressure signals are detected by an ultrasonic transducer with 1-MHz center frequency and 1-inch diameter. The transducer rotated in 600 steps around the head and the pressure signals were averaged 250 times at each scanning step, and then applied to MTT image reconstruction. A darkened circle in the reconstructed image clearly reveals the agar column enclosed inside the skull in Fig. 10.29.

The potential of MTT imaging as a low-power and compact technique for capturing biometric features from subcutaneous vasculature was explored using the common experimental setup and commercial acoustic transducers [Aliroteh and Arbabian, 2018]. Proof-of-concept MTT imaging of synthetic phantoms, plant

Fig. 10.28 MTT imaging of Rhesus monkey skull filled with phantom material: (**a**) Diagram showing the axial imaging plane for MTT imaging; (**b**) head image where the rear part of the head is at top

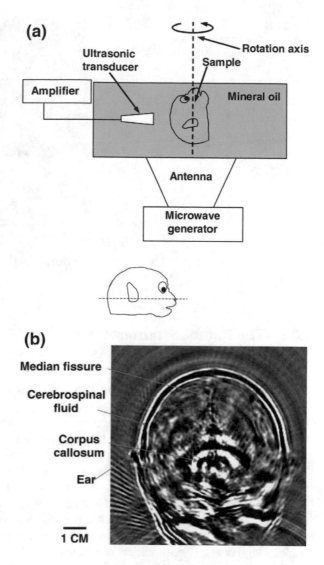

vasculature, and earthworm blood vessels with 50 W of peak power for a submillimeter spatial resolution has been reported. The well-known microwave contrast advantage makes it possible to differentiate blood vessels.

In summary, MTT continues to be a subject of vigorous research both from a systems development perspective and as a dual imaging modality amenable to greater utility in a wide range of application scenarios. The well-known microwave contrast and ultrasound acoustic resolution advantages continue to drive innovations. While MTT has not yet been put in use clinically, it may well be found in some biomedical diagnostic situations soon.

Fig. 10.29 Transcranial
MTT image of an agar
column in a human skull

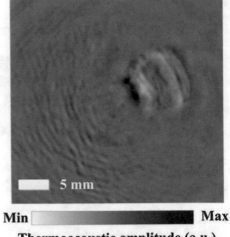

5 mm

Min ▭ Max

Thermoacoustic amplitude (a.u.)

10.2   The Havana Syndrome: Reports of Diplomatic Personnel's Sonic Attacks

As described in Chap. 1, from 2016 to 2018, there have been multiple reports that personnel of some foreign services have been experiencing health issues associated with hearing loud buzzing or what was described as bursts of sound. Staff members were reporting symptoms of hearing loss, ringing in the ears, headaches, nausea, and problems with balance or vertigo, which are suggestive of a connection to the inner ear apparatus within the human head. Officials had difficulty to pin down the source of sound. There were speculations that they may have been attacked with an advanced weapon. Assuming reported accounts are reliable, it is plausible that the loud buzzing, burst of sound, or acoustic pressure waves may have been covertly delivered using high-power pulsed microwave radiation, rather than blasting the subjects with conventional sonic sources [Lin, 2017; 2018a, b].

The U.S. National Academies of Sciences, Engineering, and Medicine (NASEM) recently released its study report "An Assessment of Illness in U.S. Government Employees and Their Families at Overseas Embassies" [Relman and Pavlin, 2020]. The NASEM report makes the point that, among the mechanisms the study committee considered in examining plausible causes of the described illnesses, "directed, pulsed radio frequency (microwave) energy appears to be the most plausible mechanism" in explaining these cases, especially in individuals with the distinct early symptoms.

Absorption of a single microwave pulse impinging on the head and conversion of microwave pulse to acoustic pressure wave by soft tissues inside the head may be perceived as an acoustic click or knocking sound, depending on incident microwave power density. A train of microwave pulses to the head may be sensed as an audible tune, chirp, or buzz. Depending on the power of the impinging microwave pulses, the level of induced sound pressure could be considerably above the threshold of auditory perception. Indeed, they may approach or exceed levels of discomfort and even tissue injury, including reported headaches, ringing in the ears, nausea, and problems with balance or vertigo. Furthermore, compared with individuals not experiencing the loud bursts of sound, brain magnetic resonance imaging (MRI) revealed significant differences in whole brain white matter volume, regional gray and white matter volumes, cerebellar tissue microstructural integrity, and functional connectivity in the auditory and visuospatial subnetworks but not in the executive control subnetwork [Verma et al., 2019]. However, the clinical importance of these differences is not definitive.

Apparently, while the clinical symptoms presented are concussion-like, the MRI images did not resemble usual presentations of traumatic brain injury or concussion. However, clinical experiences with concussion are mostly derived from externally inflicted impact injuries such as a hit to the head against the ground or other rigid bodies, which may set brain tissues into violent motion against the skull. A high-power microwave pulse-generated acoustic pressure wave could be initiated in the brain and then reverberate inside the head (see results of computer simulations in Chap. 9) and potentially reinforcing the initial pressure to cause injury of brain matters. Thus, it is conceivable that the MRI images from high-power microwave-induced pressure or shock waves may have entirely different manifestations of brain injury or concussion. The clinical importance of these differences is uncertain at present and may command further study for clarification.

Nevertheless, it appears that the U.S. military has maintained some interest on the topic of microwave auditory effect and has awarded a research contract entitled, "Remote Personnel Incapacitation System" through the navy's small business innovative research program [Navy SBIR, 2004]. The initial goal of the project was to design and build a prototype non-lethal weapon based on the microwave auditory effect. The transient personnel incapacitation system is dubbed MEDUSA (Mob Excess Deterrent Using Silent Audio). The weapon relies on a combination of pulse parameters and pulse power to raise the auditory sensation to the "discomfort" level to deter personnel from entering a protected perimeter. While the status or outcome of this project maybe privileged, there are indications that hardware was built, and power measurements were taken to confirm the required pulse parameters enabling observation of the microwave auditory effect, which is an expected situation that was never in doubt.

As mentioned in Chapter 1, the requisite microwave technology is mature and in general commercially available. Longer distances and higher power scenarios would require more bulky equipment and sophisticated aiming devices, but packable equipment is possible for closer range nonlethal applications. Also, existing hardware may need to be optimized to meet some specific requirements in defined or finely targeted operations.

Note that the known near-zone thresholds determined under controlled laboratory conditions for peak microwave power density of auditory perception in human subjects with normal hearing are given in Table 6.5. While there are greater variations in measured threshold values over the entire range of 1 to 70μs of pulse widths involved, the subset of data for 10 to 32μs fall within a narrower range. Considering that the ambient noise levels in all three experiments were essentially the same, it may be reasonable to conclude that the averaged threshold power densities of 2.1 to 40 kW/m² of 14 kW/m² as a realistic threshold peak power density for induction of the microwave auditory effect in the near field of 1250 to 3000 MHz microwaves with pulse widths between 10 and 30μs. In other words, the 14 kW/m² per pulse peak power density generates a barely audible sound level of 0 dB. To generate sound at 60 dB, or the audible level for normal conversation, requires 1000-fold higher power density per pulse. To generating tissue injuring level of sound at 120 dB would take another 1000-fold increase in required peak power density, or 14 GW/m² per pulse. The corresponding theoretical temperature elevation would be about 1 °C, which is deemed "safe" by currently promulgated protection guidelines.

For plane-wave equivalent exposures, the computations from Chap. 9 provide two sets of data that are suitable for comparison with the results described above. In one case the reported threshold peak incident power density for an anatomical head model is 3 kW/m² for 20-μs pulses at 915 MHz [Watanabe et al., 2000]. For the other, the threshold is about 50 kW/m² for 20-μs pulses at 2450 MHz [Yitzhak et al. 2014]. The corresponding peak incident power densities at the 120 dB injury level are therefore between 3 and 50 GW/m² per pulse, which bracket the 14 GW/m² per pulse from the above calculation for near-zone exposures. These peak power densities are close to and encompass the 23.8 GW/m² value for dielectric breakdown of air. As the dielectric permittivity of all biological and physical materials is greater than that of the air or free space (Chap. 4), the intrinsic impedances are always smaller than that of air (Chap. 2). The breakdown peak power density in skin, muscle, and brain tissues, for example, would be a factor of 6 to 7 higher or 142 to 166 GW/m² for a microwave pulse at 1000 to 3000 MHz. Thus, if the microwave auditory effect is weaponized, lethal, or nonlethal, at sufficiently high powers, it is likely for the microwave pulses to cause auditory pathway nervous tissue injury or damages to brain tissues by reverberating sonic shock waves first. It would not be by microwave pulse-induced hyperthermia through excessive temperature elevation in the brain, nor by dielectric breakdown of brain, muscle, or skin tissues. Note that the units of measure of kW/m² or GW/m² per pulse refers to power density (power per square meter) not the total output power of any source.

10.3   Directed Messaging and Mind Control

Over the years, some have expressed concerns about the capability of communicating directly with humans by pulsed microwaves. In fact, earlier while studying microwave-induced auditory effects in human subjects, it was noticed that in addition to zip, click and knock sounds from exposure to single pulses of microwave radiation, short trains of rectangular microwave pulses are heard as chirps with tones corresponding to the pulse repetition frequencies [Guy et al., 1973, 1975]. It was also found that when the pulse generator was keyed manually such that each closing and opening of a push-button switch resulted in emitting a short rectangular pulse of microwave energy-transmitted digital codes (Morse code) were received and accurately interpreted by the targeted subject [Lin, 1978].

Also, soon after a system was setup for exploring the microwave auditory effect at the Walter Reed Army Institute of Research (WRAIR) [Sharp, et al., 1974], two of the principal investigators were able to demonstrate direct communication of simple speech via appropriate modulation of microwave energy [Justesen, 1975]. They tape-recorded each of the single-syllable words for digits between one and ten. The speech waveforms of each word were then converted to digital signals in such a fashion that each time an analog speech wave crossed the zero reference in the negative direction, a short pulse of microwave energy was emitted from the transmitter. By subjecting themselves to the exposure of "speech modulated" microwave energy, the investigators reported the ability to hear, identify, and distinguish the words transmitted. Communication of more complex words and sentences was not attempted to the author's knowledge.

Nevertheless, as noted above, to generate sounds at 60 dB, or the level for hearing normal conversation, requires 14 MW/m^2 per pulse of peak microwave power density, which is about 1000-fold below the power density per pulse for tissue injury. A directional beam of microwave pulses could conceivably be used for directed messages (coded or simple words) that can only be perceived by targeted person or persons in the same microwave beam. Aside from command and control, there are worries that such systems not only may cause disruption to hearing, they could also be used for psychological torture or mind control.

10.4   Microwave Signal at the Moscow Embassy

The U.S. Department of State's dissemination of possible attacks on its foreign service staff described above is reminiscent of an event that occurred in the mid-1970s, when the biological and health effects of RF and microwave radiation were brought to the fore by the revelation of two U.S. ambassadors to Moscow dying of cancer; the ambassador at that time, Walter Stoessel, threatened to resign because of widespread staff concerns regarding potential health implications of an unusual microwave signal the Soviets were directing at the U.S. Embassy in Moscow [Lin, 2018a, b]. (Stoessel died of leukemia in 1986. He was 66 years old [NYT, 1986]).

Available reports suggest that, since the 1960s, it had been known that the Soviets were targeting the upper floors of the central wing of the Embassy in Moscow with low-level microwave radiation from near-by buildings, first from one, and then two sites. Indeed, it may have begun as soon as embassy staff had moved into the complex on Tchaikovsky Boulevard back in 1953 and persisted through at least 1988.

The levels of microwave exposure fluctuated over the years. The direction and intensity of the 0.5 to 10 GHz microwave signal changed in 1975 from 0.5 mW/m^2 for 9 h per day to 0.1 to 1.5 mW/m^2 for 18 h per day. But it was always directed toward or aimed at the upper floors of the embassy. Only background levels were detected elsewhere in the embassy complex.

It is noteworthy that these levels were actually below what was considered as safe for human exposure in the United States at the time. As a matter of fact, between 1966 and 1982, the American National Standards Institute guideline for safe human exposure to microwave radiation was 100 W/m^2. Essentially, the same standard was adopted by the U.S. Occupational Safety and Health Administration. By contrast, the Soviets' standards were 1 and 10 mW/m^2 for the general public and occupational exposures, respectively [Michaelson and Lin, 1987]. The Soviet standards were derived from observations made in laboratory experiments using small animals and surveys of people occupationally exposed to microwave radiation.

The existence of the microwave signal had been kept secret for years because, at the time, no one knew exactly why the Soviets were doing it or that there might be any health consequences. Conjectures on intent and purpose were abounded; the Soviet's motivation, however, remained a mystery. Speculations ranged from the Soviets' attempting to disable or interrupt communication signals from American electronic listening devices, to the energizing electronic snooping gadgets that may have been imbedded in the building prior to American occupancy, to influencing the psychological states of, or to brainwashing embassy staff.

At the time, Soviet publications and scientific understanding concerning micro-wave interaction with animal behavior and neurophysiology, including the central nervous system, were replete with accounts of direct effects of low-level microwave exposure on animals and humans [Petrov, 1972]. There was also broad interests in clinical and hygienic or public health effects, especially under occupational conditions. Most of the reported hygienic effects were, in essence, subjective complaints, commencing 2–5 years after the start of work involving microwave radiation. Specific categories include asthenic syndrome, characterized by depression, fatigue, headache, irritability, loss of appetite, and memory; autonomic syndrome, which features fainting spells, heart enlargement, and pulse and pressure lability; and dien-cephalic syndrome, represented by digestive abnormality, insomnia, and sexual dysfunction.

This episode harkens back to 1952, when American diplomatic security personnel discovered a tiny electronic eavesdropping device in the American ambassador's Moscow office. It was concealed inside a carved Great Seal of the United States given by the Soviets to the U.S. ambassador in 1945 and the ambassador proudly hung it in the embassy. The device was a passive RF transponder-type sensor, a predecessor of current-day RFID sensor technology, activated by RF energy launched from an external source, which allowed Soviet agents to eavesdrop on secret conversations for 7 years before detection. From that point on, the embassy was under periodic surveillance for electronic signals. Thus, there was a distinct reason for the discovery of the microwave signal at the embassy complex on Tchaikovsky Boulevard.

In any case, according to reports, when Ambassador Stoessel learned about the microwave signal, he threatened to resign unless the embassy community was informed. As a result, the existence of the microwave signal was finally made public in a press conference called by the ambassador in 1976. Still, the embassy community felt betrayed about being kept in the dark for so long.

It is conceivable that more of the embassy community was anxious about the effect the microwave signal might be having on their health, especially after the U.S. Congress enacted the Radiation Control for Health and Safety Act in 1968 [PL 90-602, 1968]. The preceding deliberations had highlighted a general lack of current scientific knowledge on biological effects and health implications of both ionizing and non-ionizing radiation exposure and revealed the considerable amount of unnecessary radiation, including microwaves people were exposed to each year.

The U.S. Congress had declared then that the public's health and safety must be protected against the dangers of radiation from electronic products. The act authorized the federal government to set radiation standards, monitor compliance, and undertake research. It directed the U.S. Department of Health, Education, and Welfare to establish and carry out an electronic product radiation control program designed to protect the public's health and safety from radiation emitted by electronic products, including microwaves.

An immediate benefit of Stoessel press conference was the installation of metallic screens on the embassy's outer windows to provide a substantive degree of shielding against microwave penetration into the building [Brodeur, 1976]. The nominal screening efficiencies are about 1000–100,000, depending on such factors as mesh size and type of materials.

Furthermore, while apparently unknown to the embassy staff, other activities were already underway. The U.S. government had initiated a research program, code-named Operation Pandora, at the Walter Reed Army Institute of Research with Herb Pollack, a physician, Joseph Sharp, a behavioral psychologist, and Mark Grove, a microwave electronics engineer, as principals. The program subjected trained Rhesus monkeys to microwave exposures, mimicking characteristics gathered by monitors at the Moscow embassy. Microwave-induced interruption in the monkey's performance for food was studied in a classic operant conditioning experimental protocol. The study was terminated in 1969. Reported findings included some microwave irradiation-associated "aberrant behavior." However, agreement on an unambiguous consensus for behavioral psychological changes was not reached among project personnel.

Apparently, soon after the highly publicized press conference where the existence of the Moscow embassy microwave signal was finally made public in 1976, a 2-year epidemiological research was initiated by the U.S. Department of State at Johns Hopkins University [Silverman, 1980]. The study involved 1827 employees and 3000 dependents at the Moscow Embassy, as well as 2561 employees and 5000 dependents at comparable other U.S. Foreign Service posts in the Soviet bloc (Belgrade, Budapest, Leningrad, Prague, Sophia, Warsaw, and Zagreb) during the period from 1953 to 1975 as a control population. The control or comparison group was chosen to match the study group in selection criteria and environmental factors such as climate, diet, disease, and general social milieu, except these personnel were not subjected to microwave irradiation.

The purpose was to assess any differences in morbidity and mortality between the Moscow and comparison groups. Extensive efforts were devoted to identifying and tracing the populations. Information on illness, conditions, or symptoms was gathered and validated. Death certificates were used to ascertain mortality. Standardized mortality ratios and morbidity indices for various groups were developed for the study. At its completion in 1978, the investigators' conclusion was that the Moscow and comparison groups did not differ significantly in overall and specific mortality, and no compelling evidence was observed to implicate the Moscow microwave signal in any adverse health effect. Nevertheless, investigators had noted that the study population was relatively young, and it may be too early to detect long-term health and mortality outcomes.

The most perplexing question remains: What was the Soviets' purpose in microwaving the Moscow embassy? There was an indication, at a later point, that the Soviets had intimated it was a jamming signal calculated to thwart U.S. electronic spying devices. It is a plausible but simplistic rationalization; the signal strengths were rather weak for jamming maneuvers.

One possibility is that the Soviets were using microwaves to activate the numerous snooping devices they had implanted in the building prior to American occupancy. This obviously is an expected scenario given the success of the bug concealed in the carved Great Seal, especially because electronic fabrication capability had advanced by leaps and bounds since 1945. However, ever since the discovery of the Great Seal bug, surveillance for electronic signals had become routine at the embassy complex.

Another intriguing proposition was that the Soviets were bouncing the microwave signal off the embassy's window glass in an attempt to eavesdrop on conversations taking place inside the office. The theory is as follows. Conversation-generated sound waves would create tiny vibrations in the glass windowpane (vibrations that do occur, in principle). The reflection of the microwave signal impinging on the windowpane would be modulated by the vibrations in amplitude and phase, which might then be demodulated electronically to reproduce conversations taking place inside the office. While feasible, there are technological challenges in converting the faint electronic signals to voice, but doing so is not impossible. At best, the reflected microwave intensity would be down approximately 90% in a potentially noisy electronic operating environment. Nevertheless, given the demonstrated sophistication of the Great Seal bug from an earlier period, it could have been well within the Soviets' capability.

Finally, it might have been a designed exploitation of the Soviet understanding of how prolonged exposure to low-level microwaves affects the mental state of exposed subjects. If that was the intent, the Moscow microwave signal indeed may have accomplished, in part, its intended purpose. The embassy staff clearly showed anxiety after learning about the existence of the microwave signal. They were concerned about the potential health effects the microwave signal might be having on them or their children. It became somewhat of a morale problem for a time.

10.5   Pulsed Radar Exposure Induced Fighter Pilot Disorientation

Military pilots often report cognitive performance challenges during flight operations. Many have reported experiences with spatial disorientation (SD), in which the pilot's perception of aircraft position, motion, altitude, or attitude does not correspond to reality [Bellenkes et al, 1992; Lyons et al, 2006; Poisson and Miller, 2014; Takada et al, 2009]. Spatial disorientation has posed a significant problem to the military pilots and continues to be a challenge today. These reports indicate that spatial disorientation mishaps occur in fighter/attack aircraft at more than five times the rate of nonfighter/attack, fixed-wing aircrafts. Indeed, the rate of spatial disorientation related accidents has been estimated to be 11-12% of military aircraft crashes [Lyons et al, 2006; Takada et al, 2009].

The number of spatial disorientation accidents at night are higher than at daytime. The visual illusions at night because of the degraded visual environment are well documented. During night flights, pilots sometimes confuse ground lights with stars and unlighted areas of the Earth as night sky. From 1993 to 2013, spatial disorientation accounted for 72 Class A mishaps, 101 deaths, and 65 aircraft lost in US Air Force pilots [Poisson and Miller, 2014]. Mishaps are defined as Class A if they result in death or permanent total disability, or destruction of aircraft. It is noteworthy that mishap rates for F-16 fighter/attack aircraft were found to be marginally higher than for other fighter/attack aircraft.

Spatial disorientation may be caused by several human factors, such as the visual, vestibular, and the somatosensory systems involved in cognitive performance. Aside from potential effects on cognitive functions, there may be potential responses of more abrupt and distractive nature resulting from exposure to high-power pulsed RF and microwave radiation [Lin, 2020, 2021b].

The fighter/attack aircraft's cockpits are flooded with RF signals from on-board emissions, communication links, and navigation electronics, including strong electromagnetic (EM) fields from audio headsets and helmet tracking technologies. It has been hypothesized that the cockpit RF and EM fields, especially the frequencies between 9 kHz and 1 GHz, may influence cognitive performance including spatial disorientation, task saturation, and mis-prioritization. However, RF and EM fields in cockpits are not currently monitored; little effort has been made to shield pilots from these fields, and the potential impacts of these fields on a pilot's cognition have not been assessed.

It is reasonable to assume that the fighter cockpits are subjected to strong impinging RF and/or microwave radar pulses under some operational conditions. Common characteristics of these radar pulses are high peak power (GW), short pulse width (μs), and fast pulse rise time (ns). Depending on specific materials and designs of the helmets, the RF and microwave radiation could penetrate and reverberate inside the pilot's helmet and head to generate even higher RF and EM fields inside the pilot's head under these circumstances. The resulting exposure to high-power pulsed RF radiation and associated microwave energy deposition in head tissues may not produce overt tissue heating but can elicit sensitive biological responses. The prime examples include microwave pulse-induced acoustic pressure waves in the head (microwave auditory effect) [Lin, 1980, 2018a, b; Lin and Wang, 2007] and the startle reflex and motor reaction behaviors observed in laboratory animals [Brown et al, 1994; Lu et al, 2000; Seaman et al, 1994]. Indeed, there may be similarities between these pulsed microwave induced responses. The microwave auditory effect has been implicated in the sonic attacks on diplomats in Havana [Lin, 2018a, b, 2021a; Relman and Pavlin, 2020; also Sect. 10.2]. However, as discussed below, the microwave power threshold needed for startle reflex and motor reaction is higher than that required to induce the microwave auditory effect. Indeed, the microwave auditory effect may be the cause for the startle reaction resulting from a much louder microwave-induced sound. Thus, it is plausible that an unexpected, sudden, and intense auditory stimulus inside the head from a high-power radar could elicit a classic acoustic startle reflex that can cause abrupt changes in pilot's head position, orientation, or attention [Lin, 2020, 2021b].

The startle reflex and motor reaction behaviors are innate responses of mammals to an unexpected and sudden occurrence, such as intense auditory, visual, somato-sensory, or tactile stimulus that interrupts ongoing behavior, distracts from attentional function, initiates actions, and prepares the individual against a potential threat. Concurrently, it activates a protective stance to prevent injury and may alert the person or animal to instigate evasive behaviors [Fleshler, 1965; Gómez-Nieto et al, 2020].

For 1.25 GHz microwave radar pulses (10 μs, 80 Hz), the threshold for startle reflex was found to be 0.29 kJ/kg. The microwave energy was associated with less than 0.1 °C potential rise in the bulk tissue temperature in mice [Brown et al, 1994]. Clearly, the response is not associated with microwave induced tissue heating. Furthermore, it has been shown that a single 1.25 GHz microwave pulse (0.8 to 1.0 μs) to the head of rats at 22 - 43 mJ/kg per pulse (peak SAR, 2 3 - 48 kW/kg) could modify the acoustic startle response [Seaman and Beblo, 1992]. In another investigation, 1.25 GHz microwave pulses averaging 0.96 μs wide and 35.5 - 86.0 kW/kg peak power (SAR of 66.6 - 141.8 mJ/kg in absorbed energy) were reported to modify the startle response in rats [Seaman et al, 1994].

In summary, the ambient RF-EM field levels in a typical fighter/attack aircraft's cockpit is unclear now. Quantitative surveys and measurements are necessary to allow the proper assessment of the RF-EM field's potential effects on pilot's brain activity, neurophysiology, and behavioral responses. Noticeably, the fighter cockpits are subjected to strong impinging RF and/or microwave radar pulses under some operational conditions. There are two pulsed microwave induced auditory responses from humans and mammals when the head is exposed to high-power microwave pulses, which could impact a pilot's cognitive performance and behavioral response [Lin, 2020, 2021b]. These are microwave pulse-induced acoustic pressure waves in the head: the microwave auditory effect and the acoustically induced startle reflex and motor reaction from a sudden, unexpected, and intense auditory stimulus inside the pilot's head. The startle reaction from a sudden unexpected auditory stimulus may cause the pilot to experience spatial disorientation, in which the pilot's perception of aircraft position, motion, altitude, or attitude does not correspond to actuality.

The above hypothesis is formulated based on theoretical considerations along with available but limited experimental evidence. The kind of confirmational studies that would be useful are neurophysiological and psychophysical investigations of pulsed microwave exposed animals, including observations of subhuman primate's behavioral and performance responses. Furthermore, a set of human factors experiments may be appropriately designed to investigate the operability of pulsed radar exposure-induced startle on fighter pilot disorientation.

References

Aliroteh M, Arbabian A (2018) Microwave-induced thermoacoustic imaging of subcutaneous vasculature with near-field RF excitation. IEEE Trans Microw Theory Tech 66:577–588

Bellenkes A, Bason R, Yacavone DW (1992) Spatial disorientation in naval aviation mishaps: a review of class A incidents from 1980 through 1989. Aviat Space Environ Med. 63(2):128–131

Brodeur P (1976) Microwave I, New Yorker, Dec. 13, 1976, pp 50–110

Brown DO, Lu ST, Elson EC (1994) Characteristics of microwave evoked body movements in mice. Bioelectromagnetics, 15:143–161

Cao C, Nie L, Lou C, Xing D (2010) The feasibility of using microwave-induced thermoacoustic tomography for detection and evaluation of renal calculi. Phys Med Biol 55:5203–5212

Chan KH, Lin JC (1988) Microwave induced thermoelastic tissue imaging. In: Proceedings of the IEEE/EMBS annual international conference, New Orleans, pp 445–446

Chan KH, Lin JC, Olsen RG (1984) Microwave thermoelastic tissue imaging. In: Application of optical instrumentation in medicine XII conference, San Diego, CA, February 1984

Chi Z, Zhao Y, Huang L, Zheng Z, Jiang H (2016) Thermoacoustic imaging of rabbit knee joints. Med Phys 43:6226–6233

Chi Z, Zhao Y, Yang J, Li T, Zhang G, Jiang H (2019) Thermoacoustic tomography of in vivo human finger joints. IEEE Trans Biomed Eng 66:1598–1608

Cui Y, Yuan C, Ji Z (2017) A review of microwave-induced thermoacoustic imaging: excitation source, data acquisition system and biomedical applications. J Innov Optical Health Sci 10(4):1–18

Fleshler M (1965) Adequate acoustic stimulus for startle reaction in the rat. J Comp Physiol Psychol. 60:200–207

Fu Y, Ji Z, Ding W, Ye F, Lou C (2014) Thermoacoustic imaging over large field of view for three-dimensional breast tumor localization: a phantom study. Med Phys 41(11):110701

Gómez-Nieto R, Hormigo S, López DE (2020) Prepulse inhibition of the auditory startle reflex assessment as a hallmark of brainstem sensorimotor gating mechanisms. Brain Sci. 10: 639–654

Guy AW, Taylor EM, Ashleman B, Lin JC (1973) Microwave Interaction with Auditory Systems of Human and Cats. In: Proc IEEE International Microwave Symp, Boulder, Colo, pp 231–232

Guy AW, Lin JC, Chou CK (1975a) Electrophysiological effects of electromagnetic fields on animals. In: Fundamentals and applied aspects of nonionizing radiation. Plenum Press, pp 167–211

Guy AW, Chou CK, Lin JC, Christensen D (1975b) Microwave-induced acoustic effects in mammalian auditory systems and physical materials. Ann NY Acad Sci 247:194–218

Huang L, Ge S, Zheng Z, Jiang H (2019) Technical Note: Design of a handheld dipole antenna for a compact thermoacoustic imaging system. Med Phys 46(2):851–856

Ji Z, Lou C, Yang S, Xing D (2012) Three-dimensional thermoacoustic imaging for early breast cancer detection. Med Phys 39:6738–6744

Jin X, Li C, Wang LV (2008) Effects of acoustic heterogeneities on transcranial brain imaging with microwave-induced thermoacoustic tomography. Med Phys 35(7):3205–3214. https://doi.org/10.1118/1.2938731

Justesen DR (1975) Microwaves and behavior. Am Psychol 30:391–401

Kruger RA, Kiser WL Jr (2001) Thermoacoustic CT of the breast: pilot study observations. Proc SPIE 4256:1–5

Kruger RA, Kopecky KK, Aisen AM, Reinecke DR, Kruger GA, Kiser WL (1999) Thermoacoustic CT with radio waves: a medical imaging paradigm. Radiology 211:275–278

Kruger RA, Miller KD, Reynolds HE, Kiser WL Jr, Reinecke DR, Kruger GA (2000) Breast cancer in vivo: contrast enhancement with thermoacoustic CT at 434 MHz-feasibility study. Radiology 216:279–283

Kruger RA, Stantz K, Kiser JWL (2002) Thermoacoustic CT of the breast. Proc SPIE 4682:521–525

Kruger RA, Kiser WL Jr, Reinecke DR, Kruger GA (2003) Thermoacoustic computed tomography using a conventional linear transducer array. Med Phys 30:856–860

Ku G, Wang LV (2000a) Combining microwave and ultrasound: scanning thermoacoustic tomography. In: Proceedings of the 22nd annual EMBS international conference, Chicago, IL, pp 2321–2323, July 23–28.

Ku G, Wang LV (2000b) Scanning thermoacoustic tomography in biological tissue. Med Phys 27:1195–1202

Ku G, Wang LV (2001) Scanning microwave-induced thermoacoustic tomography: signal, resolution, and contrast. Med Phys 28:4–10

Lin JC (1976a) Microwave auditory effect-a comparison of some possible transduction mechanisms. J Microw Power 11:77–81

Lin JC (1976b) Microwave induced hearing sensation: some preliminary theoretical observations. J Microw Power 11:295–298

Lin JC (1977a) On microwave-induced hearing sensation. IEEE Trans Microw Theory Tech 25:605–613

Lin JC (1977b) Further studies on the microwave auditory effect. IEEE Trans Microw Theory Tech 25:936–941

Lin JC (1978) Microwave auditory effects and applications. CC Thomas, Springfield

Lin JC (1980) The Microwave Auditory Phenomenon. Proceedings of IEEE 68(1):67–73

Lin JC (1982) Electromagnetic instrumentation for medical imaging. International workshop on biomedical instrumentation and measurement, NIH/CWRU, Cleveland, OH, March 1982

Lin JC (1985) Frequency optimization for microwave imaging of biological tissues. Proc IEEE 72:374–375

Lin JC (1986) Microwave propagation in biological dielectrics with application to cardiopulmonary interrogation. In: Medical applications of microwave imaging, Larsen LE, Jacobi H (eds) IEEE Press, New York, pp 47–58

Lin JC (2005) Microwave thermoelastic tomography and imaging. In: Advances in electromagnetic fields in living systems, vol 4. Springer, New York, pp 41–76

Lin JC (2017) Mystery of sonic health attacks on havana-based diplomats. URSI radio science bulletin no. 362, September, pp 102–103

Lin JC (2018a) Strange reports of weaponized sound in Cuba. IEEE Microw Mag 19(1):18–19

Lin JC (2018b) Microwave signal at the embassy in Moscow. IEEE Microw Magazine 19(2):20–22

Lin JC (2020) Fighter pilot disorientation and pulsed microwave radiation from radars. URSI Radio Science Bulletin 373:86–88

Lin JC (2021a) Sonic health attacks by pulsed microwaves in Havana revisited. IEEE Microwave Magazine 22(3):71–73

Lin JC (2021b) New ICEMAN project seeks answers to fighter pilot disorientation. IEEE Microwave Magazine 22(4):16–18

Lin JC, Chan KH (1983) Microwave thermoelastic tissue imaging. In: IEEE conference on engineering in medicine and biology society, Columbus, OH, September 1983

Lin JC, Chan KH (1984) Microwave thermoelastic tissue imaging – system design. IEEE Trans Microwave Theory Tech 32(8):854–860

Lin JC, Wang JW (2007) Hearing of microwave pulses by humans and animals: effects, mechanism, and thresholds. Health Physics 92(6):621–628

Lou C, Yang S, Ji Z, Chen Q, Xing D (2012) Ultrashort microwave induced thermoacoustic imaging: a breakthrough in excitation efficiency and spatial resolution. Phys Rev Lett 109:218101

Lu ST, DeLorge JO (2000) Biological effects of high peak power radiofrequency pulses. In:Advances in electromagnetic felds in living systems. Lin J (ed). Kluwer Academic/ Plenum Publishers, New York, pp 207–264

Lyons TJ, Ercoline W, O'Toole K, Grayson K (2006) Aircraft and related factors in crashes involv ing spatial disorientation: 15 years of U.S. Air Force data. Aviat Space Environ Med 77(7):720–723

Manohar S, Razansky D (2016) Photoacoustics: a historical review. Adv Opt Photonics 8(4):586–617

Michaelson SM, Lin JC (1987) Biological effects and health implications of radiofrequency radiation. Chapter 20. Plenum, New York

Nan H, Arbabian A (2017) Peak-power-limited frequency-domain microwave-induced thermoacoustic imaging for handheld diagnostic and screening tools. IEEE Trans Microw Theory Tech 65(7):2607–2616

Nan H, Liu S, Dolatsha N, Arbabian A (2015) A 16-element wideband microwave applicator for breast cancer detection using thermoacoustic imaging. In: Proceedings of the PIERS, pp 243–247.

Nasoni RL, Evanoff Jr. GA, Halverson PG, Bowen T (1984) Thermoacoustic emission by deeply penetrating microwave radiation. In Proceedings IEEE ultrasonics symposium, pp 633–638

Navy SBIR, Remote Personnel Incapacitation System, http://www.navysbirprogram.com/NavySearch/Summary/summary.aspx?pk=F5B07D68-1B19-4235-B140-950CE2E19D08. Last accessed in August 2020.

Nie L, Xing D, Zhou Q, Yang D, Guo H (2008) Microwave-induced thermoacoustic scanning CT for high-contrast and noninvasive breast cancer imaging. Med Phys 35:4026–4032

NYT (1986) Walter J. Stoessel Jr. dies at 66; a former ambassador to Moscow. New York Times, 11 December 1986

Olsen RG (1982) Generation of acoustical images from the absorption of pulsed microwave energy. In: Powers JP (ed) Acoustical imaging, vol 11. Plenum, New York, pp 53–59

Olsen RG, Lin JC (1983) Acoustical imaging of a model of a human hand using pulsed microwave irradiation. Bioelectromagnetics 4:397–400

Patch SK, Hull D, See WA, Hanson GW (2016) Toward quantitative whole organ thermoacoustics with a clinical array plus one very low-frequency channel applied to prostate cancer imaging. IEEE Trans Ultrason Ferroelectr Freq Control 63:245–255

Petrov IR (ed) (1972) Influence of microwave on the organisms of man and animals. NASA Washington, DC

Poisson RJ III, Miller ME (2014) Spatial disorientation mishap trends in the U.S. Air Force 1993–2013. Aviat Space Environ Med 85(9):919–924

Public Law 90-602 (1968) 90th Congress, H.R. 10790, October 18, 1968: An Act, to Amend the Public Health Service Act to provide for the protection of the public health from radiation emissions from electronic products, Washington, DC

Relman RA, Pavlin JA (2020) An assessment of illness in U.S. government employees and their families at overseas embassies. National Academies Press, Washington, D.C.

Seaman RL, Beblo DA (1992) Modification of acoustic startle by microwave pulses in the rat: a pre liminary report. Bioelectromagnetics 13(4):323–328

Seaman RL, Beblo DA, Raslear TG (1994) Modifcation of acoustic and tactile startle by single microwave pulses. Physiol Behav 55(3):587–595

Sharp JC, Grove HM, Gandhi OP (1974) Generation of acoustic signals by pulsed microwave energy. IEEE Trans Microw Theory Tech 22:583–584

Silverman C (1980) Epidemiologic studies of microwave effects. Proc IEEE 68(1):78–84

Soltani M, Rahpeima R, Kashkooli FM (2019) Breast cancer diagnosis with a microwave thermoacoustic imaging technique-a numerical approach. Med Biol Eng Comput 57(7):1497–1513

Song J, Zhao Z, Wang J, Zhu X, Wu J, Nie Z-P, Liu QH (2013) An integrated simulation approach and experimental research on microwave induced thermo-acoustic tomography system. Prog Electromagn Res 140:385–400

Su JL, Chen DS (1992) Improvement of reconstruction algorithm in computerized thermoelastic wave tomography. In: Proceedings National Science Council, ROC Part B: life sciences, vol. 16, pp 31–38

Su JL, Lin JC (1985) Acoustic imaging of induced thermal expansion of biological tissue. In: IEEE engineering in medicine biology conference, Chicago, IL, September 1985

Su J-L, Lin JC (1987) Thermoelastic signatures of tissue phantom absorption and thermal expansion. IEEE Trans Biomed Eng BME-34(2):179–182

Su JL, Lin JC (1990) Thermoelastic waves computerized tomography system. J ROC Biomed Engg Soc 10:4–9

Su JL, Lin JC (1991) Computerized thermoelastic wave tomography. In: World Congress medical physics and biomedical engineering, Kyoto, Japan, July 1991

Takada Y, Hisada T, Kuwada N, Sakai M, Akamatsu T (2009) Survey of severe spatial disorientation episodes in Japan Air Self-Defense Force fighter pilots showing increased severity in night fight. Mil Med. 174(6):626–630

Verma R, Swanson RL, Parker D, Ould Ismail AA, Shinohara RT, Alappatt JA, Doshi J, Davatzikos C, Gallaway M, Duda D, Chen HI, Kim JJ, Gur RC, Wolf RL, Grady MS, Hampton S, Diaz-Arrastia R, Smith DH (2019) Neuroimaging findings in US Government personnel with possible exposure to directional phenomena in Havana, Cuba. JAMA 322(4):336–347

Wang LV, Zhao X, Sun H, Ku G (1999) Microwave-induced acoustic imaging of biological tissues. Rev Sci Instrum 70:3744–3748

Wang X, Bauer DR, Witte R, Xin H (2012) Microwave-induced thermoacoustic imaging model for potential breast cancer detection. IEEE Trans Biomed Eng 59(10):2782–2791

Watanabe Y, Tanaka T, Taki M, Watanabe SI (2000) FDTD analysis of microwave hearing effect. IEEE Trans Microw Theory Tech 48:2126–2132

Wen L, Yang S, Zhong J, Zhou Q, Xing D (2017) Thermoacoustic imaging and therapy guidance based on ultra-short pulsed microwave pumped thermoelastic effect induced with superparamagnetic iron oxide nanoparticles. Theranostics 7(7):1976–1989

Xu M, Wang LV (2002) Time-domain reconstruction for thermoacoustic tomography in a spherical geometry. IEEE Trans Med Imaging 21:814–822

Xu Y, Wang L (2006) Rhesus monkey brain imaging through intact skull with thermoacoustic tomography. IEEE Trans Ultrason Ferroelectr Freq Control 53:542–548

Xu MH, Ku G, Wang LHV (2001) Microwave-induced thermoacoustic tomography using multi-sector scanning. Med Phys 28(9):1958–1963

Yan A, Lin L, Liu C, Shi J, Na S, Wang LV (2019) Microwave-induced thermoacoustic tomography through an adult human skull. Med Phys 46:1793–1797. https://doi.org/10.1002/mp.13439

Yitzhak NM, Ruppin R, Hareuveny R (2014) Numerical simulation of pressure waves in the cochlea induced by a microwave pulse. Bioelectromagnetics 35:491–496

Zhao ZQ, Song J, Zhu XZ, Wang JG, Wu JN, Liu YL, Nie ZP, Liu QH (2013) System development of microwave induced thermoacoustic tomography and experiments on breast tumor. Prog Electromagn Res 134:323–336

Appendix

A.1. Table of Standard Prefixes Used with Units of the International System (SI) of Measurements

Prefix Name	Abbreviation	Magnitude
Atto	a	10^{-18}
Femto	f	10^{-15}
Pico	p	10^{-12}
Nano	n	10^{-9}
Micro	μ	10^{-6}
Milli	m	10^{-3}
Centi	c	10^{-2}
Deci	d	10^{-1}
Deka	da	10^{1}
Hecto	h	10^{2}
Kilo	k	10^{3}
Mega	M	10^{6}
Giga	G	10^{9}
Tera	T	10^{12}

© Springer Nature Switzerland AG 2021
J. C. Lin, *Auditory Effects of Microwave Radiation*,
https://doi.org/10.1007/978-3-030-64544-1

Index

© Springer Nature Switzerland AG 2021
J. C. Lin, *Auditory Effects of Microwave Radiation*,
https://doi.org/10.1007/978-3-030-64544-1

Printed in the United States
by Baker & Taylor Publisher Services